Microbiology Monographs
Volume 21

Series Editor: Alexander Steinbüchel
Münster, Germany

For further volumes:
www.springer.com/series/7171

Min-Tze Liong
Editor

Probiotics

Biology, Genetics and Health Aspects

Springer

Editor
Associate Professor Dr. Min-Tze Liong
School of Industrial Technology
Universiti Sains Malaysia
11800 Penang
Malaysia
mintze.liong@usm.my

Series Editor
Professor Dr. Alexander Steinbüchel
Institut für Molekulare Mikrobiologie und Biotechnology
Westfälische Wilhelms-Universität
Corrensstr. 3
48149 Münster
Germany
steinbu@uni-muenster.de

ISSN 1862-5576 e-ISSN 1862-5584
ISBN 978-3-642-20837-9 e-ISBN 978-3-642-20838-6
DOI 10.1007/978-3-642-20838-6
Springer Heidelberg Dordrecht London New York

Library of Congress Control Number: 2011932712

Cover design: SPi Publisher Services

Printed on acid-free paper

Springer is part of Springer Science+Business Media (www.springer.com)

Preface

Probiotics have a long history of safe use and have been well-documented for their health benefits on hosts. This volume is a collection of various topics on probiotics, ranging from the microbiological aspects of both prokaryotic and eukaryotic probiotic microorganisms to genetic modifications, maintenance, health benefits and claims, detection, genetic modifications, safety and market trends.

The chapter by Daniel O. Otieno discusses prokaryotic probiotics and their classification based on morphology, ability to form spores, method of energy production, nutritional requirements and reaction to the Gram stain. Prokaryotic probiotics are mainly from the genera of *Lactobacillus* and *Bifidobacterium*. *Lactobacillus* has 106 validly described species, out of which 56 species have probiotic potential. On the other hand, *Bifidobacterium* currently has 30 species validly described, with eight having probiotic capabilities. A close study of these microorganisms revealed that probiotic microorganisms are likely to be Gram positive, mostly rod shaped but with fewer spherically shaped ones, nonspore forming and nonflagellated bacteria.

The chapter by Sukanta K. Nayak covers the microbiological aspects of eukaryotic probiotics. Several eukaryotic microorganisms have been identified with probiotic properties and have been consumed as single cell protein and/or as components of food starters for human and animal consumption. A limited number of these eukaryotic probiotics originate from fungi/moulds/yeasts. Among the eukaryotic probiotics, yeasts especially *Saccharomyces* species are dominant and routinely used in a broad range of hosts. This chapter deals with the occurrence, distribution, taxonomic characterization and detail modes of action of eukaryotic probiotics with special reference to yeasts in human and other animals.

The chapter by Alexander G. Haslberger et al. provides an overview of probiotic strain characterization, gut metagenomics and the analytical methods (FISH, PCR, RAPD, DGGE, repPCR, PFGE, RFLP, microarray, high throughput sequencing) required for their study. Molecular microbiological analysis has increased the understanding of the diversity and phylogeny of beneficial strains and their functions. These modern techniques, including genotyping methods, become increasingly important for species identification and for the differentiation of probiotic strains. The precise classification and identification of probiotic strains give a strong indication of its typical habitat and origin, safety and technical applicability and provides possibilities for monitoring and product quality.

The chapter by Zhibing Zhang and Benjamin D. Huckle discusses the importance and types of protection techniques for probiotic microorganisms, with a particular focus on encapsulation and compression coating. Beneficial effects of probiotic bacteria depend on the viability of cells once delivered to the intestines. Cells tend to lose their viability with time during storage and the passage through acidic gastric fluids. Enteric coating materials are discussed as suitable for compression coating, while also improving cell storage stability and ensuring cell survival during exposure to harsh acidic gastric fluids. Techniques for controlled release of cells to the colon, including the use of hydrophobic disintegrants such as pectin, are also described.

The chapter by Chathuranga T. Bamunuarachchige et al. addresses the issues of genetically engineered probiotic microorganisms. Improved technologies in genomics and proteomics have led to greater understanding on the beneficial characteristics of probiotics that are enhanced via genetic modification (GM). GM probiotics are mainly associated with improved gut survival and persistence, tolerance of packing and storage conditions and as successful delivery vehicles for therapeutics. These are attributed to increased tolerance to osmolarity, bile salt and reduced water activity. With the expression of various molecules such as antigens, enzymes and molecules of immunological importance within probiotic microbes, the use of GM probiotics in the field of therapeutics looks promising. Safety issues of GM probiotics are also discussed.

The chapter by Wai-Yee Fung et al. documents the strong in vivo and in vitro evidences of probiotics on gut health, emphasizing on reestablishing the intestinal ecosystem balance, and alleviating gut and malabsorption disorders such as diarrhea, lactose intolerance and irritable bowel syndrome. Probiotics are also therapeutic against postoperative complications and inflammatory bowel diseases, in addition to exerting antibacterial and anticancer properties in the gut, due to their ability to attenuate the immune system. Mechanisms involved include competitive exclusion of pathogenic bacteria for nutrients and adhesion sites, production of antimicrobial bacteriocins and metabolites, and gut immunomodulation.

The chapter by Narayan C. Mandal and Vivekananda Mandal documents the different strains or species of orally consumed probiotics in conferring various new health benefits beyond gut well-being. New roles include modulation of immunological parameters, allergy and lung emphysema. Probiotics fight invading pathogens by various mechanisms, such as competitive inhibition, production of active principles such as bacteriocins, hydrogen peroxide and organic acids. In addition to inhibiting pathogens, probiotics also contribute to improving the immunological and physiological state of the host by interfering with metabolic processes. As they colonize the vital parts of the human intestines, probiotics are intricately involved with different systems of the human body and alleviate the problems associated with them.

The chapter by Siok-Koon Yeo et al. discusses nondairy new carriers for probiotics. Despite being an ideal substrate for probiotics, the growth of probiotics in dairy products is often inhibited by excessive acidification, antagonistic effect of starter cultures and the presence of oxygen during processing. The drawbacks of milk-based carrier associated with cholesterol contents, and lactose intolerance has prompted the development of alternative carriers for probiotics. Currently, soy-based, cereal-based, fruits, vegetables and meat products are developed as potential probiotic carriers. These nondairy products contain reasonable amounts of carbohydrates, fibers, proteins and vitamins that support the growth of probiotics, and protective components that are able to protect probiotics during gut transit, processing and storage. The challenges of these new carriers are also discussed.

The chapter by Maria G. Cifone et al. documents the clinical and experimental evidences of probiotic benefits at the skin. Scientific and evidence-based reports strengthen the assumption that certain probiotics can contribute to modulate cutaneous microflora, lipid barrier, skin immune system, leading to the preservation of skin homeostasis. In this chapter, recent evidences available from scientific literature as well registered patents have been summarized in relation to actual or potential topical applications of probiotics in the field of dermatology. Altogether the evidences reported in this review afford the possibility of designing new strategies based on a topical approach for the prevention and treatment of cutaneous disorders.

The chapter by Istan Siro addresses the challenges of the various probiotic health claims. Substantiation of claims should be based on scientific evidences, which requires a long and expensive procedure. Different in vitro and in vivo methods are applied for screening and characterizing the putative probiotic strains. Although useful, these assessment tools must be validated by properly designed human clinical studies. Poor prior selection, limited capacity of in vitro tests, and unsuitable animal models often contribute to contradictions between in vitro findings and in vivo feasibility. This chapter reviews the crucial steps of substantiation of health claims associated with probiotics with special emphasis on the related challenges.

The chapter by Fumiaki Abe documents the safety aspects of probiotics, rising from issues including bacterial translocation causing sepsis and horizontal transfer of acquired antibiotic resistance gene. To resolve these concerns, manufacturers have to demonstrate safety of probiotics on a strain by strain basis because not all probiotics are the same. Also, probiotics harboring acquired antibiotic resistance genes should not be used to avoid the possibility of gene transfer. A high hygienic standard to prevent contamination by pathogenic bacteria or allergen during the production of probiotics is another requirement to assure the safety of probiotics. Safety regulation of various countries including the United States, Canada, Australia, New Zealand and Japan are also highlighted.

The chapter by Carlos R. Soccol provides an insight on the current probiotic market trends and future directions. Current trends in the consumption of probiotics are associated with increased levels of health-consciousness, and the availability of probiotics in the form of dietary supplements. Several companies have profited by marketing these products in different forms, with different purposes, and with recommendation for all ages. Important aspects in maintaining the viability and bioactivity of probiotic strains during processing and storage are also discussed in this chapter. The probiotic consumption by infants and the elderly has been supported by scientific evidences and represents a new niche market.

Penang, Malaysia Min-Tze Liong

Contents

Biology of Prokaryotic Probiotics 1
Daniel Obed Otieno

Biology of Eukaryotic Probiotics 29
Sukanta Kumar Nayak

Detection and Identification of Probiotic Microorganisms and Other Beneficial Organisms from the Human GI Tract 57
Berit Hippe, Jutta Zwielehner, Angelika Kolle, William M. Smith, and Alexander G. Haslberger

Maintenance and Protection of Probiotics 87
Benjamin D. Huckle and Zhibing Zhang

Genetic Engineering of Probiotic Microorganisms 109
T.C. Bamunuarachchige, H.A.M. Wickramasinghe, D.M.J.C. Dissanayaka, and N.A.D. Wickramarathna

Roles of Probiotic on Gut Health 139
Wai-Yee Fung, Huey-Shi Lye, Ting-Jin Lim, Chiu-Yin Kuan, and Min-Tze Liong

New Health Potentials of Orally Consumed Probiotic Microorganisms 167
Vivekananda Mandal and Narayan C. Mandal

Carriers of Probiotic Microorganisms 191
Siok-Koon Yeo, Joo-Ann Ewe, Celestine Sau-Chan Tham, and Min-Tze Liong

Use of Probiotics for Dermal Applications 221
Benedetta Cinque, Cristina La Torre, Esterina Melchiorre, Giuseppe Marchesani, Giovanni Zoccali, Paola Palumbo, Luisa Di Marzio, Alessandra Masci, Luciana Mosca, Paola Mastromarino, Maurizio Giuliani, and Maria Grazia Cifone

Challenges of Beneficial Health Claims 243
István Siró

Safety and Regulation of Probiotic Foods and Supplements 269
Fumiaki Abe

Current Market Trends and Future Directions 299
Caroline T. Yamaguishi, Michele R. Spier, Juliano De Dea Lindner, Vanete T. Soccol, and Carlos R. Soccol

Index 321

Biology of Prokaryotic Probiotics

Daniel Obed Otieno

Contents

1 Introduction 2
2 The Phylogenic Tree of Prokaryotic Probiotic Microorganisms 3
 2.1 Bacteria 3
 2.2 The Archaea 4
3 Morphology of Prokaryotic Probiotic Bacteria 6
 3.1 Coccus 7
 3.2 Bacilli 7
 3.3 Spirochaete 8
 3.4 Vibrio 8
4 Reaction of Prokaryotic Bacteria to the Gram Stain 8
5 The Cellular Structures of Prokaryotic Bacteria 9
 5.1 The Structures External to the Cell Wall 10
 5.2 The Structures Internal to the Cell Wall 12
6 Reproduction of Prokaryotic Probiotic Bacteria 16
7 The Taxonomy of Prokaryotic Probiotic Bacteria 17
 7.1 The Genus *Lactobacillus* 18
 7.2 The Genus *Bifidobacterium* 20
 7.3 Other Genera 22
8 Prokaryotic Bacteria and Interaction with the Environment 22
9 Relationship Between Prokaryotes and Eukaryotes 25
References 25

Abstract Bacteria and archaea are two distinct phyla of the prokaryotic kingdom containing many different species of microorganisms. Prokaryotic probiotics are single-celled nonnucleated organisms which when consumed live in adequate numbers confer a health benefit to the host. They can be classified based on

D.O. Otieno (✉)
Washington State University, Center for Bioenergy and Bioproduct – BSEL, Washington State University TriCities, 2710 University Drive, Richland, WA 99354, USA
e-mail: dotieno@tricity.wsu.edu

M.-T. Liong (ed.), *Probiotics*, Microbiology Monographs 21,
DOI 10.1007/978-3-642-20838-6_1, © Springer-Verlag Berlin Heidelberg 2011

morphology, ability to form spores, method of energy production, nutritional requirements, and reaction to the Gram stain. Currently, there are no known probiotic archaea but they have an important potential in the synthesis of prebiotics and other bioproducts due to their unique characteristics. There are however, probiotic bacteria mainly coming from the genera of *Lactobacillus* and bifidobacteria. *Lactobacillus* has 106 validly described species, out of which 56 species have probiotic potential. On the other hand, Bifidobacteria currently has 30 species validly described, with 8 having probiotic capabilities. A close study of these microorganisms reveal that probiotic bacteria are likely to be Gram positive, mostly rod shaped but with fewer spherically shaped ones, nonspore forming and nonflagellated bacteria.

1 Introduction

Prokaryotes are a group of microorganisms that do not possess a cell nucleus or any other membrane-bound organelles. An *organelle* refers to structures within the cell membrane that perform functions analogous to what an organ does in the body. Previously, prokaryotes were thought not to have organelles; however, some examples have recently been identified (Kerfeld et al. 2005) that possess these structures. They are not as structurally complex and for a long time were thought not to have any internal structures enclosed by lipid membranes. They were also previously thought to have little internal organization. Some prokaryotes have been found to have microcompartments such as carboxysomes, which are the subcellular compartments measuring 100–200 nm in diameter and are enclosed by a shell of proteins (Kerfeld et al. 2005). In addition, there have been descriptions of membrane-bound magnetosomes in bacteria (Komeili et al. 2006; Scheffel et al. 2006) as well as the nucleus-like structures of the *Planctomycetes* that are surrounded by lipid membranes (Fuerst 2005). These studies have revealed that a typical prokaryotic cell have important major organelles and cellular structures including nucleoid, ribosome, vesicles, rough endoplasmic reticulum, golgi apparatus, cytoskeleton, smooth endoplasmic reticulum, mitochondria, vacuole, cytosol, lysosome, and centriole. These structures determine the cell's structure, function, growth, origin, survival, and distribution as well as the taxonomic classification of the microorganism. Thus, despite their apparent simplicity compared to eukaryotes, they pretty much can exist in any given optimal environment. Table 1 presents a summary of the most important organelles in prokaryotic bacteria, their corresponding functions, structure component, and the microorganisms they are found in.

There remains two major domains of prokaryotic microorganisms, namely, the bacteria and the archea. This chapter will focus on prokaryotic probiotic microorganisms, which admittedly, only form a very small fraction of the entire population of prokaryotic microorganisms. In spite of this reality, the relatively few prokaryotes, mostly from bacteria and a few archaea are very important in their probiotic functions. Although archaea have an important niche of

Table 1 A summary of prokaryotic cell organelles, components, functions, structure, and the organisms they are found in

Prokaryotic organelles and cell components			
Organelle/ macromolecule	Main function	Structure	Organisms
Carboxysome	Carbon fixation	Protein-shell compartment	Some bacteria
Chlorosome	Photosynthesis	Light harvesting complex	Green sulfur bacteria
Flagellum	Movement in external medium	Protein filament	Some prokaryotes and eukaryotes
Magnetosome	Magnetic orientation	Inorganic crystal, lipid membrane	Magnetotactic bacteria
Nucleoid	DNA maintenance, transcription to RNA	DNA–protein	Prokaryotes
Plasmid	DNA exchange	Circular DNA	Some bacteria
Ribosome	Translation of RNA into proteins	RNA–protein	Eukaryotes, prokaryotes
Thylakoid	Photosynthesis	Photosystem proteins and pigments	Mostly cyanobacteria

applications, most of the well-known probiotics are of the bacterial domain. The highlights in this chapter will be on probiotic prokaryotic bacteria. According to the currently adopted definition by FAO/WHO, probiotic bacteria are: "Live microorganisms which when administered in adequate amounts confer a health benefit on the host" (FAO/WHO, 2001). Lactic acid bacteria (LAB) and bifidobacteria are the most common types of microbes used as probiotics, but certain yeasts and bacilli may also be helpful. Probiotics are commonly consumed as part of fermented foods with specially added active live cultures, such as in yogurt, soy yogurt, or as dietary supplements. Archaea are increasingly finding important application in bioproducts, pharmaceutical, food, and biofuel industry.

2 The Phylogenic Tree of Prokaryotic Probiotic Microorganisms

2.1 *Bacteria*

The bacteria are a large group of single-celled, prokaryote organisms that are microscopic, typically measuring a few micrometers (10^{-6}), hence referred to as microorganisms. They do not contain a nucleus and rarely harbor membrane-bound organelles. Although the term *bacteria* traditionally included all prokaryotes, the scientific classification changed after the discovery in the 1990s that prokaryotes consist of two very different groups of organisms that evolved independently from an ancient common ancestor, namely, Bacteria and Archaea (Woese et al. 1990). Therefore both bacteria and archaea are part of the domain considered prokaryotic. The classification of bacteria is generally based on five main criteria, namely, morphology, ability to form spores, method of energy production, nutritional requirements, and reaction to the Gram stain. These aspects will be discussed in detail in the various subsections of this chapter.

2.2 *The Archaea*

The archaea are also a group of single-celled microorganisms and forms an important component of prokaryotic microorganisms. A single individual or species from this domain is called an *archaeon* (sometimes spelled "archeon"). Like bacteria they also have no cell nucleus or any other membrane-bound organelles within their cells. In the past they were regarded as an unusual group of bacteria and named archaebacteria; however, after studies of the sequencing of their ribosomal RNA, it became obvious that they bore no close relationship to the bacteria (http://users.rcn.com 2010). Because of their independent evolutionary history (Zuckerkandl and Pauling 1965) and differences in their biochemistry from other forms of life, they are now classified phylogenetically as a separate domain in the three-domain system consisting of Archaea, Bacteria, and Eukarya as shown in Fig. 1. The archaea are divided into four recognized phyla, but it is well established that many more phyla may exist. Of these groups, the Crenarchaeota and the Euryarchaeota are most intensively studied, and to date, more than 250 species have been discovered. Classification of archaea has only been made possible through analysis of their nucleic acids in samples from the environment. Based

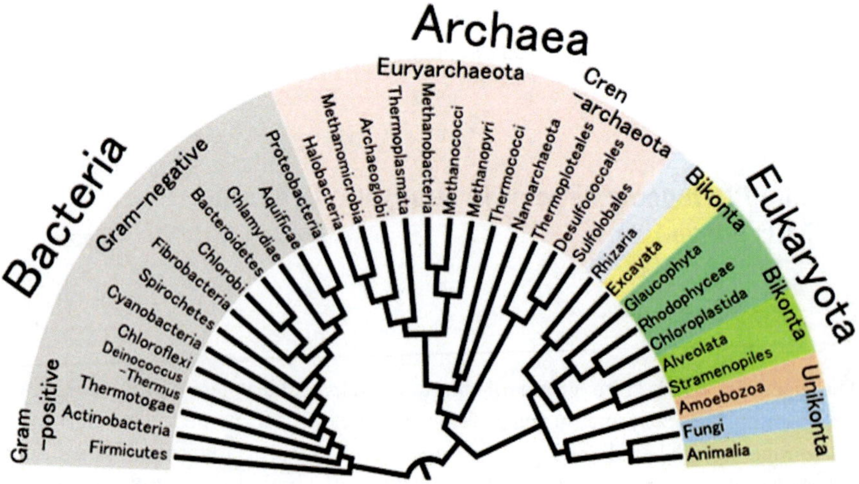

Fig. 1 Phylogenetic tree showing the diversity of prokaryotes, compared to eukaryotes (Woese 1998)

on these studies, more evidence of their differences from bacteria was established (Pace 2006).

Archaea and bacteria are visually similar in size and shape, although a few archaea have very unusual shapes, such as the flat and square-shaped cells such as those of *Haloquadra walsbyi*. Despite this similarity to bacteria, archaea possess genes and several metabolic pathways that are more closely related to those of eukaryotes such as the enzymes involved in transcription and translation, as well as reliance on ether lipids in their cell membranes. The archaea exploit a much greater variety of sources of energy for metabolic processes than bacteria and eukaryotes; ranging from familiar organic compounds such as sugars, to inorganic ones including ammonia, metal ions, sun light, and even hydrogen gas as nutrients. There are also species of archaea that fix carbon into the soil. Their mode of reproduction is asexual and divides by binary fission, fragmentation, or budding. In contrast to some bacteria, they do not form spores.

Initially, archaea were regarded as extremophiles that lived in harsh environments, such as hot springs, salt lakes, and other extremes of temperature, pH, and radiation, but they have since been found in a broad spectra of habitats, including soils, oceans, and marshlands. Archaea are particularly numerous in the oceans, and the archaea in plankton may be one of the most abundant groups of organisms on the planet. Archaea are mostly recognized for their major role in both the carbon cycle and nitrogen cycle. There are currently no known examples of archaeal pathogens or parasites, but they are often mutualists or commensals. The only known archaea with probiotic potential are from the methanogens, which inhabit the gut of humans and ruminants, where their vast numbers aid digestion. Figure 2 shows branches of archaea highlighting those increasingly useful in the biotech industry such as methanogens and hyperthermophiles. If functions by intestinal microflora such as maturation of the immune system, protection against cell injury, and regulation of energy balance (O'Hara and Shanahan 2006) are considered probiotic, then the complex ecosystem which also includes archaea as constituent of the 400–1,000 gastrointestinal species from at least nine different phyla (Rajilic-Stojanovic et al. 2007) is probiotic. Most of these species especially of archaea are still uncultured and our knowledge of the microbial diversity still remains incomplete. In humans, the archaeal methanogens have been found in the anaerobic population of the oral cavity, the vagina, and the large intestine (Chaban et al. 2006). Gut methanogenic archaea that have been well determined are two species belonging to Methanobacteriales which are *Methanobrevibacter smithii* and *Methanosphaera stadtmanae* (Fricke et al. 2006). Methanogens are also useful in biotechnological processes such as in industrial catalyzation of thermochemical process such as the application of archaeal β-galactosidases in the synthesis of prebiotics such as galactooligosaccharides (GOS) from lactose at elevated temperatures. They are also useful in biotechnological processes such as biogas production and sewage treatment. Due to superior stable properties of the glycosidases from thermophilic methanogens, they have become an attractive interest for transgalactosyl reactions for GOS synthesis at high temperatures (Otieno 2010). Examples of archaea currently being used in this respect include

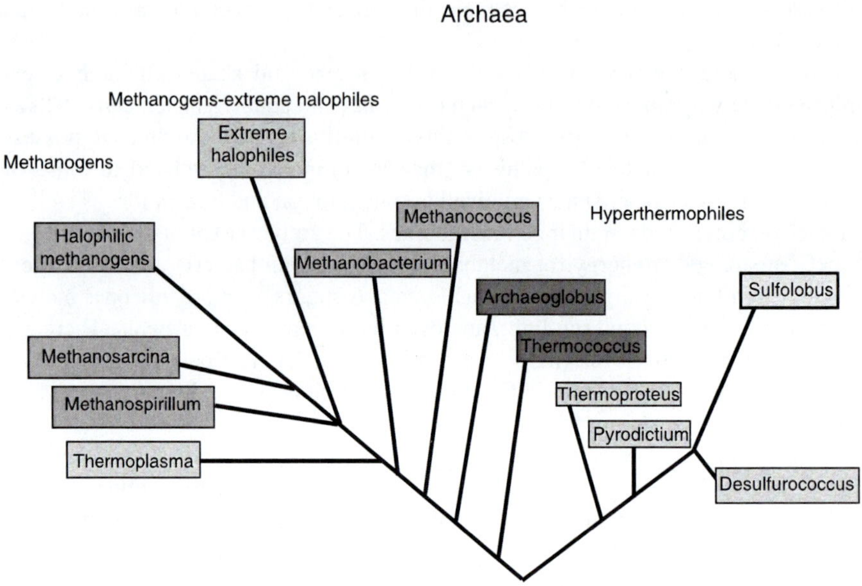

Fig. 2 Branches of archaea showing methanogens and hyperthermophiles useful in biotechnology (Baross and Holden 1996)

Thermus sp. Z-1 (Akiyama et al. 2001), *Pyrococcus furiosus* (Petzelbauer et al. 2000; Bruins et al. 2003), and *Sulfolobus solfataricus* (Reuter et al. 1999; Hansson et al. 2001) in search of GOS production at high temperatures. The rationale of using thermostable enzymes is that transgalactosyl reactions proceed more efficiently at high substrate concentrations due to the improved solubility of lactose at higher temperatures. As a result of their ability to exist and adapt in different extreme environments, they can also be grouped in various classes such as thermophiles (high temperatures), hyperthermophiles (very high temperatures such as 121°C), psychrophiles (in cold conditions and may grow best at 4°C), halophiles (in very saline environments), acidophiles (at low pH, i.e., as low as pH 1 and die at pH 7), alkaliphiles (at a high pH).

3 Morphology of Prokaryotic Probiotic Bacteria

Prokaryotic bacteria have various shapes and morphological features that are important in their classification and identification. The four basic shapes are cocci (spherical), bacilli (rod shaped), spirocheate (spiral shaped), and vibrio (comma shaped). Probiotic prokaryotes well known in the genera such as *Lactobacillus* and bifidobacteria are mostly rod shaped with fewer strains having spherical ones.

3.1 Coccus

The coccus (plural cocci) can be used to describe any bacterium that has a circular shape. They can be solitary in nature, but also group together to form different patterns in clusters. *Diplococci* are two-celled pairs of bacteria, while *Streptococci* are those arranged in a chain formation. The *Sarcina* are those having a cuboidal cell arrangement, while the tetrad ones are arranged in a square formation. The *Staphylococcus* genus usually contains cells arranged in large clusters of bacterial cells. Figure 3 shows the arrangements of Cocci into different patterns from which they derive their names. Whereas the arrangements of the bacterial cells take different forms, the bacterial cells themselves remain circular. Most of the bacteria having spherical shapes are known to be disease causing or pathogenic in nature, e.g., *Staphylococcus aureus*. However, there are a few well-known prokaryotic bacteria that have probiotic function such as *Streptococcus thermophilus* and *Enterococcus faecium*. *S. thermophilus* is a probiotic well known for its use in yogurt production. It uses its endogenous lactase to break down lactose into glucose and galactose. This enables a wider consumption of dairy products which also includes consumers who are lactose intolerant. Other Streptococcus strains which are probiotic include *Streptococcus cremoris* and *Streptococcus infantis*. *E. faecium* has shown in studies to be helpful for diarrhea through shortening the duration of disease symptoms. It also inhibits pathogenic microbes, such as rotavirus and lower LDL or bad cholesterol. Although *E. faecium* is considered a transient guest, i.e., does not occur naturally in the GIT, it is a welcome natural resident in the human body.

3.2 Bacilli

These are bacteria that are rod shaped and the singular form is referred to as *bacillus*. They are found in many different taxonomic groups of bacteria; however,

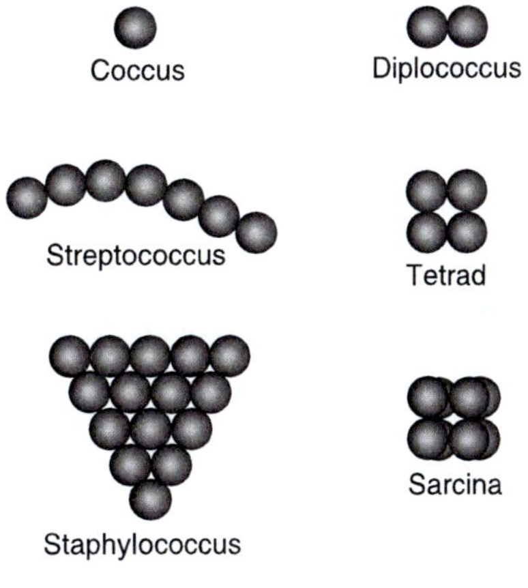

Fig. 3 Different patterns of spherically shaped prokaryotic bacteria. (Kaiser 1999) http://student.ccbcmd.edu/~gkaiser/goshp.html

the name *Bacillus*, capitalized and italicized, refers to a specific genus of bacteria. The name Bacilli, capitalized but not italicized, can also refer to a more specific taxonomic class of bacteria that includes two orders, one of which contains the genus *Bacillus*. They are also usually solitary, but can combine in clusters to form diplobacilli, streptobacilli, and palisades (http://www2.nemcc.edu). The most studied and well-known probiotic prokaryotic bacteria are from the genera of lactobacilli and bifidobacteria. A majority of the currently used probiotic prokaryotes are the Gram-positive rod-shaped bacteria.

3.3 Spirochaete

They are also spelled spirochetes, and belong to a phylum of Gram-negative bacteria, which have long, helically coiled (spiral-shaped) cells known as the flagella which is sometimes referred to as axial filament (Ryan and Ray 2004). They are not known to be probiotic, hence will not be discussed into greater detail in this chapter.

3.4 Vibrio

Vibrio is also a genus of Gram-negative bacteria that distinctly possess a curved rod shape, (Faruque and Nair 2008). They too are not known to be probiotic.

4 Reaction of Prokaryotic Bacteria to the Gram Stain

Gram staining is important in the identification of bacteria and is named after the nineteenth-century Danish bacteriologist named Hans Christian Gram who developed it. In the Gram staining process as shown in Fig. 4, the bacterial cells are first stained with a purple dye known as crystal violet. The cells are then treated with alcohol or acetone, followed by counterstaining with a dye of a different color (such as safarin or fuchsine) which is red or pink in color. Gram-positive bacteria are those that are stained dark blue or violet by Gram staining. This is in contrast to Gram-negative bacteria, which cannot retain the crystal violet stain, instead taking up the counterstain (safarin or fuchsine) and appearing red or pink. Gram-positive organisms are able to retain the crystal violet stain because of the high amount of peptidoglycan in the cell wall. Gram-positive cell walls typically lack the outer membrane found in Gram-negative bacteria. Figure 5 shows the marked difference in the wall structure of Gram-positive and Gram-negative bacteria, specifically indicating a thicker peptidoglycan layer among the Gram-positive bacteria, which influences their ability to retain or lose the crystal violet stain.

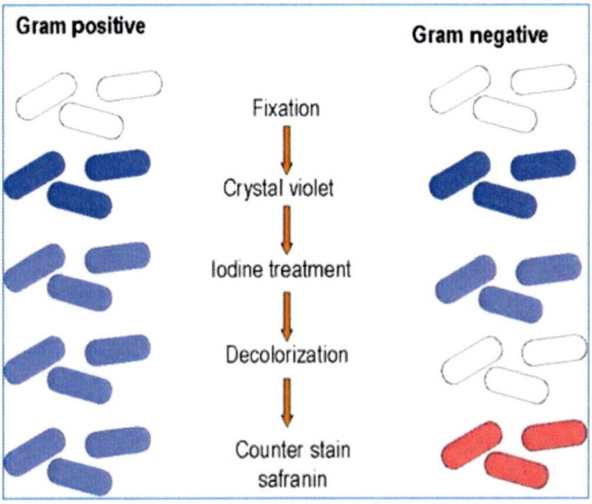

Fig. 4 Gram staining procedure in the identification of bacteria (Fox 2010). Generated by Alvin Fox, USC School of Medicine http://pathmicro.med.sc.edu/fox/culture

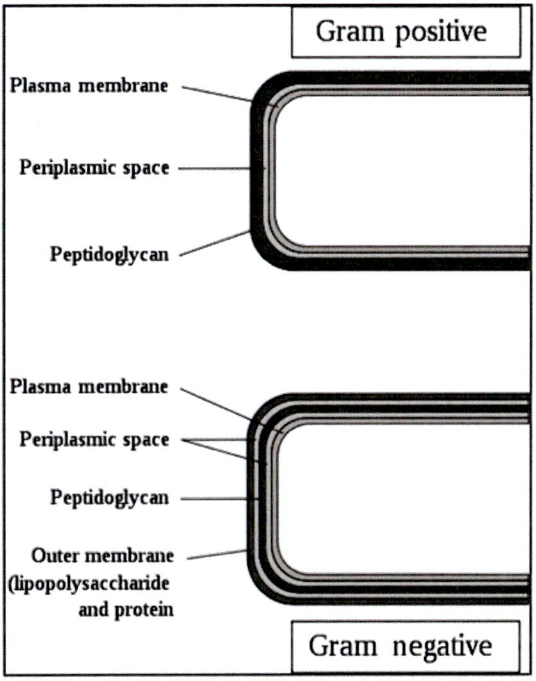

Fig. 5 Gram-positive and Gram-negative cell wall structure (Madigan and Martinko 2005)

5 The Cellular Structures of Prokaryotic Bacteria

The cellular structures are important for specific biological functions and are divided into two main classes, namely, (a) the structures external to the cell wall and (b) the structures internal to the cell wall.

5.1 The Structures External to the Cell Wall

The capsule is a very large structure of Gram-negative prokaryotic cells which lies outside the cell wall of the bacteria. It is basically composed of polysaccharides but may also include polypeptides among certain bacterial strains. Probiotic prokaryotes do not possess a bacterial capsule.

A flagellum is a tail-like projection that protrudes from the cell body of Gram-negative prokaryotic cells and helps in locomotion. Probiotic prokaryotic bacteria do not have a flagellum.

Fimbriae and pili are structures made of the protein pilin. The fimbriae are shorter and thinner than flagella and are used for attachment and not movement. They are found on some Gram-negative pathogenic bacteria and they enable the bacteria to attach to surfaces, such as mucous membranes. The pili on the other hand are comparatively longer than the fimbriae and much fewer in number, i.e., only one or two per cell. Some are used to join bacterial cells in preparation for the process of conjugation (transfer of DNA), hence sometimes called conjugation pili. There are no known probiotic prokaryotic bacteria possessing fimbriae and pili.

The cell wall is also located outside the cell membrane and it is a tough, usually flexible but sometimes fairly rigid layer that surrounds some types of cells. Its main functions are to provide the cells with structural support and protection, while also acting as a filtering mechanism. Perhaps its most important function is to act as a pressure vessel, preventing overexpansion when water enters the cell. Prokaryotic probiotic bacteria are known to have cell walls. There are known strains within species of probiotic bacteria having cell walls are *Bifidobacterium longum, Bifidobacterium animalis, Lactococcus lactis, Lactobacillus johnsonii, Lactobacillus planetarium, Lactobacillus acidophilus,* and *Lactobacillus casei* amongst others.

Prokaryotic cell walls are mainly made of the peptidoglycan material. However, the archaea do not contain peptidoglycan, instead, their cell wall contains pseudomurein. Peptidoglycan consists of disaccharide units connected by polypeptides to form a lattice. The many layers of peptidoglycan form a thick rigid structure that provides shape and protection. There are also teichoic acids in the cell walls, which consist of an alcohol and phosphate and the exact structure of the acids varies among species of bacteria. This is the part that often acts as an antigen and is used for identification. Some cell walls contain large amounts of mycolic acid thus making the bacteria to be regarded as acid-fast bacteria. The mycolic acid is in a layer outside the peptidoglycan.

As shown in Fig. 6, the peptidoglycan layer in the bacterial cell wall is a crystal lattice structure formed from linear chains of two alternating amino sugars, namely, *N*-acetylglucosamine (GlcNAc or NAG) and *N*-acetylmuramic acid (MurNAc or NAM). The alternating sugars are connected by a β-(1, 4)-glycosidic bond. Attached to the *N*-acetylmuramic acid is a peptide chain of three to five amino acids. Peptidoglycan layer is important in serving as a structural role in the bacterial

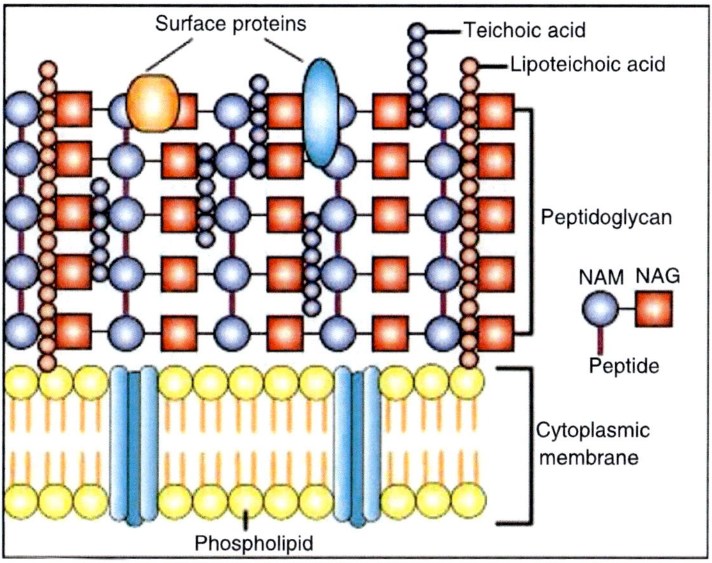

Fig. 6 The structure of Gram-positive prokaryotic cell wall (Kaiser 1999) http://student.ccbcmd.edu/~gkaiser/goshp.html

cell wall, giving structural strength, as well as counteracting the osmotic pressure of the cytoplasm. The peptidoglycan layer is substantially thicker in Gram-positive bacteria (20–80 nm) than in Gram-negative bacteria (7–8 nm), with the attachment of the S-layer. Peptidoglycan forms around 90% of the dry weight of Gram-positive bacteria but only 10% of Gram-negative strains. In Gram-positive strains, it is important in attachment roles and stereotyping purposes (Salton and Kim 1996). For both Gram-positive and Gram-negative bacteria, particles of approximately 2 nm can pass through the peptidoglycan (Demchick and Koch 1996). The functions of peptidoglycan (glycocalyx) include protecting the probiotic bacteria from phagocytosis, making them protected from digestive enzymes along the gastrointestinal track, allowing the bacterial cell to attach to a surface, for example, the intestinal linings, formation of a biofilm which protects the bacterial cells, often making body defenses, antiseptics, and antimicrobials less effective and may serve as a source of nutrients and protect against cell dehydration. The surface proteins on the peptidoglycan have different roles which include acting as enzymes for metabolic processes within the cell, adhesions, to enable the bacterium to attach intimately to the host and other surfaces in order to colonize and resist flushing, and for resisting phagocytic destruction.

A few prokaryocytes such as members of the genus *Mycoplasma* do not have a cell wall. As a result, they are also extremely tiny and their plasma membranes need extra strength which is provided through incorporation of sterols. Bacteria known to have atypical cell walls are pathogenic or potentially pathogenic.

5.2 *The Structures Internal to the Cell Wall*

Various internal structures to the cell wall exist for cellular physiological functions and maintenance. These include the cytoplasmic membrane, cytoplasm, nucleiod, ribosomes, plasmids, and endospores.

5.2.1 Cytoplasmic Membrane (Cell Membrane)

It refers to selectively permeable membrane that encloses the cytoplasm of a cell. It is composed of a bilayer of phospholipids embedded with proteins which determines what enters and leaves the cell. It is also known as a plasma membrane or cell membrane.

As shown in Fig. 7, the prokaryotic cell membrane separates the interior of prokaryotic cells from the outside environment. The cell membrane is selectively permeable to ions and organic molecules such as sugars and lipids, thus facilitating the transport of materials needed for survival. The movement of compounds i.e nutrients across the membrane can be passive, that is, occurring without the input of cellular energy, or active, requiring the cell to expend energy in moving it. The cell membrane also helps the cell to maintain the cell potential. It consists of the phospholipid bilayer with embedded proteins, which are involved in a variety of cellular processes such as cell adhesion, ion conductivity, and cell signaling. The plasma membrane also serves as the attachment surface for the extracellular

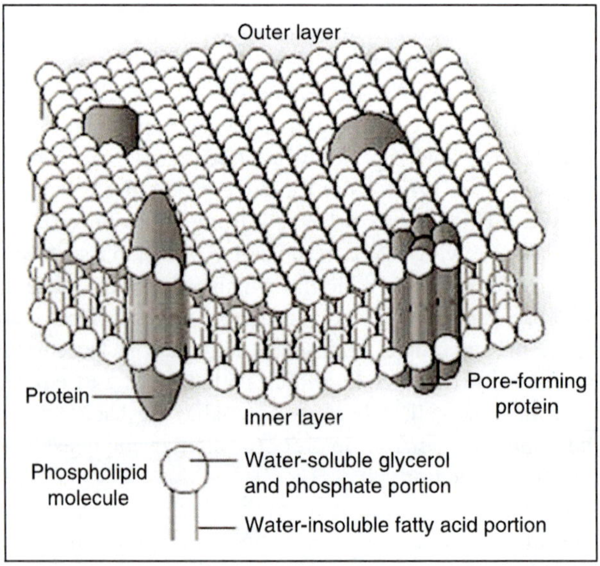

Fig. 7 Details of the phospholipid bilayer and phospholipid in a prokaryotic cell membrane (Kaiser 1999) http://student.ccbcmd.edu/~gkaiser/goshp.html

glycocalyx and cell wall and intracellular cytoskeleton. Cell signaling is part of a complex system of communication that determines basic cellular activities and coordinates cell actions. The ability of bacterial cells to perceive and correctly respond to their microenvironment is the basis of development, tissue repair, and immunity as well as normal tissue homeostasis. Communication also occurs between the cells. In the human gastrointestinal tract, bacteria exchange signals with each other and also with human epithelial and immune system cells. Many cell signals are carried by molecules that are released by one cell and move to make contact with another cell. In bacteria, the plasma membrane is also associated with the breakdown of nutrients and the synthesis of ATP.

5.2.2 Cytoplasm

The cytoplasm is a thick, semitransparent and gel-like liquid residing inside the plasma membrane and holds membrane organelles. It contains dissolved and suspended materials such as proteins (mostly enzymes), carbohydrates, lipids, and organic ions. It is the site of most cellular activities such as metabolic pathways including glycolysis, and processes such as cell division. The inner, granular mass is called the endoplasm and the outer, clear and glassy layer is called the cell cortex or the ectoplasm. Species of probiotic prokaryotes like any other cell do posses a cytoplasm and include some of the best known strains used such as *B. animalis* ssp. *lactis* BB12 and *L. acidophilus* NCFM.

Constituents of cell cytoplasm – The cytoplasm has three major elements: the cytosol, organelles, and inclusions. The cytosol is the portion within membrane-bound organelles, making up about 70% of the cell volume and is composed of water, salts, and organic molecules. The cytosol is a translucent fluid in which the plasmic elements are suspended. It also contains the protein filaments that make up the cytoskeleton, as well as soluble proteins and small structures such as ribosomes, proteasomes, and the mysterious vault complexes (van Zon et al. 2003). The inner, granular and more fluid portion of the cytoplasm is referred to as endoplasm. Some major organelles that are suspended in the cytosol are the mitochondria, the endoplasmic reticulum, the Golgi apparatus, vacuoles, and lysosomes. The cytoplasmic inclusions are small particles of insoluble substances suspended in the cytosol. A huge range of inclusions exist in different cell types but the most prominent in prokaryotic cells are the lipid droplets, also known as adipocytes, where lipids (fatty acids), proteins, and sterols are stored (Murphy 2001). The lipid droplets make up much of the volume of adipocytes, which are the specialized lipid-storage cells. All prokaryotic probiotic bacteria do possess cell cytoplasm from which all metabolic processes take place.

Cytoplasmic inclusions – These are reserve deposits of chemicals within the cytoplasm that tend to accumulate when nutrients are plentiful. They include:

(a) Metachromatic granules (volutin granules) – they are phosphate reserves that can be used in synthesizing ATP for metabolic processes within the cell.

(b) Polysaccharide granules – they are glycogen granules and starch granules.
(c) Lipid inclusions – the most common is poly-β-hydroxybutyric acid (PHB) which is found in *Bifidobacterium, Bacillus,* and others.
(d) Sulfur granules – these are found in bacteria whose metabolism involves oxidation of sulfur thus depositing these as energy reserve. Some *Bacilli* are an example of sulfur-containing granules.
(e) Carboxysomes – are found in bacteria that use carbon dioxide as their only carbon source from these granules, requiring an enzyme to perform this process.
(f) Gas vacuoles – these contain gases within the bacterial cell. These may aid floatation.
(g) Magnetosomes – inclusions of iron oxide that act like magnets. Function uncertain, but it is thought they can decompose hydrogen peroxide.

5.2.3 Nucleiod

Prokaryotic cells do not have a nucleus. Instead, they have a *nucleus-like* structure in the irregularly shaped region of the cytoplasm known as nucleoid. It is therefore the nuclear body of a prokaryotic (bacterial) cell, which is usually composed of a single molecule of circular, chromosomal DNA. As shown in Fig. 8, the nucleoid has a nuclear material without a nuclear membrane and it is where the genetic material is localized (Thanbichler et al. 2005).

The nucleoid is largely composed of about 60% DNA while the rest are proteins and a small amount of RNA. The latter two constituents are likely to be mainly messenger RNA and the transcription factor proteins found regulating the bacterial genome. Proteins helping to maintain the super-coiled structure of the nucleic acid are known as nucleoid proteins or nucleoid-associated proteins. The chromosome's DNA twists into a tight helix and is connected to the plasma membrane. Proteins of the plasma membrane are believed to be responsible for the replication of DNA. The DNA-binding proteins often use other mechanisms to promote compaction which can lead to the occurrence of DNA bending.

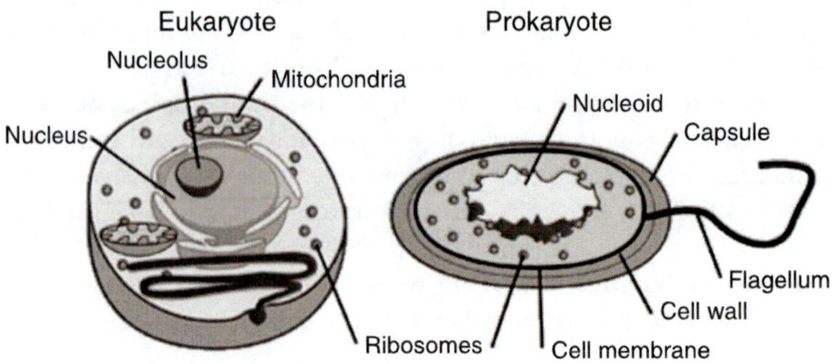

Fig. 8 Prokaryotic cell showing the nucleoid (Browning et al. 2010)

5.2.4 Ribosomes

Ribosomes are the components of cells where the protein synthesis takes place. In prokaryotic cells, the ribosomes are mostly scattered in the cytoplasm. The process involves utilization of DNA to make RNA, and subsequently used to make protein. The DNA sequence in genes is copied into a messenger RNA (mRNA). The information in the mRNA is then read and used to create proteins through a process known as translation in which the mRNA is used as a template for the correct sequence of amino acids in a particular protein. The amino acids are attached to transfer RNA (tRNA) molecules, which enter one part of the ribosome and bind to the messenger RNA sequence. The attached amino acids are then joined together by another part of the ribosome, which then moves along the mRNA, "reading" its sequence and producing a chain of amino acids. Ribosomes are divided into two subunits, one larger than the other and each subunit consist of a protein and a ribosomal RNA. The smaller subunit binds to the mRNA, while the larger subunit binds to the tRNA and the amino acids. When a ribosome finishes reading an mRNA, these two subunits split apart.

5.2.5 Plasmids

These are gene carrying, circular DNA molecules separate from the chromosome which are not involved in reproduction. They are found in most bacteria including those considered probiotic. They usually contain 5–100 genes, which code for proteins that may give the cell resistance to certain antibiotics, to toxic metals, ability to produce certain toxins, or to synthesize enzymes not included in the chromosomal genes. Proteins coded on plasmid DNA are usually not required for cells under ideal conditions, but they can give the cell advantages for survival in a given environment, e.g., in the gut or in the intestinal environment. Copies of plasmids can be transferred from one cell to another and once a cell has a plasmid, it will replicate the plasmid and pass a copy on to each daughter cell during cell division.

5.2.6 Endospores

These are dormant cells, formed when certain genera (mainly *Bacillus* which include species and strains considered probiotic) find themselves facing adverse conditions such as lack of nutrients, drying, presence of toxic materials, or any unfavorable change in the environment. Endospores once formed can withstand extreme conditions and survive for apparently unlimited periods of time. They can perform normal cell functions such as taking in nutrients, carrying out metabolic activities, and reproduction. They too are affected by pH variations, heat, drying, lack of nutrients, and most of them are relatively easy to kill. The process of formation of endospores is referred to as sporulation or sporogenesis.

6 Reproduction of Prokaryotic Probiotic Bacteria

Bacteria and archaea as prokaryotes reproduce through asexual reproduction, usually by binary fission or budding. The genetic exchange and recombination still occur, but through a form of horizontal gene transfer. This is different from the replicative process in which the DNA material is transferred between two cells through bacterial conjugation.

Asexual reproduction is therefore the reproduction which does not involve meiosis, ploidy reduction, or fertilization. Only one parent is involved in asexual reproduction. A more stringent definition is agamogenesis which refers to reproduction without the fusion of gametes. Asexual reproduction is the primary form of reproduction for single-celled organisms such as the archaea, bacteria, and protists. While all prokaryotes reproduce asexually (without the formation and fusion of gametes), mechanisms for lateral gene transfer such as conjugation, transformation, and transduction are sometimes likened to sexual reproduction (Narra and Ochman 2006).

Binary fission, also referred to as prokaryotic fission, is the form of asexual reproduction and cell division used by all prokaryotes. This process results in the reproduction of a living prokaryotic cell by division into two parts both of which have the potential to grow to the size of the original cell as shown in Fig. 9.

Binary fission begins with DNA replication which opens up into a replication bubble (note: prokaryotic DNA replication usually has only one origin of replication, whereas eukaryotes have multiple origins of replication). The replication bubble separates the DNA double strand, each strand acting as template for synthesis of a daughter strand by semiconservative replication, until the entire prokaryotic DNA is duplicated. After this process, cell growth occurs where each circular DNA strand attaches to the cell membrane. The cell then elongates, causing

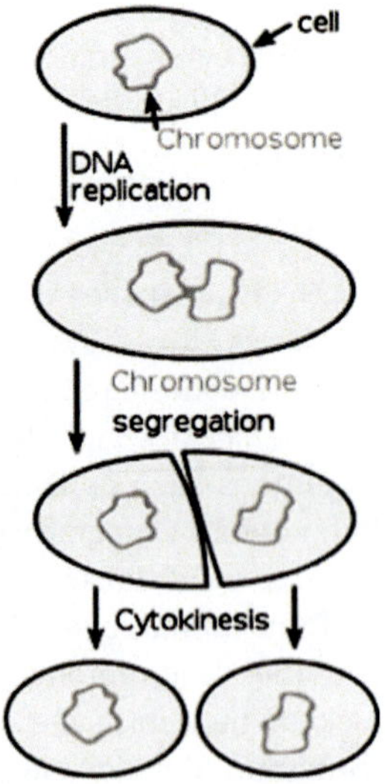

Fig. 9 Process of prokaryotic binary fission (Harry et al. 2006)

the two chromosomes to separate. Cell division in bacteria is controlled by dozen proteins that collect around the site of division known as the FtsZ. There, these proteins direct assembly of the division septum. The cell wall and plasma membrane starts growing transversely from near the middle of the dividing cell and the parent cell separates into two nearly equal daughter cells, each having a nuclear body (Weiss 2004). The cell membrane then invaginates (grows inwards) and splits the cell into two daughter cells, separated by a newly grown cell plate. There are some other forms of asexual reproduction such as budding, vegetative reproduction, sporogenesis, fragmentation, parthenogenesis, agamogenesis, apomixes, and nucellar embryony which does not apply to prokaryotes.

7 The Taxonomy of Prokaryotic Probiotic Bacteria

Taxonomy may be defined as the process of cataloguing biodiversity, as it is the scientific study of the diversity of organisms with the ultimate goal of characterizing and arranging them in an orderly manner (Schleifer and Ludwig 1994). Classification, identification, and nomenclature are the three separate but related subdisciplines of taxonomy. *Classification* is the process of clustering organisms into taxonomic groups (*taxa*) on the basis of similarities or relationships. *Nomenclature* is the assignment of names to the taxonomic groups according to international rules. Finally, *identification* is the process of determining the belonging of a new isolate to one of the established and named taxa (Staley and Krieg 1989). Bacterial taxonomy is driven by the innate human desire to recognize and understand the world around them and this requires a logical ordering of items (Rosselló-Mora 2005).

In the hierarchical way of classification, only two prokaryotic domains are presently recognized, namely, the Archaea and Bacteria. Domains are divided into phyla and the levels below the phylum are classes, orders (or subdivisions, depending on the group), families, genera, and species as shown in Table 1 of chapter "New health potentials of orally-consumed probiotic microorganisms." Different taxonomic levels are characterized by different suffixes in the taxon names. As microorganisms have too simple a structure and too few informative characters compared with higher organisms (e.g., morphology), progress in bacterial taxonomy has always been dependent on advances in technology such as *molecular* data. These data have been made possible with the discovery of DNA as the depositary material of genetic information and the improvement of techniques suitable to deal with the smallest components of cells. However, different methods of analysis have different resolution power, therefore the complete investigation of the identity of a microorganism could be carried out only through the comparison of results of a large number of techniques referred to as the *polyphasic* approach to bacterial systematic. This approach is based on the fact that the more data one has and to compare, the more complete and accurate the identification (Colwell 1970; Vandamme et al. 1996). The definition and naming of

Table 2 An example of hierarchal classification of probiotic bacteria: *Bifidobacterium* and *Lactobacillus*

Taxon	Name	Name
Dormain	Bacteria	Bacteria
Phylum	*Actinobacteria*	*Firmicutes*
Class	*Actinobacteria*	*Bacilli*
Sub-class	*Actinobacteridae*	*Actinobacteridae*
Order	*Bifidobacterales*	*Lactobacillales*
Family	*Bifidobacteriaceae*	*Lactobacillaceae*
Genus	*Bifidobacterium*	*Lactobacillus/Paralactobacillus/Pediococcus*
Species	*Bifidobacterium animalis*	*Lactobacillus acidophilus/casei*

Source: *Taxonomic outline of the prokaryotes* (Garrity et al. 2004)

species is necessary for practical reasons; however, it should be noted that microbiologists work with *strains*, as the strain is the microbial individual. It is possible that properties can be assigned to species when a number of different strains are studied. However, when tests are performed and properties attributed, they are, first of all, characteristics of the strain under study.

Species of the genera *Lactobacillus* and *Bifidobacterium* are some of the most important taxa used in the food industry, feed production, and in human nutrition. They are mostly renowned for the probiotic properties exhibited by some of their strains.

The taxonomic orders for microorganisms are species, genus, family, order, class, phylum, and domain. Except for domain these are rarely used. Table 2 shows the hierarchal taxonomic arrangement and identification of two important and well-studied genera of prokaryotic probiotic, namely, *Bifidobacterium* and *Lactobacillus*. The most significant and well-known prokaryotic probiotic bacteria belong to the two phylums, namely, *Actinobacteria* and *Firmicutes*. Whereas Bifidobacteria, which is the most predominant microorganism in the gastrointestinal tract of humans belong to the phylum *Actinobacteria,* the other known probiotic species such as *L. acidopillus* and *L. casei*, both under the genus *Lactobacillus* belong to the phylum *Firmicutes*. *Acidophilus* is the most well-known probiotic bacteria of which there are many strains that offer health benefits.

7.1 *The Genus* Lactobacillus

The genus *Lactobacillus* belongs to the LAB, a definition which groups Gram-positive, non-spore-forming, catalase-negative bacterial species able to produce lactic acid as main end-product of the fermentation of carbohydrates. Catalase is a common enzyme found in nearly all living organisms that are exposed to oxygen, where it decomposes hydrogen peroxide to water and oxygen (Chelikani et al. 2004). Catalase has one of the highest turnover numbers of all enzymes, i.e., one molecule of catalase can convert 40 million molecules of hydrogen peroxide to water and oxygen each second (Goodsell 2004).

Morphologically, they present as rod-shaped microorganisms although they can appear circular hence sometimes referred to as coccobacilli. They are fermentative, microaerophylic, and chemo-organotrophic, requiring rich media to grow. Considering DNA base composition of the genome, they usually show a GC content (guanine–cytosine content – which is the percentage of nitrogenous bases on a DNA molecule, from a possibility of four different ones, also including adenine and thymine) of between 32 and 51 mol%. They are almost ubiquitous: they are found in environments where carbohydrates are available, such as food (dairy products, fermented meat, sour doughs, vegetables, fruits, beverages), respiratory, GI and genital tracts of humans and animals, and in sewage and plant material. Based on the *Taxonomic Outline of the Prokaryotes*, given in Table 1 of chapter "New health potentials of orally-consumed probiotic microorganisms," the genus *Lactobacillus* belongs to the phylum *Firmicutes,* class *Bacilli,* order *Lactobacillales,* family *Lactobacillaceae* and its closest relatives, being grouped within the same family, are the genera *Paralactobacillus* and *Pediococcus.* Currently, the genus *Lactobacillus* includes 106 validly described species, making it, the most numerous genus of the order *Lactobacillales.* Moreover, one more species (*Lactobacillus tucceti,* Chenoll et al. 2006) has been described but it has not been validated yet, while few other species have description but are yet to be published including *Lactobacillus composti, Lactobacillus farraginis*, and *Lactobacillus parafarraginis* (Endo and Okada 2007a, b) and *Lactobacillus secaliphilus* (Ehrmann et al. 2007).

Table 3 presents a summary of seven already validated species in the genus *Lactobacillus,* comprising of two subspecies or more. There are other species such as *L. salivarius* which have not been substantially separated in two subspecies, therefore leading to the amendment of the species description (Li et al. 2006).

Table 3 Summary of the genus *Lactobacillus* and its species and subspecies

Genus	Species	Subspecies
		L. aviaries subsp. *aviarius*
	Lactobacillus aviarius	*L. aviarius* subsp. *araffinosus*
		L. coryniformis subsp. *coryniformis*
	Lactobacillus coryniformis	*L. coryniformis* subsp. *torquens*
		L. delbrueckii subsp. *delbrueckii*
		L. delbrueckii subsp. *bulgaricus*
		L. delbrueckii subsp. *indicus*
	Lactobacillus delbrueckii	*L. delbrueckii* subsp. *lactis*
		L. kefiranofaciens subsp. *kefiranofaciens*
	Lactobacillus kefiranofaciens	*L. kefiranofaciens* subsp. *kefirgranum*
		L. paracasei subsp. *paracasei*
	Lactobacillus paracasei	*L. paracasei* subsp. *tolerans*
		L. plantarum subsp. *plantarum*
	Lactobacillus plantarum	*L. plantarum* subsp. *argentoratensis*
		L. sakei subsp. *sakei*
Lactobacillus	*Lactobacillus sakei*	*L. sakei* subsp. *carnosus*

Genera *Paralactobacillus* and *Pediococcus* are constituted by one and ten species, respectively, and no subspecies has yet been identified.

Out of the 106 validly described species, only 56 species of the genus *Lactobacillus* have been identified to date. From this group, the Lactobacilli species widely being used as probiotics include *L. acidophilus, Lactobacillus brevis, Lactobacillus bulgaricus, L. casei, Lactobacillus cellobiosus, Lactobacillus crispatus, Lactobacillus curvatus, Lactobacillus fermentum, Lactobacillus GG (Lactobacillus rhamnosus* or *L. casei* subspecies *rhamnosus), Lactobacillus gasseri, Lactobacillus johnsonii, Lactobacillus plantarum,* and *Lactobacillus salivarus.* Other probiotic strains of *Lactobacillus* are *L. acidophilus* BG2FO4, *L. acidophilus* INT-9, *Lactobacillus plantarum* ST31, *Lactobacillus reuteri, L. johnsonii* LA1, *L. acidophilus* NCFB 1748, *L. casei* Shirota, *L. acidophilus* NCFM, *L. acidophilus* DDS-1, *Lactobacillus delbrueckii* subspecies *delbrueckii, L. delbrueckii* subspecies *bulgaricus* type 2038, *L. acidophilus* SBT-2062, *Lactobacillus brevis, L. salivarius* UCC 118, and *Lactobacillus paracasei* subsp *paracasei F19.* There are many strains of *Lactobacillus* and bifidobacteria in different culture collection centers worldwide that are used in probiotic and prebiotic research.

7.2 The Genus Bifidobacterium

Bifidobacteria are Gram-positive polymorphic branched rods that occur singly, in chains or clumps. They have various shapes, including short, curved rods, club-shaped rods, and bifurcated Y-shaped rods. Their name is derived from the observation that they often exist in a Y-shaped or bifid form. They are nonspore forming, nonmotile, and nonfilamentous. They are anaerobic and chemoorganotrophs, having a fermentative type of metabolism. They produce acid but not gas from a variety of carbohydrates. They are catalase negative, with some exceptions such as *Bifidobacterium indicum* and *Bifidobacterium asteroides* when grown in presence of air. Their genome GC content varies from 42 to 67 mol% (Biavati and Mattarelli 2001). They are *saccharolytic* organisms that produce acetic and lactic acids without generation of CO_2, except during degradation of gluconate. They are also classified as LAB. For a long time bifidobacteria were regarded as relatives to lactobacilli based on the correlation supported by the analysis of the murine/ peptodoglycan structure (Kandler and Lauer 1974). Bifidobacteria degrade hexoses using a unique metabolic pathway referred to as the fructose-6-phosphate pathway, also known as the bifid shunt. The key enzyme involved in this pathway is fructose-6-phosphoketolase, which was considered a taxonomic character for the identification on the genus level (Biavati and Mattarelli 2001). However, due to the reclassification of *Bifidobacterium* species into new genera, it can now be considered a taxonomic marker for the Family *Bifidobacteriaceae.* A complete survey of identification methods for bifidobacteria has been compiled by Ventura et al. (2004).

They occur in animal and human habitats and in particular they have been isolated from feces, rumen of cattle, sewage, human vagina, dental caries, and honey bee intestine. In the gastrointestinal track, they are the most predominant bacteria. Newborns, especially those that are breast-fed, are colonized with bifidobacteria within days after birth. As shown too in Table 3 of chapter "New health potentials of orally-consumed probiotic microorganisms," the genus *Bifidobacterium* belongs to the phylum *Actinobacteria*, class *Actinobacteria*, subclass *Actinobacteridae*, order *Bifidobacteriales*, family *Bifidobacteriaceae*. There are other genera which belong to this family, but are considered probiotic and include *Aeriscardovia, Falcivibrio, Gardnerella, Parascardovia,* and *Scardovia.*

Currently, 30 species of bifidobacteria given in Table 4 have been isolated, validated, and identified. Out of these, eight species currently being used as probiotics include *Bifidobacterium adolescentis, Bifidobacterium bifidum, B. animalis, Bifidobacterium thermophilum, Bifidobacterium breve, B. longum, Bifidobacterium*

Table 4 Summary of the genus *Bifidobacterium* and its species and subspecies

Genus	Species	Subspecies
	Bifidobacterium adolescentis	
	Bifidobacterium angulatum	
		B. animalis subsp. *animalis*
	Bifidobacterium animalis	*B. animalis* subsp. *lactis*
	Bifidobacterium asteroides	
	Bifidobacterium bifidum	
	Bifidobacterium boum	
	Bifidobacterium breve	
	Bifidobacterium catenulatum	
	Bifidobacterium choerinum	
	Bifidobacterium coryneforme	
	Bifidobacterium cuniculi	
	Bifidobacterium dentium	
	Bifidobacterium longum	
	Bifidobacterium magnum	
	Bifidobacterium merycicum	
	Bifidobacterium minimum	
	Bifidobacterium pseudocatenulatum	
		B. pseudolongum subsp. *pseudolongum*
	Bifidobacterium pseudolongum	*B. pseudolongum* subsp. *globosum*
	Bifidobacterium psychraerophilum	
	Bifidobacterium pullorum	
	Bifidobacterium ruminantium	
	Bifidobacterium saeculare	
	Bifidobacterium scardovii	
	Bifidobacterium subtile	
		B. thermacidophilum subsp. *thermacidophilum*
	Bifidobacterium thermacidophilum	*B. thermacidophilum* subsp. *porcinum*
Bifidobacterium	*Bifidobacterium thermophilum*	

infantis, and *Bifidobacterium lactis*. Specific strains of bifidobacteria used as probiotics include *B. breve strain Yakult*, *B. breve* RO7O, *B. lactis* Bb12, *B. longum* RO23, *B. bifidum* RO71, *B. infantis* RO33, *B. longum* BB536, and *B. longum* SBT-2928. These species with probiotic functions are not phylogenetically related, hence underlining the strain specificity of those characteristics.

7.3 Other Genera

Lactococcus – *Lactococci* are Gram-positive facultative anaerobes. They are also classified as LAB. *L. lactis* (formerly known as *Streptococcus lactis*) is found in dairy products and is commonly responsible for the souring of milk. Lactococci that are used or are being developed as probiotics include *L. lactis*, subspecies *cremoris* (*S. cremoris*), *L. lactis* subspecies *lactis* NCDO 712, *L. lactis* subspecies *lactis* NIAI 527, *L. lactis* subspecies *lactis* NIAI 1061, *L. lactis* subspecies *lactis* biovar diacetylactis NIAI 8W, and *L. lactis* subspecies *lactis* biovar diacetylactis ATCC 13675.

S. thermophilus – *Streptococcus thermophilus* is a Gram-positive facultative anaerobe. It is a cytochrome-, oxidase-, and catalase-negative organism that is nonmotile, nonspore forming, and homofermentative. *S. thermophilus* is an alpha-hemolytic species of the viridans group. It is also classified as LAB. *S. thermophilus* is found in milk and milk products. It is a probiotic and used in the production of yogurt. *Streptococcus salivarus* subspecies *thermophilus* type 1131 is a probiotic strain of *S. thermophilus*.

Enterococcus – *Enterococci* are Gram-positive, facultative anaerobic cocci of the *Streptococcaceae* family. They are spherical to ovoid and occur in pairs or short chains. Enterococci are catalase-negative, nonspore forming and usually nonmotile. Enterococci are part of the intestinal microflora of humans and animals. *E. faecium* SF68 is a probiotic strain that has been used in the management of diarrheal illnesses.

8 Prokaryotic Bacteria and Interaction with the Environment

The bacterial cell like any other organism can be vulnerable and yet has to adapt in any given environment. For prokaryotic probiotic bacteria which only have a cell wall but with a thicker peptidoglycan layer, the lack of a cell capsule would appear to make them more vulnerable. It is therefore suggested that Gram-positive bacteria are more susceptible to the external factors such as lysozymes than Gram-negative bacteria. In addition, if the Gram-positive cell loses its wall and is in hypotonic surroundings, it will undergo lysis (burst) due to entrance of water. If the surroundings are isotonic, the Gram-positive cell can live with no cell wall as a protoplast. Lysozyme does not affect Gram-negative cell walls to the same extent.

A Gram-negative cell treated with lysozyme becomes a spheroplast. It retains its outer membrane and remnants of the cell wall, thus become being able to adapt better to lysozyme attack compared to Gram positive. Gram-negative organisms are not usually highly susceptible to antimicrobials like penicillin because the outer membrane gives them some protection. The plasma membrane (cytoplasmic) is also vital to the survival of the bacterial cell. Some chemical compounds such as aldohols and quaternary ammoniums as well as antibiotics can damage the plasma membrane, thus destroying the bacterial cell.

Prokaryotic probiotic cells obtain their nutrients for metabolism by allowing movement of materials (nutrients) across cell membranes as determined by their energy requirements. Consequently, there are passive processes and active processes. A passive process does not require energy and the chemicals can only move from an area with greater concentration to lesser concentration, thus following the concentration gradient. There are two types of passive processes.

(a) Simple Diffusion – in which there is a net movement of molecules or ions from an area of higher concentration to an area of lower concentration. This movement continues until the two concentrations become equal. Many small molecules enter or leave cells this way.
(b) Osmosis – this is really still simple diffusion but this time water (solvent) molecules do the moving instead of the solutes (dissolved material). Osmosis occurs across a membrane that is permeable to water but relatively impermeable to the solutes. Water molecules move from an area of higher water concentration to an area of lower water concentration. There are three ways osmosis can affect a bacterial cell depending on the surroundings.

i Isotonic surroundings – Refers to an environment where water and solute concentration are equal inside the cell and in the solution that surrounds the cell. Since concentrations are already equal, no net movement occurs.

$$\text{Water} = \text{Water}$$
$$\text{Solutes} = \text{Solutes}$$

ii Hypotonic surroundings – Refers to an environment where the water concentration is higher and solute concentration is lower outside the cell than inside the cell. Bacteria often find themselves in these surroundings. In these conditions, water would enter the cell, even enough to cause the cell to burst, if not for the cell wall.

$$\text{Lower Water} \leftarrow \text{Higher Water}$$
$$\text{Higher Solute} \rightarrow \text{Lower Solute}$$

iii Hypertonic surroundings – Refers to when water concentration is lower and solute concentration higher outside the cell than inside. Water would tend to leave the cell. If enough leaves, this could kill the cell. This is the principal being used in food preservation such as in salted meats and jellies to prevent food spoilage by bacteria.

$$\text{Higher Water} \leftarrow \text{Lower Water}$$
$$\text{Lower Solute} \rightarrow \text{Higher Solute.}$$

(c) Facilitated diffusion – Refers to when the substance to be transported combines with a plasma membrane carrier protein called a transporter. Transporters being protein are specific to the substance being transported. As the substance molecule moves into the open end of the transporter molecule, its presence causes the carrier molecule to change shape, the open end of the carrier allows the substance to move in. The carrier then transports the substance across the cell membrane into the cytoplasmic membrane and releases the substance inside the cell. The carrier protein then returns to its original shape, ready to repeat the process. Movement can continue as long as there is a need to move the nutrient from higher concentration (outside the cell) to lower concentration (inside the cell). However, passive processes are less important in prokaryotic bacterial cells than in eukaryotic cells since these cells frequently find themselves in surroundings where nutrients are in low concentration, so passive processes are unable to meet their needs. Another reason is that only relatively small molecules can enter cells by passive processes. In some cases, bacteria produce enzymes and release them out into their surroundings (extracellular enzymes) to break down larger molecules into a size that can enter into the cell passively. Example of probiotic bacteria is *L. lactis* which uses its lactase to convert lactose into glucose and galactose for metabolic processes.

On the other hand, there can be an active transport process across the bacterial cell walls. Active process, however, requires energy in the form of adenosine triphosphate (ATP). As a result of this, they can move substances from lower concentration to higher concentration (against the gradient). There are two types of active processes.

i *Active transport* – this is quite similar to facilitated diffusion in that a specific transporter protein molecule in the plasma membrane is required. The difference is that ATP energy is required, and that substances can be moved from lower concentrations to higher concentrations. In this process, the substance is not changed as it enters the cell.

ii *Group translocation* – this occurs only in prokaryocytes. It is a special form of active transport in which the substance being transported is chemically modified while it is transported. One example is glucose, which is changed to glucose phosphate as it is delivered from outside to inside the cell. The plasma membrane is completely impermeable to glucose phosphate, so it cannot leave the cell, even if the concentration inside the cell becomes quite high. Glucose then acts as a carbon source to meet the metabolic requirements of the cell.

Table 5 Characteristic differences between prokaryotes and eukaryotes

Characteristic	Prokaryotic	Eukaryotic
Size	Smaller – typical size is 0.2–2 μm in diameter	Larger – typical size is 10–100 μm in diameter
Nucleus	No nuclear membrane or nucleoli (nucleoid)	True nucleus with nuclear membrane and nucleoli
Membrane-enclosed organelles	Absent	Many, including lysosomes, Golgi complex, ER, mitochondria, and chloroplasts
Flagella	Consist of two protein building blocks	Complex; consist of multiple microtubules
Phagocytosis	None	Some can carry on phagocytosis
Glycocalyx	Present as a capsule or slime layer	Present in some cells that lack a cell wall
Cell wall	Usually present, chemically complex (peptidoglycan)	When present, usually simple, NO peptidoglycan
Plasma membrane	No carbohydrates, almost all lack sterols	Sterols and carbohydrates incorporated
Cytoplasm	No cytoskeleton	Cytoskeleton
Ribosomes	Smaller size (70S)	Larger size (80S) except those within organelles
Chromosome (DNA) arrangement	Single circular chromosome, proteins associated with DNA are not histones	Multiple linear chromosomes with histone proteins
Cell division	Binary fission No meiosis; transfer fragments of DNA only	Mitosis Meiosis

9 Relationship Between Prokaryotes and Eukaryotes

The cells of prokaryotic and eukaryotic share numerous characteristics as shown in Table 5. However, the important distinctions among them are in terms of size, existence of a true nucleus or not, occurrence of membrane enclosed organelles, flagella, ability to initiate phagocytosis, ribosome sizes and manner of cell division amongst others.

Prokaryotes have a larger surface-area-to-volume ratio giving them a higher metabolic rate, a higher growth rate and consequently a shorter generation time compared to Eukaryotes (Campbell 2003).

References

Akiyama K, Takase M, Horikoshi K, Okonogi S (2001) Production of galactooligosaccharides from lactose using β-glucosidase from *Thermus sp. Z-1*. Biosci Biotechnol Biochem 65:438–441

Baross JA, Holden JF (1996) Enzymes and proteins from hyperthermophilic microorganisms. Adv Protein Chem 48:1–34

Biavati B, Mattarelli P (2001) The family *Bifidobacteriaceae*. In: The prokaryotes, Release 3.7

Browning DF, Grainger DC, Busby SJW (2010) Effects of nucleoid-associated proteins on bacterial chromosome structure and gene expression. Curr Opin Microbiol 13(6):773–780

Bruins ME, Strubel M, Lieshout JFT, Janssen AEM, Boom RM (2003) Oligosaccharide synthesis by the hyperthermostable beta-glucosidase from *Pyrococcus furiosus*: kinetics and modelling. Enzyme Microb Technol 33:3–11

Campbell N (2003) Biology: concepts and connections. Pearson Education, San Francisco

Chaban B, Ng SY, Jarrell KF (2006) Archaeal habitats from the extreme to the ordinary. Can J Microbiol 52:73–116

Chelikani P, Fita I, Loewen PC (2004) Diversity of structures and properties among catalases. Cell Mol Life Sci 61(2):192–208

Chenoll E, Carmen MM, Aznar R (2006) *Lactobacillus tucceti* sp. nov., a new lactic acid bacterium isolated from sausage. Syst Appl Microbiol 29:389–395

Colwell RR (1970) Polyphasic taxonomy of the genus *Vibrio*: numerical taxonomy of *Vibrio cholerae, Vibrio parahaemolyticus,* and related *Vibrio* species. J Bacteriol 104:410–433

Demchick PH, Koch AL (1996) The permeability of the wall fabric of *Escherichia coli* and *Bacillus subtilis*. J Bacteriol 178 (3): 768–73. PMID 8550511. PMC 177723. http://jb.asm.org/cgi/reprint/178/3/768. Accessed 10 Dec 2010

Ehrmann MA, Brandt M, Stolz P, Vogel RF, Korakli M (2007) *Lactobacillus secaliphilus* sp. nov., isolated from type II sourdough fermentation. Int J Syst Evol Microbiol 57:745–750

Endo A, Okada S (2007a) *Lactobacillus farraginis* sp. nov. and *Lactobacillus parafarraginis* sp. nov., heterofermentative lactobacilli isolated from a compost of distilled shochu residue. Int J Syst Evol Microbiol 57:708–712

Endo A, Okada S (2007b) *Lactobacillus composti* sp. nov., a lactic acid bacterium isolated from a compost of distilled shochu residue. Int J Syst Evol Microbiol 57:870–872

Faruque SM, Nair GB (ed) (2008). *Vibrio cholerae*: genomics and molecular biology. Caister Academic Press. ISBN 978-1-904455-33-2. http://www.horizonpress.com/vib. Accessed 10 Dec 2010

Fox A (2010) Culture and identification of infectious agents. http://pathmicro.med.sc.edu/fox/culture. Accessed 13 Dec 2010

Fricke WF, Seedorf H, Henne A, Kruer M, Liesegang H, Hedderich R, Gottschalk G, Thauer RK (2006) The genome sequence of *Methanosphaera stadtmanae* reveals why this human intestinal archaeon is restricted to methanol and H_2 for methane formation and ATP synthesis. J Bacteriol 188:642–658

Fuerst JA (2005) Intracellular compartmentation in planctomycetes. Annu Rev Microbiol 59:299–328

Garrity GM, Bell JA, Lilburn TG (2004) Taxonomic outline of the prokaryotes. Bergey's manual of systematic bacteriology, 2nd edn, Release 5.0, Springer, New York. DOI: http://dx.doi.org/10.1007/bergeysoutline200405. Accessed 10 Dec 2010

Goodsell DS (2004). "Catalase". Molecule of the Month. RCSB Protein Data Bank. http://www.rcsb.org/pdb/static.do?p=education_discussion/molecule of the month/pdb571.html. Accessed 10 Dec 2010

Hansson T, Kaper T, van der Oost J, de Vos WM, Adlercreutz P (2001) Improved oligosaccharide synthesis by protein engineering of β-glucosidase CelB from hyperthermophilic *Pyrococcus furiosus*. Biotechnol Bioengr 73:203–210

Harry E, Monahan L, Thompson L (2006) Bacterial cell division: the mechanism and its precision. Int Rev Cytol 253:27–94

http://users.rcn.com/jkimball.ma.ultranet/BiologyPages/A/Archaea.html. Accessed 03 Dec 2010

http://www2.nemcc.edu/bkirk/Template%201/MICROCHAPTER4NOTES.htm. Accessed 01 Dec 2010

Kaiser GE (1999) Arrangements of Cocci. Dr. Kaiser's Microbiology Homepage. http://student.ccbcmd.edu/~gkaiser/goshp.html. Accessed 01 Dec 2010

Kandler O, Lauer E (1974) Neuere Vorstellungenzur Taxonomic der Bifidobacterien. Zentralbl. Bakteriol.Parasitenkd. Infektionskr. Hyg Abt 1 Orig Reihe A228: 29–45

Kerfeld CA, Sawaya M, Tanaka S, Nguyen CV, Phillips M, Beeby M, Yeates T (2005) Protein structures forming the shell of primitive bacterial organelles. Science 309(5736):936–938

Komeili A, Li Z, Newman DK, Jensen GJ (2006) Magnetosomes are cell membrane invaginations organized by the actin-like protein MamK. Science 311(5758):242–245

Li Y, Raftis E, Canchaya C, Fitzgerald GF, van Sinderen D, O'Toole PW (2006) Polyphasic analysis indicates that *Lactobacillus salivarius* subsp. *salivarius* and *Lactobacillus salivarius* subsp. *Salicinius* do not merit separate subspecies status. Int J Syst Evol Microbiol 56:2397–2403

Madigan M, Martinko J (eds) (2005) Brock biology of microorganisms, 11th edn. New York, Prentice Hall, ISBN 0-13-144329-1

Murphy DJ (2001) The biogenesis and functions of lipid bodies in animals, plants and microorganisms. Prog Lipid Res 40(5):325–438

Narra HP, Ochman H (2006) Of what use is sex to bacteria. Curr Biol 16(17):R705–R710

O'Hara AM, Shanahan F (2006) The gut flora as a forgotten organ. EMBO Rep 7:688–693

Otieno DO (2010) Synthesis of β-galactooligosaccharides from lactose using microbial β-galactosidases. Compr Rev Food Sci F 9:471–482

Pace NR (2006) Time for a change. Nature 441(7091):289. doi:10.1038/441289a DOI:dx.doi.org. PMID 16710401

Petzelbauer I, Reiter A, Splechtna B, Kosma P, Nidetzky B (2000) Transgalactosylation by thermostable *β*-glycosidases from *Pyrococcus furiosus* and *Sulfolobus solfataricus.* Binding interactions of nucleophiles with the galactosylated enzyme intermediate makes major contributions to the formation of new & beta-glycosides during lactose conversion. Eur J Biochem 267:5055–5066

Rajilic-Stojanovic M, Smidt H, de Vos WM (2007) Diversity of the human gastrointestinal tract microbiota revisited. Environ Microbiol 9:2125–2136

Report of a Joint FAO/WHO Expert Consultation on Evaluation of Health and Nutritional Properties of Probiotics in Food Including Powder Milk with Live Lactic Acid Bacteria (October 2001). "Health and Nutritional Properties of Probiotics in Food including Powder Milk with Live Lactic Acid Bacteria". Food and Agriculture Organization of the United Nations, World Health Organization. http://www.who.int/entity/foodsafety/publications/fs_management/en/probiotics.pdf. Accessed 01 Dec 2010

Reuter S, Rusborg NA, Zimmermann W (1999) β-Galactooligosaccharide synthesis with β-galactosidases from *Sulfolobus solfataricus*, *Aspergillus oryzae*, and *Escherichia coli.* Enzyme Microb Technol 25:509–516

Rosselló-Mora R (2005) Updating prokaryotic taxonomy. J Bacteriol 187:6255–6257

Ryan KJ, Ray CG (eds) (2004) Sherris medical microbiology (4th edn). McGraw Hill, New York, ISBN 0-8385-8529-9

Salton MRJ, Kim KS (1996) Structure. In: Barron S et al. (eds) Baron's Medical Microbiology, 4th edn. University of Texas Medical Branch. ISBN 0-9631172-1-1. http://www.ncbi.nlm.nih.gov/books/bv.fcgi?rid=mmed.section.289#297. Accessed 25 Nov 2010

Scheffel A, Gruska M, Faivre D, Linaroudis A, Plitzko JM, Schüler D (2006) An acidic protein aligns magnetosomes along a filamentous structure in magnetotactic bacteria. Nature 440 (7080):110–114

Schleifer KH, Ludwig W (1994) Molecular taxonomy: classification and identification. In: Priest FG et al (eds) Bacterial diversity and systematics. Plenum, New York

Schwan T (1996) Ticks and Borrelia: model systems for investigating pathogen-arthropod interactions. Infect Agents Dis 5(3):167–181

Staley JT, Krieg NR (1989) Classification of prokaryotic organisms: an overview. In: Staley JT, Bryant MP, Pfennig N, Holt JG (eds) Bergey's manual of systematic bacteriology, vol 3. Williams & Wilkins, Baltimore, pp 1601–1603

Thanbichler M, Viollier PH, Shapiro L (2005) The structure and function of the bacterial chromosome. Curr Opin Genet Dev 15:153–162
van Zon A, Mossink MH, Scheper RJ, Sonneveld P, Wiemer EA (2003) The vault complex. Cell Mol Life Sci 60(9):1828–1837
Vandamme P, Pot B, Gillis M, De Vos P, Kersters K, Swings J (1996) Polyphasic taxonomy, a consensus approach to bacterial systematics. Microbiol Rev 60:407–438
Ventura M, van Sinderen D, Fitzgerald GF, Zink R (2004) Insights into the taxonomy, genetics and physiology of bifidobacteria. Antonie Van Leeuwenhoek 86:205–223
Weiss DS (2004) Bacterial cell division and the septal ring. Mol Microbiol 54(3):588–597
Woese C (1998) The universal ancestor. Proc Natl Acad Sci USA 95(12):6854–6859
Woese CR, Kandler O, Wheelis ML (1990) Towards a natural system of organisms: proposal for the domains Archaea, Bacteria, and Eucarya. Proc Natl Acad Sci USA 87(12):4576–4579
Zuckerkandl E, Pauling L (1965) Molecules as documents of evolutionary history. J Theor Biol 8(2):357–366

Biology of Eukaryotic Probiotics

Sukanta Kumar Nayak

Contents

1 Introduction 30
 1.1 Eukaryotic Probiotics: An Overview 31
2 Eukaryotic Probiotics Versus Prokaryotic Probiotics 32
3 Sources of Eukaryotic Probiotics 34
4 Taxonomic Characterization of Eukaryotic Probiotics 34
5 Properties of Eukaryotic Probiotics 35
 5.1 Ability to Tolerate Stress 36
 5.2 Ability to Survive in the Gastrointestinal Tract 37
 5.3 Ability to Adhere to the Gastrointestinal Tract 38
6 Beneficial Effects of Eukaryotic Probiotics 38
 6.1 Nutritional Benefits 39
 6.2 Disease Protection 40
7 Modes of Action of Eukaryotic Probiotics 41
 7.1 Trophic Effects on the Gastrointestinal Tract 41
 7.2 Stimulation of Immunity 41
 7.3 Synergistic Activity 42
 7.4 Antagonistic Activity 43
 7.5 Toxin Neutralization Activity 43
8 Safety Issues Related to Eukaryotic Probiotics 44
9 Genetic Manipulation of Eukaryotic Probiotics 44
10 Conclusion 45
References 45

Abstract Probiotics are viable microorganisms which upon ingestion confer health benefits to the host. Any microorganism irrespective of its origin, capable of surviving in the digestive tract of host and exerting such effects can be a candidate.

S.K. Nayak (✉)
Laboratory of Fish Pathology, Department of Veterinary Medicine, Nihon University, 1866 Kameino, Fujisawa, Kanagawa 252-8510, Japan
e-mail: sukantanayak@rediffmail.com

M.-T. Liong (ed.), *Probiotics*, Microbiology Monographs 21,
DOI 10.1007/978-3-642-20838-6_2,

Most of the currently used probiotics belong to prokaryotic origin. Unlike prokaryotes, several eukaryotic microorganisms can also be very useful to animal's health. Since a long time, eukaryotes are used as single cell protein and/or as components of food starters for human and animal consumption throughout the world. Apart from these uses, certain eukaryotic microorganisms are also used as probiotics since they can withstand the harsh milieu of gut and execute beneficial effects in host. While bacterial probiotics are common, only limited eukaryotic probiotics belonging to fungi/moulds/yeasts are used in human and animal practices. Nowadays interest in eukaryotic probiotics is on the rise and in most of the cases, their efficacy and usefulness has been confirmed by firm scientific evidences. Among the eukaryotic probiotics, yeasts especially *Saccharomyces* species are dominant and routinely used in a broad range of hosts. This chapter deals with the occurrence, distribution, taxonomic characterization, and detail modes of action of eukaryotic probiotics with special reference to yeasts in human and other animals.

1 Introduction

The concept of using microorganisms as promoter of life perhaps dates back to almost a century when the Russian scientist Elie Metchnikoff proposed the, then revolutionary, concept of consuming live microorganisms for better health. As the knowledge on such types of microorganisms increases and their beneficial effects are scientifically established, the area of what is now known as "probiotics" has made tremendous progress. The term "probiotic" was originated from the Greek words "pro" and "bios" which mean "for life" (Gismondo et al. 1999) and are often called as promoter of life that help in a natural way to improve the overall health status of host. Probiotics are living microorganisms which upon ingestion in adequate amounts confer health benefits to host (FAO/WHO 2002). Therefore, any nonpathogenic microorganism, irrespective of its origin, capable of surviving inside the gut of host and exerting beneficial effects can be a candidate for probiotic use (Ouwehand et al. 2002). Over the years, probiotics are gaining scientific and commercial interest and are now quite commonplace in our daily life starting from health promoting functional foods to therapeutic, prophylactic, and growth supplements (O'Sullivan 2001; Ouwehand et al. 2002; Boyle et al. 2006). Apart from human beings, probiotics have also profound effect on a broad range of animals ranging from terrestrial to aquatic species. Probiotics, which encompass numerous prokaryotic and eukaryotic microorganisms, have become increasingly popular in the past few decades. Most of the currently used probiotics belong to prokaryotes and lactic acid bacteria, bifidobacteria, enterococci, and several other bacteria are the common examples of prokaryotic probiotics (Tuohy et al. 2003). On the contrary, the use of eukaryotic as probiotics is limited and only a few probiotics of eukaryotic origin are commercially available for human and animal practices (Czerucka et al. 2007; Martins et al. 2007, 2009).

1.1 Eukaryotic Probiotics: An Overview

Eukaryotic microorganisms can be very useful to animal's health as probiotics. There are several food/(feed) grade eukaryotes, such as algae (e.g., *Chlorella, Spirulina* species), fungi (e.g., *Aspergillus*, *Penicillium* species)/yeasts (e.g., *Saccharomyces, Candida, Kluyveromyces, Pichia, Torulopsis* species), which are being consumed by human and animals throughout the world since a very long time. These organisms are mostly used as single cell protein and/or as components of food starters. However, there are certain eukaryotes when supplemented in live conditions through diet are found to execute probiotics like beneficial effects in host. Therefore, it is believed to be a very crucial event in the field of probiotics for the developing new candidate species beyond prokaryotic origin. Nowadays significant interest in eukaryotic probiotics is on the rise and in most of the cases, their efficacy and usefulness has been confirmed by firm scientific evidences. Most of the eukaryotic probiotics used in human and animal practices belong to fungi/yeasts/moulds with yeasts to be the dominant group. The common examples of eukaryotic microorganisms with probiotic properties include the genus *Saccharomyces, Pichia, Metschnikowia, Yarrowia, Candida, Debaryomyces, Isaatchenkia, Kluyveromyces,* and *Aspergillus* (Table 1).

Historically, yeast has been used for fermentation purposes since 1,550 BC. However, nowadays yeasts are a part of nutritional supplements and health food realms due to their established beneficial probiotic effects. Yeast especially the genus *Saccharomyces* is widely used probiotics in human and animals throughout the world (Jakobsen and Narvhus 1996; Lourens and Viljoen 2001;

Table 1 Systemic position of some of the eukaryotic microorganisms showing probiotic activities in different animals including human beings

Phylum	Class	Order	Family	Genus	Species
Basidiomycota	*Urediniomycetes*	*Sporidiales*	*Sporidiobolaceae*	*Cryptococcus*	*mujuensis*
					cuniculi
Ascomycota	*Saccharomycetes*	*Saccharomycetales*	*Dipodascaceae*	*Yarrowia*	*lipolytica*
			Metschnikowiaceae	*Metschnikowia*	*lochheadii*
			Saccharomycetaceae	*Candida*	*humilis*
					pintolepsii
					saitoana
					utilis
					pararugosa
				Debaryomyces	*hansenii*
					cereviae
				Kluyveromyces	*Marxianus*
					lodderae
					lctis
				Pichia	*anomala*
				Saccharomyces	*cerevisiae*
				Isaatchenkia	*orientalis*
	Euromycetes	*Eurotiales*	*Trichocomaceae*	*Aspergillus*	*oryzae*
					niger

Buchl et al. 2010; Moslehi-Jenabian et al. 2010). *Saccharomyces cerevisiae* and *Saccharomyces cerevisiae* var *boulardii* are the only yeast strains commercialized for human uses (Sargent and Wickens 2004; Czerucka et al. 2007; McFarland 2010). *S. cerevisiae* possess the generally recognized as safe (GRAS) status from the Food and Drug Administration (FDA, USA). Similarly, the European Union has also approved five probiotic strains of *S. cerevisiae* (NCYC SC 47, NCYC 1026, CNCM I-1077, CNCM I-1079, and MUCL 39885) for application in animal feed (Buchl et al. 2010). Besides, several other yeasts such as *Candida pintolopesii*, *Candida utilis*, and *Candida saitoana* (Bovill et al. 2001; Leuschner et al. 2004) and many filamentous fungi belonging to *Aspergillus* species (*A. niger*, *A. oryzae*) are also used as probiotics in animal feeds (Lee et al. 2006b). Like *S. cerevisiae*, *A. oryzae* has the GRAS status from FDA and its safety is also supported by the World Health Organization (FAO/WHO 1987).

While various strains belonging to these species are already been used as probiotics, several other yeasts belonging to *Kluyveromyces*, *Isaatchenkia*, and *Debaryomyces* species are emerging as new generation of eukaryotic probiotics. Strains of these species possess all the desired features of an ideal probiotic including antioxidative, antifungal, antibacterial, antiinflammatory, and antitumoral properties (Oh et al. 2002; Diniz et al. 2003; Lopitz-Otsoa et al. 2006; Lee et al. 2008; Chen et al. 2010).

2 Eukaryotic Probiotics Versus Prokaryotic Probiotics

Bacteria are the natural and predominant colonizers ($\approx$99%) in the gastrointestinal tract of an organism and hence most of the currently used probiotics belong to prokaryotes. On contrary to prokaryotes, eukaryotes are not dominant and/or even part of the naturally occurring microbiota of gastrointestinal tract of many animals and till date only limited strains are available for human and animal uses. So far, tremendous efforts have been given on prokaryotic probiotics, whereas limited emphasis has been placed on eukaryotic probiotics. Both the groups differ greatly in their properties and modes of action and a comparative account of eukaryotic probiotic (yeast) and prokaryotic probiotics (bacteria) is given in Table 2.

It is already established that probiotics can exert beneficial metabolic and immunological effects in host. However, there are also chances of adverse metabolic and immunological responses like deconjugation and dehydroxylation of bile salts, transformation of conjugated primary bile salts into free secondary bile salts, and excessive degradation of host intestinal mucus layer due to manipulation of gut microbiota with the use of probiotics. Transformation of conjugated primary bile salts into toxic free secondary bile salts in the small bowel due to ingestion of fermented dairy products containing *Lactobacillus acidophilus* and *Bifidobacterium* species has been reported in healthy humans with a terminal ileostomy (Marteau et al. 1995). However, no such effects are observed for any of the eukaryotic probiotics.

Table 2 Comparison between prokaryotic and eukaryotic probiotics with reference to yeast

Sl. No.	Parameters	Prokaryotic (bacteria)	Eukaryotic (yeast)
1	Size	Small (≈0.5 × 5 μm)	Large (≈10 × 5 μm)
2	Cell wall composition	Peptidoglycan, lipoteicheic acid, lipopolysaccharide	Chitin, glucan, mannose, phospohopeptidomannan, phospholipomannan
	Optimum growth conditions		
	pH	6.5–7.5	4.5–6.5
	Temperature	10–80°C	20–30°C
	Tolerance to gastric acid	Yes	Yes
3	Tolerance to bile salts	Yes	Yes
4	Resistant to antibiotics	No	Yes
5	Ability to transfer genetic materials	Yes	No
6	Natural occurrence in gut	Predominant (up to 99% of gut microbes)	Sporadic occurrence (less than 1%)
7	Ability to colonize in the gut	High	Low to moderate
8	Synergistic effects on other microbes	No	Yes
9	Application as probiotics	Wide range of animals	Limited application
	Effect on host		
	Growth and nutrition	Yes	Yes
	Immunostimulation	Yes	Yes
10	Protection	Yes	Yes
11	Ability to produce antagonistic compounds	High	Low (occasionally)
12	Ability to neutralize enterotoxin	No	Yes

Likewise, the transfer of antibiotic resistance gene which is now believed to be a major concern for bacterial probiotics is not reported for any of the eukaryotic probiotics (Czerucka et al. 2007). There exists the possibility of antibiotic resistant gene transfer from bacterial probiotics to pathogens in the gut of host. However, most of the probiotic yeasts lack such plasmid encoded genes (Kourelis et al. 2010a) and till date no such transfer of genetic material from yeast to bacteria is reported. Finally, the hallmark of probiotic yeasts is that they can be used during antibiotic treatment.

Further, probiotic strains of *S. cerevisiae* and *S. cerevisiae* var *boulardii* are sensitive to nonabsorbable antimycotics such as nystatine but can be used with reabsorbable antifungal agents such as fluconazole (Dixit and Gandhi 2010). On the other hand, bacterial probiotics cannot be used during antibiotics treatment.

3 Sources of Eukaryotic Probiotics

The isolation and characterization of suitable eukaryotic probiotics from natural sources warrants special considerations. Over the years, there is a great interest in developing suitable eukaryotic probiotics in general and probiotic yeasts in particular for human and animal practices. Yeasts may not be as ubiquitous as bacteria but they can thrive in diverse niches such as plants, animals, soil, water, and atmosphere. They are also associated with the skin and gastrointestinal tract of human and animals including aquatic animals (Suh et al. 2005; Gatesoupe 2007; Scanlan and Marchesi 2008; Urubschurov et al. 2008). Yeasts are predominantly associated with many food items especially dairy products (Jakobsen and Narvhus 1996; Fleet 2006; Chen et al. 2010). A scan of the literature suggests that dairy and dairy-related products could be a good source of many potential probiotics such as *Candida* (*C. humilis*), *Debaryomyces* (*D. hansenii*, *D. occidentalis*), *Kluyveromyces* (*K. lactis*, *K. lodderae*, *K. marxianus*), *Yarrowia* (*Y. lipolytica*), and several other species (Fleet 1990; Lopez-Daz et al. 1995; Kumura et al. 2004). Many times the gastrointestinal tract of human and animals is a good source of probiotic yeasts. Recently, Kourelis et al. (2010a, b) have succeeded in developing suitable probiotic strains belonging to *Saccharomyces* and *Kluyveromyces* species isolated from feta cheese and human gastrointestinal tract. Apart from these sources, several yeasts from marine sources such as *Yarrowia lipolitica* and *Candida tropicalis* are not only found to colonize in the gut of animals but also exert nutritional and other beneficial probiotic effects in host (Hirimuthugoda et al. 2007; Chi et al. 2010).

4 Taxonomic Characterization of Eukaryotic Probiotics

One of the most important aspects of assessing the efficacy of probiotics requires the understanding of individual strains, each of which is unique and different. Characterization of a strain is very important which in turn provides exact information on the nomenclature of the strain, its origin, and even presumed safety (Salminen et al. 2001). Detailed information on the taxonomic position of an organism is therefore very important in the selection process of probiotics. As per joint FAO/WHO guidelines (http://www.who.int/foodsafety/fs management/en/probiotic guidelines.pdf), the presumptive probiotics should be identified up to genus and strain level by internationally accepted methods such as DNA–DNA hybridization or sequencing of DNA encoding 16SrRNA and strain identification with pulsed field gel electrophoresis or randomly amplified polymorphic DNA.

Over the years, researchers have succeeded in identifying and characterizing several probiotic yeast strains from different sources by using molecular techniques (Posteraro et al. 2005; MacKenzie et al. 2008; Buchl et al. 2010; Chen et al. 2010). The ITS regions situated between the small and large subunits of rDNA, and NTS that separate the ribosomal gene clusters present high degrees of inter- and intraspecies variability (Baleiras Couto et al. 1996), in contrast to the rRNA genes, which are subject to greater evolutionary constraints and are thus more highly conserved (Molina et al. 1993). Biolog YT microplate and chromosomal PFGE karyotyping are often used for identification of yeasts (Baleiras Couto et al. 1996; Valente et al. 1996). As an alternative to sequencing, restriction fragment length polymorphism analysis of amplified rDNA regions has also been used. However, the sequencing of the D1/D2 fragment of the 26 S rRNA gene is suitable for identification of yeasts from different sources since strains of one yeast species differ in less than 1% of D2 nucleotide sequence (Peterson and Kurtzman 1991; Kurtzman and Robnett 1997, 1998). Similarly, the taxonomic characterization of *Aspergillus* species is very important due to their economic importance as both pathogenic and beneficial activities. Nowadays, several useful strains of *Aspergillus* species have been characterized by various molecular techniques (Klich and Mullaney 1987; Klich et al. 1995; Kumeda and Asao 2001; Montiel et al. 2003; Lee et al. 2006a).

5 Properties of Eukaryotic Probiotics

Eukaryotic microorganisms often exhibit potentially exploitable physiological and metabolic characteristics of an ideal probiotic. Several features which contribute to the success of eukaryotic microorganisms like yeasts as probiotics include their robust size, morphological diversity (budding, pseudomycelial), nutritional flexibility (ability to utilize a broad range of nitrogen, carbon, and phosphorous sources), stress tolerance ability (to low pH/oxygen/water activity, high osmotic pressure), enzyme secreting potency (secrete a broad range of enzymes such as lipase, peptidase, amylase, invertase, phytase, etc.), antioxidative/antitumor/antimicrobial activity (effective against a wide range of pathogens), and ability to produce several other useful metabolites. Furthermore, most of the presumptive probiotic yeasts are nonmycotoxic, nonallergenic, and nonpathogenic in nature (Fredlund et al. 2002).

All these characteristics not only contribute to their ability to resist the harsh milieu of the digestive tract of host but also are indispensable for their beneficial probiotic effects in host. However, the biological properties of various yeasts show considerable intraspecies variability and their beneficial properties are considered to be strain specific (Posteraro et al. 2005). *S. cerevisiae* var *boulardii* is genetically identical to *S. cereviceae* but often exhibits very distinct behavior particularly in relation to growth pattern, resistance to stress and temperature variations (Fietto et al. 2004; Nicoli and Castro 2004; Graff et al. 2008). *S. cerevisiae* var *boulardii* strains are better tolerant to acidic stress and grow faster at 37°C than *S. cereviceae* (Fietto et al. 2004). Edwards-Ingram et al. (2007) also recorded similar type of viability of *S. boulardii* to that of the control at pH 2 while laboratory *S. cerevisiae* strains was less than 4% of the control at same pH.

Yeasts can tolerate a wide range of temperature, salt concentration, and pH depending upon the strain. Temperature can affect the growth and metabolic activities of yeast. Generally, yeasts grow best at a temperature range from 20 to 30°C. The lower temperature limit is around 20°C while the higher limit for certain probiotic yeasts can be as high as 50°C. Koedrith et al. (2008) isolated a thermotolerant strain of *S. cereviceae* capable of growing rapidly at both high (40/41°C) and low temperatures from banana leaves. With regard to pH, most of the yeasts can grow very well in between the pH 4.5–6.5 but nearly all species can survive at low pH (up to 2.5). Similarly, the salt tolerance potency varies greatly among different probiotic yeasts. Probiotic strains of *S. cerevisiae* can resist up to 1.5 M NaCl while several other natural probiotic strains like *D. hansenii* which can resist NaCl up to 2.5 M with optimum growth at 0.5 M NaCl.

5.1 Ability to Tolerate Stress

Probiotics irrespective of origin are expected to undergo various stress conditions during industrial processing such as drying, heating, and freezing. Therefore, the stress tolerance nature of any probiotics is of paramount importance. Yeasts can survive the adverse environmental fluctuations by rapidly adopting their internal systems to meet challenges of new environment. However, it depends upon the strains, cell age, and type. The ability of any probiotics to tolerate heat stress is very crucial since they have to undergo the industrial processing. *S. cerevisiae* usually respond to heat stress in two phases. The initial phase of the response is the gaining of thermotolerance with accumulation of trehalose followed by the induction of heat shock proteins when it begins to grow at the increased temperature. Besides, antioxidants enzymes and plasma membrane ATPase are also involved in this process (Piper 1993). Trehalose protects the membranes and proteins from stress (Hounsa et al. 1998) and its accumulation/depletion is closely correlated with changes during the induction of thermotolerance in yeast (Hottiger et al. 1987, 1992; de Virgillio et al. 1994; Martins et al. 2008). Further, Hsp104 is believed to play a role not only in thermotolerance but also in other kinds of stress in *S. cerevisiae* (Sanchez et al. 1992; Holubarova et al. 2000). Li et al. (2006) observed that heat stress (39°C) rapidly lead to a notable increase in the binding of HSF to ScSSA1 and HSP104 genes with a peak after 10 min of heat stress followed by decline with continuous exposure to heat stress for up to 1 h.

Yeasts also exhibit broad range of tolerance towards osmotic stress and studies indicate that KCl and sorbitol are less inhibitory to yeast growth than NaCl (Gaxiola et al. 1992). While KCl and sorbitol only cause osmotic stress, NaCl in addition to osmotic stress also contributes to Na^+ toxicity (Serrano 1996).

Usually yeasts adopt several mechanisms to avoid depletion of water in the cell during osmotic stress that could lead to the cell shrinkage and eventually death. Glycerol plays a critical role in overcoming the stress to balance the osmotic pressure across the cell membrane and adjusting the external water pressure (Blomberg and Adler 1989, 1992; Blomberg 2000). Studies indicate that among the two genes that are responsible for the production of glycerol, namely glycerol-3-phosphate dehydrogenase (GPD1 and GPD2), only GPD1 is induced by stress in yeast (Eriksson et al. 1995).There is also an indication of accumulation of intracellular trehalose during osmotic stress (MacKenzie et al. 1988; Sharma 1997; Kofli et al. 2006). Further, increase of intracellular K^+ and decrease of intracellular Na^+ by overexpression of either HAL1 or HAL3 is crucial for NaCl tolerance in *S. cerevisiae* (Gaxiola et al. 1992; Rios et al. 1997).

While thermo- and osmotolerance nature of probiotics is necessary to withstand the processing stress, their ability to tolerate low temperature is equally important for maintenance. The ability of probiotics to resist low temperature is also very crucial since decreased efficiency of several probiotics is often reported when maintained at low temperature for long duration. Yeasts cells show distinct resistance to freezing injury and the growth conditions of yeasts significantly modify the cellular responses to freezing injury (Morris et al. 1986, 1988). The cells in their early exponential phase or stationary phase of culture show more resistant to freezing injury (Morris et al. 1988; Pardo et al. 2009). Morris et al. (1988) observed a clear relationship between the morphology during freezing of *S. cereviceae* cells from the late exponential phase of culture and viability upon thawing. Recently, Pardo et al. (2009) also reported that −20°C maintained *S. cereviceae* var *boulardii* cells to show similar types of specific growth rate and biomass as that of normal cells with survival percentage of 0.31 and 11.5 when freezed the cells in their exponential phase and stationary phase of growth, respectively. Furthermore, they have found pretreatment of *S. cerevisiae* var *boulardii* cells in media with water activity 0.98 can lead to a tenfold enhancement in survival at −20°C for 2 months.

However, the entire process of tolerance of yeasts to various stresses is very complex which need detail information from genome, proteome, and metabolome studies.

5.2 Ability to Survive in the Gastrointestinal Tract

Probiotics need to survive the inevitable biological barriers of gut (Saarela et al. 2009). It is apparent from the previous in vitro studies that eukaryotic probiotics can tolerate the simulated environmental conditions of the digestive tract. This fact is also supported by several in vivo studies where eukaryotic probiotics are not only found in viable conditions but also execute positive effects. In vitro studies indicate that yeasts belonging to *Saccharomyces, Debaryomyces,* and *Kluyveromyces* species are extremely tolerant to bile salt (Kumura et al. 2004; van der Aa kuhle et al. 2005; Pardo et al. 2009; Chen et al. 2010; Kourelis et al. 2010a). van der Aa kuhle et al. (2005) recorded that probiotic *S. cerevisiae* can withstand 0.3% Oxgall at a pH of 2.5 while certain strains of *S. cereviceae* and *K. lactis* can tolerate 0.15% bile salt at low pH of 3.0 (Kourelis et al. 2010a). The high tolerance potency of *Torulaspora delbrueckii, D. hansenii, K. lactis, K. marxianus,* and *K. lodderae* towards simulated gastrointestinal conditions, further accentuates their possible use as probiotics in near future (Kumura et al. 2004; Psani and Kotzekidou 2006).

In vivo studies also support this fact and probiotic strains of *S. cerevisiae*/*S. cerevisiae* var *boulardii* are not only able to survive in the digestive tract of host but also are found to persist at high populations in the gut of host. The pharmacokinetics study indicates that probiotic *S. cerevisiae* var *boulardii* when ingested orally achieved a steady state in the gut within 3 days but eliminated within 2–5 days after it is discontinued (Blehaut et al. 1989). Such elimination from the digestive tract is believed to be due to the barrier effect of the complex established resident microbiota of the gut (Fiems et al. 1993; Chaucheyras et al. 1998).

5.3 Ability to Adhere to the Gastrointestinal Tract

The necessity of probiotics adherence and subsequent colonization to the gut of host for exerting beneficial effects remains unsolved (Fedorak and Madsen 2004). However, adhering potency of probiotics to the intestinal tract of host is believed to be crucial in order to ensure their maintenance in the intestinal tract for a longer period of time (Ouwehand et al. 1999a). Probiotic strains of *Saccharomyces, Debaryomyces, Candida, Isaatchenkia,* and *Kluyveromyces* species possess noteworthy adhesive potency as evidenced from their in vitro adhesive capabilities to various cells (Kumura et al. 2004; Kourelis et al. 2010a). However, these yeasts differ greatly in the attachment with respect to sources, species, and strains. Ouwehand et al. (2000) reported that *S. cerevisiae* var *boulardii* from Precosa R exhibits good adhesive tendency which is congruent with their earlier study (Ouwehand et al. 1999b). Similar type of high adhesive potency of *S. cerevisiae* isolates from blue veined cheeses is also reported by van der Aa kuhle et al. (2005). On contrary to these findings, several other strains of *S. cerevisiae* var *boulardii* are found to possess poor to low in vitro adherence potency to various cells such as Vero cells (monkey kidney cells), Caco-2 (Tasteyre et al. 2002; Kumura et al. 2004). In a comparative study, Kumura et al. (2004) recorded best adhesive potency of *K. lactis* to human enterocyte-like Caco-2 cells followed by *K. marxianus, K. lodderae,* and *D. hansenii* while *C. humilis, D. occidentalis, S. cerevisiae,* and *Y. lipolytica* strains exhibited least adhesive potency.

6 Beneficial Effects of Eukaryotic Probiotics

Probiotics are now becoming a popular and important tool in the health management strategy of human and animals. Eukaryotic probiotics, when ingested orally as food/feed supplements exert several types of nutritional benefits in hosts. Some of the nutritional and growth benefits which are mainly documented in animals are as follows.

6.1 Nutritional Benefits

Probiotics have profound effect on nutrition of host and can influence on various digestive processes, especially cellulolysis and synthesis of microbial protein and increase in the absorption of nutrients. *S. cerevisiae* is considered as one of the probiotics that, when administered through the digestive tract, have a positive impact on the hosts health (Patterson and Burkholder 2003). In various animals, *S. cerevisiae* supplementation often lead to enhanced body weight gain with increase in feed intake (Churchil and Mohan 2000; Sharma et al. 2001; Kim et al. 2002; Nilson et al. 2004; Kabir 2009; Shareef and Al-Dabbagh 2009). Apart from *S. cerevisiae*, *A. oryzae* is also very useful to animals from nutritional point of view (Lee et al. 2006a, b, c). *A. oryzae* often secretes various digestive (amylolytic and proteolytic) enzymes which are indispensable to the host nutrition. Dietary supplementation of *A. oryzae* can significantly enhance the weight gain (Wallentine et al. 1986) and the dry matter digestibility in cows (Wiedmeier et al. 1987; Gomez-Alarcon et al. 1990). In broilers, *A. oryzae* is not only reported to enhance the feed intake and weight gain but also reduce the ammonia and serum cholesterol level (Kim et al. 2003). Similarly, another probiotic belonging to *Aspergillus* species, *A. niger* is also reported to improve the blood quality with low cholesterol and glucose level in broilers (Al-Kassie et al. 2008).

Probiotics can influence the production as well as the quality of milk. Yeasts either live or fermented have been used in dairy cattle for more than 60 years (Schingoethe et al. 2004). Eukaryotic probiotics help in improving the dry matter intake (Williams et al. 1991; Wohlt et al. 1991, 1998; Yu et al. 1997; Dann et al. 2000), the percentage of milk fat, Solids-Not-Fat, protein (Putman et al. 1997; Wang et al. 2001a), and production level (Harris and Lobo 1988; Arambel and Kent 1990; Piva et al. 1993; Robinson and Garrett 1999; Masek et al. 2008). In dairy cows, supplementation of *S. cerevisiae* can increase the production of acetate, propionate, and total volatile fatty acids (Nisbet and Martin 1991; Piva et al. 1993; Miller-Webster et al. 2002). Likewise, other eukaryotic probiotic like *A. oryzae* is also found to enhance the milk yields (Marcus et al. 1986; Wallentine et al. 1986; Kellems et al. 1987).

Probiotics can influence the egg quality and production percentage and inclusion of live yeasts into laying hen diets significantly improve the egg production percentage (Kim et al. 2002; Shivani et al. 2003), egg weight/mass (Han et al. 1999; Park et al. 2001, 2002; Omar 2006), egg shell breaking strength (Park et al. 2002), as well as reduce the percentage of soft or broken eggs (Park et al. 2001). On the other hand, Dizaji and Pirmohammadi (2009) could not observe any substantial influence of dietary supplementation of eukaryotic-based probiotics "Biosaf SC 47" (*S. cerevisiae*, strain NCYC SC 47) on the overall egg performance of laying hens. Earlier, Yousefi and Karkoodi (2007) also failed to record any significant effect on egg mass, weight, and production, except improvement in yolk weight and Haugh unit of egg by supplementing *S. cerevisiae* diets of laying hens.

The involvement of probiotics especially prokaryotic in improving the overall meat quality has been questioned due to several contradictory findings. However, not much work has been done on these aspects of eukaryotic probiotics. Preliminary reports indicate that whole *S. cerevisiae* as well as its extract can improve the meat tenderness and oxidative stability of broiler meat. It is assumed that certain antioxidant factors present in *S. cerevisiae* are involved in shifting the oxidative factor of fatty acid profile in the meat (Zhang et al. 2005).

6.2 Disease Protection

Over the years, several studies have been conducted to establish the disease protecting ability of yeasts. The therapeutic potency of yeasts especially *Saccharomyces* species and their mechanisms, pharmacokinetics, and pharmacodynamics is well documented in animal models as well as in clinical trials (Rodrigues et al. 1996; Periti and Tonelli 2001; Girard et al. 2003; Dalmasso et al. 2006; Dixit and Gandhi 2010). Out of 16 species of *Saccharomyces*, specific strains of *S. cerevisiae* / *S. cerevisiae* var *boulardii* are only used as biotherapeutic agents. Their biotherapeutic effects especially in the treatment of general digestive problems, diarrhea, amebiasis, irritable bowel syndrome, inflammatory bowel syndrome, bacterial overgrowth in short bowel syndrome, Crohn's disease, ulcerative colitis, and lyme disease are noteworthy (McFarland et al. 1994, 1995; WHO 1995; Bleichner et al. 1997; Guslandi et al. 2000). Probiotic strains of *S. cerevisiae* and *S. cerevisiae* var *boulardii* are extremely effective in treatment and prevention of various types of diarrhea, including infectious types such as acute diarrhea in children, diarrhea caused by pathogens in adults, traveler's diarrhea, AIDS-associated diarrhea, antibiotic-associated diarrhea, diarrhea in ill tube-fed patients, and *Clostridium difficile*-associated diarrhea (Surawicz et al. 1989a, b; Cetina and Sierra 1994; Bleichner et al. 1997; Aloysins et al. 2005; Martins et al. 2005; Czerucka et al. 2007).

Furthermore, *S. cerevisiae* var *boulardii* is also effective for lactose intolerance, urinary tract infections, vaginal yeast infections, high cholesterol levels, hives, fever blisters, canker sores, and teen-age acne. Besides, their usefulness against food allergies, yeast (candida) infection (Murzyn et al. 2010), and parasitic infestation and in reestablishing the normal gut functions after long-term antibiotic therapy in hosts (McFarland et al. 1994) are also recorded. In a recent study, Murzyn et al. (2010) found that *S. cerevisiae* var *boulardii* as well as its extract can inhibit the expression of various genes associated with the virulence of *C. albicans*. They have established the involvement of capric acid as an active compound responsible for preventing the growth, hyphae formation, partly adhesion, and biofilm formation of *C. albicans*. Nonetheless, strains of *S. cerevisiae* are also effective in preventing the adhesion of *Entamoeba histolytica* trophozoites (Elliot et al. 1991; Rigothier et al. 1994).

Further, *S. cerevisiae* can reduce the *C. albicans*, *Candida krusei*, and *Candida pseudotropicalis* in the digestive tract of normal and antibiotic treated rats/mice (Seguela et al. 1978; Ducluzeau and Bensaada 1982) as well as other pathogens such as *Salmonella typhimurium* and *Shigella flexneri* (Rodrigues et al. 1996). *S. cereviceae* var *boulardii* is also reported to decrease the inflammatory reaction and colonization of mouse intestine by the *C. albicans* infection (Jawhara and Poulain 2007).

7 Modes of Action of Eukaryotic Probiotics

Probiotics often exert beneficial effects through a variety of disparate and overlapping mechanisms. Eukaryotic probiotics adopt several mechanisms like trophic effect on gut for maintaining the intestinal homeostasis, stimulating effect on both local and systematic immunity, antimicrobial and toxin neutralizing activities.

7.1 Trophic Effects on the Gastrointestinal Tract

Eukaryotic probiotics especially yeasts exert trophic effects on gut to restore its homeostasis. Although the precise mechanisms by which yeasts exert various trophic effects are yet to be established, Buts (2009) summarized the possible mechanisms of trophic effects of *S. boulardii* in human and animals. It is well known that probiotics often improve enzymatic activity of the gut by producing several enzymes which are not produced by host to break down complex macromolecules. In human beings, *S. cerevisiae*/*S. cerevisiae* var *boulardii* can increase the lactase, glycosidase, and alkaline phosphatase activities both at the basal and apical parts of villi (Jahn et al. 1996). Oral ingestion of *S. cerevisiae* can also lead to marked increase in the specific and total activities of brush border membrane disaccharidases including sucrase, lactase, isomaltase, and maltase in human and animals (Buts et al. 1986). Therefore, *S. cerevisiae* can help to improve malabsorption in patients with sucrase-isomaltase deficiency that intentionally or unintentionally consumes sucrose and also improves some diarrheas that are associated with a decrease of the intestinal disaccharidase activities. Further, the involvement of polyamines (spermine and spermidine), present in *S. cerevisiae*/*S. cerevisiae* var *boulardii* in trophic effect on the intestinal mucosa could not be ruled out (Balasundram et al. 1994; Buts et al. 1994).

7.2 Stimulation of Immunity

Eukaryotic probiotics have also been shown to stimulate both innate and adaptive immunity which in turn may contribute to good health and disease resistance of host. Eukaryotic probiotics unlike bacterial probiotics can stimulate the immune system of diversified hosts ranging from mammalian to piscine. Yeasts are rich source of many biologically active substances which can potentially trigger various biological systems including the immune system of host (Buts et al. 1990; Rodrigues et al. 2000; Martins et al. 2007, 2009). The cell wall components of yeast are predominantly composed of complex polymers of β-glucans, α-mannans, mannoproteins, and a minor component of chitin (Smits et al. 1999). These macromolecules can eventually stimulate the immune system of host especially inflammatory response and reticuloendothelial system. Recently, Pothoulakis (2009) has elaborately discussed the anti-inflammatory mechanism of *S. cereviceae* var *boulardii*.

S. cereviceae strains are found to be better stimulator of immune system. In a comparative study, Martins et al. (2009) reported better immunostimulating activity (in terms of sIg A and IL10 production) of *S. cereviceae* var *boulardii* as compared to bacterial probiotics such as *Bifidobacterium animalis*, *E. coli*, and *Lactobacillus casei*. Probiotic strains of *S. cereviceae* var *boulardii* have been shown to enhance the phagocytic activity as well as production of different cytokines (Cuaron 1999; Rodrigues et al. 2000; Czerucka et al. 2007). Kourelis et al. (2010b) found the ability of presumptive probiotic yeast strains belonging to *S. cerevisiae* and *K. lactis* species to increase polymorphonuclear cell influx, phagocytic activity, and various cytokines in an air pouch model. However, many times, elevation of immune responses is often strain specific. Recently, Kourelis et al. (2010b) have reported significant variation in the ability of various probiotics strains in inducing the increased production of TNF-α, IFN-γ, and IL-10. However, strains like *K. latis* and *K. loddrae* showed no significant effect on the secretion of proinflammatory cytokine, IL 8 by Caco-2 cells (Kumura et al. 2004).

Saccharomyces species are found to stimulate sIgA level in different animal models. In gnotobiotic mice, it triggered the phagocytic system along with enhancement of sIg A (Rodrigues et al. 2000) while Buts et al. (1990) found significant increase in sIgA and secretory component of immunoglobulins in rats by oral administration of *S. cerevisiae*.

7.3 Synergistic Activity

Eukaryotic probiotics can exert synergistic effects on the indigenous gut microbiota of host. Supplementation of probiotic yeasts often favors the growth of certain bacteria (Newbold 1996; Chaucheyras et al. 1996, 1997), fungi, and even protozoa (Chaucheyras et al. 1995; Miranda et al. 1996). *Saccharomyces* species have been found to stimulate the growth of other microorganisms by providing essential metabolites such as pyruvate, amino acids, and vitamins (Jespersen 2003). The redox potential in the rumen which is mainly increased due to the infusion of oxygen during intake of feed and water can have detrimental effect on the anaerobes in the rumen. Yeasts can remove oxygen from the rumen and help to increase microbial population. This is further confirmed from the fact that respiration-deficient mutants of *S. cerevisiae* lack the ability to enhance microbes (Newbold et al. 1996). Recently, Hassanein and Soliman (2010) also recorded such increase in the lactobacilli population along with decrease of many pathogens in the gut of layers fed with probiotic *S. cerevisiae*.

7.4 Antagonistic Activity

Bacterial probiotics inhibit pathogens by competing for nutrients, adhesion sites, and/or by producing several antimicrobial compounds/metabolites. However, the situation may be little different for eukaryotic probiotics. There is a general consensus that eukaryotic probiotics lack the ability to produce any antimicrobial metabolites and/or inhibit pathogens (Martins et al. 2009; Kourelis et al. 2010a) but several reports indicate the successful inhibition/killing of various pathogens such as *Clostridium albicans, E. coli, Salmonella typhi, Shigella dysenteriae, Vibrio cholerae, Salmonella enteritidis,* and *Clostridium difficile* by eukaryotic probiotics (Izadnia et al. 1998; Filho-Lima et al. 2000; Czerucka and Rampal 2002).

Pathogens usually adhere to the epithelium for establishing lethal infection. But eukaryotic probiotics can prevent the pathogens to bind to intestinal cells by direct antagonistic effect and/or secreting several metabolites/enzymes. Further, *S. cerevisiae* var *boulardii* strains can compete with pathogenic microorganisms for food and mucosal receptors in the gut and thereby prevent the pathogens to colonize in the gut of host (Filho-Lima et al. 2000). Similarly, certain pathogens like *E. coli* or *Salmonella* which posses such mannose-specific type-1 fimbriae bind to mannose residues on epithelial cell membranes (Ofek et al. 1977). Yeasts like *S. cerevisiae* contain mannan in its outer layer of their cell wall and therefore can induce a protective effect against these pathogens. Such pathogens readily bind to the mannans present on the surface of yeast instead of attaching to intestinal epithelial cells (Gedek 1999; Spring et al. 2000) and then eliminated from the digestive tract of host after agglutinated by yeast.

7.5 Toxin Neutralization Activity

Lethal toxins produced by pathogens bind to specific receptors on intestinal epithelial cells and then cause mucosal damage and inflammation. Eukaryotic probiotics often destroy bacterial toxins and their receptor sides by releasing various enzymes like proteases (Castagliulo et al. 1999). In vitro studies indicate the possible involvement of protease present in both whole yeast and cell wall fraction of *S. cerevisiae* var *boulardii* in inhibiting the adhesion of *C. difficile* to Vero cells despite lacking the adhesion ability to bacteria or Vero cells (Tasteyre et al. 2002).

Yeasts not only can inhibit the production of toxins of several pathogens such as *C. difficile*, *V. cholerae*, and *E. coli* but also prevent their adverse effects (Massot et al. 1982; Corthier et al. 1986; Vidon et al. 1986). A serine protease from

S. cerevisiae effectively hydrolyzes the toxin A, one of the two potent toxins (A and B) produced by *C. difficile* (Pothoulakis and Lamont 2001). Similarly, *S. cerevisiae*/*S. cerevisiae* var *boulardii* are reported to reduce the liquid secretion and mannitol permeability caused by *C. difficile* toxin A in the rat ileum (Pothoulakis et al. 1993). A 54-kDa serine protease of *S. cerevisiae* can reduce the toxin A-induced rat ileal secretion and prevent the toxin A-mediated inflammation and villus damage by inhibiting toxin A binding to its brush border glycoprotein receptor (Pothoulakis et al. 1993; Castagliuolo et al. 1996). Nevertheless, another 120-kDa protein of *S. cerevisiae* var *boulardii* is responsible for preventing the cholera toxin action and is found to reduce the cholera toxin-mediated stimulation of cyclic adenosine monophosphate in mouse intestinal loops (Brandao et al. 1998; Czerucka et al. 2000; Czerucka and Rampal 1999; Neves et al. 2002).

8 Safety Issues Related to Eukaryotic Probiotics

Probiotics are usually safe beyond any doubt but sometimes can cause complications and side effects in susceptible individuals. Most of yeasts used as probiotic are safe and also possess Qualified Presumption of Safety Status, assigned by the European Food Safety Authority (http://www.efsa.europa.eu) and are rarely associated with outbreaks or cases of food-borne illness. Although, *S. cerevisiae* and *S. cerevisiae* var *boulardii* are routinely used in human and animal practices for several decades, few cases of infection are reported in patients. *S. cerevisiae*/*S. cerevisiae* var *boulardii* induced fungemia has been reported in humans using biotherapeutic products containing *S. cerevisiae* var *boulardii* (Zunic et al. 1991; Pletincx et al. 1995; Fredenucci et al. 1998; Niault et al. 1999; Cesaro et al. 2000; Hennequin et al. 2000; Rijnders et al. 2000; Cassone et al. 2003; Riquelme et al. 2003). However, such cases are only restricted to immunocompromised patients and/or patients being contaminated through a central venous catheter, and therefore raise some concerns over the use of live yeasts especially in the debilitated and immunosuppressed patients.

Animal model studies also indicate the low to moderate virulence nature of *S. cerevisiae*/*S. cerevisiae* var *boulardii* (McCullough et al. 1998). They can penetrate the intestinal mucosa of animals to reach other organs (Cartwright-Shamoon et al. 1996) as evident from the translocation of *S. cerevisiae* to mesenteric lymph nodes in immunosuppressed mice.

9 Genetic Manipulation of Eukaryotic Probiotics

Nowadays, much emphasis has been given on the genetic improvement of probiotics for developing suitable candidate with desirable characteristics (Steidler 2003). The genetic improvement of *Saccharomyces* and other yeasts has traditionally been relied on random mutagenesis (Wang et al. 2001b) but now protoplast fusion technique is adopted to generate fusants with desired characteristics (Martins et al. 2004). Several researchers have also suceeded in developing highly efficient biotherapeutic strains of *S. boulardii* which can resist low pH degree, high bile salts, possess high vitamin content, and exhibit antagonistic effect on several pathogens through genetic improvement (using mutation/interspecific protoplast fusion) (Nivien et al. 2006; Abosereh et al. 2007; Pasha et al. 2007; Sharaf et al. 2009). However, there are several issues associated with such manipulations and detail information on various aspects of the genetic improvement of yeast/fungi needs to be addressed before its practical application in the field of probiotics. Furthermore, issues like safety, effectiveness, impact on environment associated with such manipulations warrants further elaboration but it is beyond the scope of this review.

10 Conclusion

Although the epicenter of probiotic research has been prokaryotes, scientific and commercial interest in eukaryotic probiotics has significantly increased in recent years. There is no doubt that eukaryotic probiotics can execute a myriad of beneficial effects in a broad range of hosts. However, there are identifiable limitations to the use of eukaryotic probiotics and more in-depth studies are needed to establish scientific rationale of eukaryotic probiotics. There is a dearth of information on the diversity and function of the eukaryotic microbiota in the gut of human and other animals. Additionally, adequate knowledge on the individual strain, its source, safety, and host ranges is required for the firm establishment of eukaryotes as probiotics. More comprehensive ecological surveys are needed to determine the diversity of probiotic fungi/yeasts in nature other than *Saccharomyces* species. Successful application of modern technologies will certainly allow in developing tools for analyzing the functionality of these strains as probiotics. Integrating metagenomic and metaproteomic approaches will certainly help to develop the new generation of eukaryotic probiotics.

References

Abosereh NA, Mohamed HAA, El-Chalk ABA (2007) Genetic construction of potentially probiotic *Saccharomyces boulardii* yeast strains using intraspecific protoplast fusion. J Appl Sci Res 3(3):209–217

Al-Kassie GAM, Al-Jumaa YMF, Jameel YJ (2008) Effect of Probiotic (*Aspergillus niger*) and Prebiotic (*Taraxacum officinale*) on Blood Picture and Biochemical Properties of Broiler Chicks. Int J Poult Sci 7(12):1182–1184

Aloysins L, Souza D, Rajkumar C, Cooke J, Bulpitt CJ (2005) Probiotic in prevention of antibiotic associated diarrhea. British Med J 8:1–6

Arambel MJ, Kent BA (1990) Effect of yeast culture on nutrient digestibility and milk yield response in early- to midlactation dairy cows. J Dairy Sci 73:1560–1563

Balasundram D, Jonathan DD, Celia WT, Tabor H (1994) SPE1 and SPE2: two essential genes in the biosynthesis of polyamines that modulate +1 ribosomal frameshifting in *Saccharomyces cerevisiae*. J Bacteriol 176(22):7126–7128

Baleiras Couto MM, Eijsma B, Hofstra H, JH Huis in't Veld, van der Vossen JM (1996) Evaluation of molecular typing techniques to assign genetic diversity among *Saccharomyces cerevisiae* strains. Appl Env Microbiol 62:41–46

Blehaut H, Massot J, Elmer GW, Levy RH (1989) Disposition kinetics of *Saccharomyces boulardii* in man and rat. Biopharm Drug Dispos 10:353–364

Bleichner G, Blehaut H, Mentec H, Moyse D (1997) *Saccharomyces boulardii* prevents diarrhea in critically ill tube-fed patients. A multicenter, randomized, double-blind placebo-controlled trial. Intens Care Med 23:517–523

Blomberg A (2000) Metabolic surprises in *Saccharomyces cerevisiae* during adaptation to saline conditions: questions, some answers and a model. FEMS Microbiol Lett 182(1):1–8

Blomberg A, Adler L (1989) Roles of glycerol and glycerol-3 phosphate dehydrogenase (NAD^+) in acquired osmotolerance of *Saccharomyces cerevisiae*. J Bacteriol 171:1087–1092

Blomberg A, Adler L (1992) Physiology of osmotolerance in fungi. Adv Microbiol Physiol 33:145–212

Bovill R, Bew J, Robinson S (2001) Comparison of selective media for the recovery and enumeration of probiotic yeasts from animal feed. Int J Food Microbiol 67:55–61

Boyle RJ, Roy MRB, Tang ML (2006) Probiotic use in clinical practice: what are the risks? American J Clin Nutri 83:1256–1264

Brandao RL, Castro IM, Bambirra EA, Amaral SC, Fietto LG, Tropia MJM, Neves MJ, Santos RG, Gomes NCM, Nicoli JR (1998) Intracellular signal triggered by cholera toxin in *Saccharomyces boulardii* and *Saccharomyces cerevisiae*. Appl Environ Microbiol 64:564–568

Buchl NR, Hutzler M, Mietke-Hofmann H, Wenning M, Scherer S (2010) Differentiation of probiotic and environmental *Saccharomyces cerevisiae* strains in animal feed. J Appl Microbiol 109(3):783–791

Buts JP (2009) Twenty-five years of research on *Saccharomyces boulardii* trophic effects: updates and perspectives. Dig Dis Sci 54:15–18

Buts JP, Bernasconi P, Van Craynest MP, Maldague P, De Meyer R (1986) Response of human and rat small intestinal mucosa to oral administration of *Saccharomyces boulardii*. Pediatr Res 20:192–196

Buts JP, Bernasconi P, Vaerman JP, Dive C (1990) Stimulation of secretory IgA and secretory component of immunoglobulins in small intestine of rats treated with *Saccharomyces boulardii*. Dig Dis Sci 35:251–256

Buts JP, De Keyser N, De Raedemaeker L (1994) *Saccharomyces boulardii* enhances rat intestinal enzyme expression by endoluminal release of polyamines. Pediatr Res 36:522–527

Cartwright-Shamoon J, Dickson GR, Dodge J, Carr KE (1996) Uptake of yeast (*Saccharomyces boulardii*) in normal and rotavirus treated intestine. Gut 39:204–209

Cassone M, Serra P, Mondello F, Girolamo A, Scaletti S, Pastella E, Venditti M (2003) Outbreak of *Saccharomyces cerevisiae* subtype *boulardii* fungemia in patients neighboring those treated with a probiotic preparation of the organism. J Clin Microbiol 41:5340–5343

Castagliuolo I, Lamount JT, Nikulasson ST, Pathoulakis C (1996) *Saccharomyces boulardii* protease inhibits *Clostridium difficle* toixin: an effect in the rat ileum. Infect Immun 64:5225–5232

Castagliulo I, Riegler MF, Valenick L, LaMont JT, Pothoulakis C (1999) *Saccharomyces boulardii* protease mediates *Clostridium difficile* toxin A ad B effects in human colonic mucosa. Infect Immun 67:302–307

Cesaro S, Chinello P, Rossi L, Zanesco L (2000) *Saccharomyces cerevisiae* fungemia in a neutropenic patient treated with *Saccharomyces boulardii*. Support Care Cancer 8(6):504–505

Cetina SG, Sierra BG (1994) Evaluation thérapeutique de *Saccharomyces boulardii* chez des enfants souffrant de diarrhee aigue. Annales de Pediatrie (Paris) 41:397–400

Chaucheyras F, Fonty G, Bertin G, Gouet P (1995) Effects of live *Saccharomyces cerevisiae* cells on zoospore germination, growth and cellulolytic activity of the rumen anaerobic fungus *Neocallimastix frontalis* NCH3. Curr Microbiol 31:201–205

Chaucheyras F, Fonty G, Bertin G, Salmon JM, Gouet P (1996) Effects of a strain of *Saccharomyces cerevisiae* (Levucell SC), a microbial additive for ruminants, on lactate metabolism *in vitro*. Canadian J Microbiol 42:927–933

Chaucheyras F, Millet L, Michalet-Doreau B (1997) Effect of the addition of Levucell *Saccharomyces cerevisiae* on the rumen microflora of sheep during adaptation to high starch diets. In: Chesson A, Stewart CS, Flint HJ (eds) Proceedings of evolution of the rumen microbial ecosystem, RRI-INRA Rumen microbiologu Symposium, 20–21 Mar 1997, Aberdeen, UK

Chaucheyras F, Fonty G, Bertin G, Theveniot M, Gouet P (1998) Fate of Levucell SCI-1077 yeast additive during digestive transit in lambs. Reprod Nutr Dev 38:275–280

Chen LS, MA Y, Maubois JL, Chen LJ, Liu QH, Guo JP (2010) Identification of yeasts from raw milk and selection for some specific antioxidant properties. Int J Dairy Tech 63(1):47–54

Chi ZM, Liu G, Zhao S, Li J, Peng Y (2010) Marine yeasts as biocontrol agents and producers of bio-products. Appl Microbiol Biotechnol 86:1227–1241

Churchil R, Mohan B (2000) Effect of supplementation of broiler rations with live yeast culture. Cheiron 29(1–2):23–27

Corthier G, Dubos F, Ducluzeau R (1986) Prevention of *Clostridium difficile* induced mortality in gnotobiotic mice by *Saccharomyces boulardii*. Canadian J Microbiol 32:894–896

Cuaron P (1999) Live yeast use in growing and finishing swine. Development of a study model. In: Proceedings of the 3rd SAF-AGRI Symposium on biotechnology applied to animal nutrition, Merida, Mexico

Czerucka D, Rampal P (1999) Effect of *Saccharomyces boulardii* on cAMP and Ca^2- dependent Cl secretion in T84 cells. Dig Dis Sci 44:2359–2368

Czerucka D, Dahan S, Mograbi B, Rossi B, Rampal P (2000) *Saccharomyces boulardii* preserves the barrier function and modulates the signal transduction pathway induced in enteropathogenic *Escherichia coli* infected T84 cells. Infect Immun 68:5998–6004

Czerucka D, Rampal P (2002) Experimental effects of *Saccharomyces* on diarrhoeal pathogens. Microbes Infect 4:733–739

Czerucka D, Piche T, Rampal P (2007) Review article: yeast as probiotics – *Saccharomyces boulardii*. Aliment Pharmacol Ther 26:767–778

Dalmasso G, Cottrez F, Imbert V, Lagadec P, Peyron JF, Rampal P, Czerucka D, Groux H (2006) *Saccharomyces boulardii* inhibits inflammatory bowel disease by trapping T cells in mesenteric lymph nodes. Gastroenterology 131:1812–1825

Dann HM, Prockley JR, McCoy GC, Hutjens MF, Garret JE (2000) Effects of yeast cultures (*Saccharomyces cerevisiae*) on prepartum intake and postpartum intake and milk production of Jersey cows. J Dairy Sci 83:123–127

de Virgillio C, Hottiger T, Dominguez J, Boller T, Wiemken A (1994) The role of trehalose synthesis for the acquisition of thermotolerance in yeast. I. Genetic evidence that trehalose is a thermoprotectant. Eur J Biochem 219:179–186

Diniz RO, Garla LK, Schneedorf JM, Carvalho JCT (2003) Study of anti-inflammatory activity of Tibetan mushroom, a symbiotic culture of bacteria and fungi encapsulated into a polysaccharide matrix. Pharmacol Res 47:49–52

Dixit K, Gandhi DN (2010) Biotherapeutic properties of probiotic yeast Saccharomyces species in fermented dairy foods. Available from URL: http://www.dairyscience.info/probiotics/105-biotherapeutic-probioticyeast. Accessed 16 Apr 2010

Dizaji SB, Pirmohammadi R (2009) Effect of *Saccharomyces cerevisiae* and Bioplus 2B on performance of laying hens. Int J Agric Biol 11:495–497

Ducluzeau R, Bensaada M (1982) Effet comparé de l'administration unique ou en continu de *Saccharomyces boulardii* sur l'établissement de diverses souches de Candida dans le tractus digestif de souris gnotoxéniques. Ann Microbiol (Inst Pasteur) 133B:491–501

Edwards-Ingram L, Gitsham P, Burton N, Warhurst G, Clarke I, Hoyle D, Oliver SG, Stateva L (2007) Genotypic and physiological characterization of *Saccharomyces boulardii*, the probiotic strain of *Saccharomyces cerevisiae*. Appl Env Microbiol 73:2458–2467

Elliot DA, Katcher VB, Lowy FD (1991) A 220-kilodalton glycoprotein in yeast extract inhibits *Staphylococcus aureus* adherence to human endothelial cells. Infect Immun 59(6):2222–2223

Eriksson P, Andre L, Ansell R, Blomberg A, Adler L (1995) Cloning and characterization of GPD2, a second gene encoding sn-glycerol 3-phosphate dehydrogenase (NAD^+) in *Saccharomyces cerevisiae*, and its comparison with GPD1. Mol Microbiol 17:95–107

FAO/WHO (1987) Committee on food additives. 31. World Health Organization Technical Report Series, Geneva

FAO/WHO (2002) Guidelines for the evaluation of probiotics in food. Food and Agriculture Organization of the United Nations and World Health Organization Working Group Report, London Ontario, Canada

Fedorak RN, Madsen KM (2004) Probiotics and prebiotics in gastrointestinal disorders. Curr Opin Gastroenterol 20:146–155

Fiems LO, Cottyn BG, Dussert L, Vanacker JM (1993) Effect of a viable yeast culture on digestibility and rumen fermentation in sheep fed different types of diets. Reprod Nutr Dev 33:43–49

Fietto JL, Araujo RS, Valadao FN, Fietto LG, Brandao RL, Neves MJ, Gomes FC, Nicoli JR, Castro IM (2004) Molecular and physiological comparisons between *Saccharomyces cerevisiae* and *Saccharomyces boulardii*. Canadian J Microbiol 50:615–621

Filho-Lima JV, Vieira EC, Nicoli JR (2000) Antagonistic effect of *Lactobacillus acidophilus*. Saccharomyces boulardii and Escherichia coli combinations against experimental infections with Shigella flexneri and Salmonella enteritidis subsp. typhimurium in gnotobiotic mice. J Appl Microbiol 88:365–370

Fleet GH (1990) Yeast in dairy product a review. J Appl Bacteriol 68:199–211

Fleet GH (2006) The commercial and community significance of yeasts in food and beverage production. In: Querol A, Fleet GH (eds) Yeasts in food and beverages, vol 2. Springer, Berlin, pp 1–12

Fredenucci I, Chomarat M, Boucaud C, Flandrois JP (1998) *Saccharomyces boulardii* fungemia in a patient receiving Ultralevure therapy. Clin Infect Dis 27:222–223

Fredlund E, Druvefors U, Boysen EM, Lingsten K, Schnurer J (2002) Physiological characteristics of the biocontrol yeast *Pichia anomala* J121. FEMS Yeast Res 2:395–402

Gatesoupe FJ (2007) Live yeasts in the gut: natural occurrence, dietary introduction and their effects on fish health and development. Aquaculture 267:20–30

Gaxiola R, de Larrinoa IF, Villalba JM, Serrano R (1992) A novel and conserved salt induced protein is an important determinant of salt tolerance in yeast. EMBO J 11:3157–3164

Gedek BR (1999) Adherence of *Escherichia coli* serogroup O 157 and the *Salmonella typhimurium* mutant DT 104 to the surface of *Saccharomyces boulardii*. Mycoses 42:261–264

Girard P, Pansart Y, Lorette I, Gillardin JM (2003) Dose-response relationship and mechanism of action of *Saccharomyces boulardii* in castor-oil-induced diarrhoea in rats. Dig Dis Sci 48:770–774

Gismondo MR, Drago L, Lombardi A (1999) Review of probiotics available to modify gastrointestinal flora. Int J Antimicrob Ag 12:287–292

Gomez-Alarcon RA, Dudas D, Huber IT (1990) Influence of *Aspergillus oryzae* on rumen and total tract digestion of dietary components. J Dairy Sci 73:703

Graff S, Hussain S, Chaumeil JC, Charrueau C (2008) Increased intestinal delivery of viable *Saccharomyces boulardii* by encapsulation in microspheres. Pharmaceut Res 25(6):1290–1296

Guslandi M, Mezzi G, Sorghi M, Testoni PA (2000) *Saccharomyces boulardii* in maintenance treatment of Crohn's disease. Dig Dis Sci 45:1462–1464

Han SW, Lee KW, Lee BD, Sung CG (1999) Effect of feeding *Aspergillus oryzae* culture on fecal microflora, egg qualities and nutrient metabolizabilities in laying hens. Asian Australian J Anim Sci 12:417–421

Harris B, Lobo R (1988) Feeding yeast culture to lactating dairy cows. J Dairy Sci 71(suppl 1):276
Hassanein SM, Soliman NK (2010) Effect of Probiotic (*Saccharomyces cerevisiae*) adding to diets on intestinal microflora and performance of hy-line layers hens. J American Sci 6(11):159–169
Hennequin C, Kauffmann-Lacroix C, Jobert A, Viard JP, Ricour C, Jacquemin JL, Berche P (2000) Possible role of catheters in *Saccharomyces boulardii* fungemia. European J Clin Microbiol Infect Dis 19:16–20
Hirimuthugoda NY, Chi Z, Wu L (2007) Probiotic yeasts with phytase activity identified from the gastrointestinal tract of sea cucumbers. SPC Beche de Mer Information Bulletin 26:31–33
Holubarova A, Muller P, Svoboda A (2000) A response of yeast cells to heat stress: cell viability and the stability of cytoskeletal structures. Scripta Medica (Brno) 73(6):381–392
Hottiger T, Boller T, Wiemken A (1987) Rapid changes of heat and desiccation tolerance correlated with changes of trehalose content in *Saccharomyces cerevisiae* cells subjected to temperature shifts. FEBS Lett 220:113–115
Hottiger T, De Virgilio C, Bell W, Boller T, Wiemken A (1992) The 70-kilodalton heat-shock proteins of the SSA subfamily negatively modulate heat-shock-induced accumulation of trehalose and promote recovery from heat stress in the yeast, *Saccbaromyces cerevisiae*. Eur J Biocbem 210:125–132
Hounsa CG, Brandt EV, Thevelein J, Hohman S, Prior BA (1998) Role of trehalose in survival of *Saccharomyces cerevisiae* under osmotic stress. Microbiology 144:671–680
Izadnia F, Wong CT, Kocoshis SA (1998) Brewer's yeast and *Saccharomyces boulardii* both attenuate *Clostridium difficile* induced colonic secretion in the rat. Dig Dis Sci 43:2055–2060
Jahn HU, Ullrich R, Scheneider T, Lichr RM, Schieferdecker HL, Holst H, Zeitz M (1996) Immunological and tropical effect of *Saccharomyces cesevisceae* and *Saccharomyces boulardii* on the small intestine in healthy human volunteers. Digestion 57:95–104
Jakobsen M, Narvhus J (1996) Yeasts and their possible beneficial and negative effects on the quality of dairy products. Int Dairy J 6:755–768
Jawhara S, Poulain D (2007) *Saccharomyces boulardii* decreases inflammation and intestinal colonization by *Candida albicans* in a mouse model of chemically induced colitis. Med Mycol 45:691–700
Jespersen L (2003) Occurrence and taxonomic characteristics of strains of *Saccharomyces cerevisiae* predominant in African indigenous fermented foods and beverages. FEMS Yeast Res 3:191–200
Kabir SML (2009) The role of probiotics in the poultry industry. Int J Mol Sci 10:3531–3546
Kellems RO, Johnston NP, Wallentine MV, Lagerstedt A, Andrus D, Jones R, Huber JT (1987) Effect of feeding *Aspergillus oryzae on* performance of cows during early lactation. J Dairy Sci 70(Suppl 1):219
Kim SH, Dj YU, Park SY, Lee SJ, Ryu KS (2002) Effects of single or mixed feeding of lactobacillus and yeast on performance, nutrient digestibility intestinal microflora, and fecal NH_3 gas emission in laying hens. Korean J Poultry Sci 29:225–231
Kim SH, Park SY, Yu DJ, Lee SJ, Ryu KS, Lee DG (2003) Effect of feeding *Aspergillus oryzae* ferments on performance, intestinal microflora, blood serum components and environmental factors in broiler. Korean J Poult Sci 30:151–159
Klich MA, Mullaney EJ (1987) DNA restriction enzyme fragment polymorphism as a tool for rapid differentiation of *Aspergillus flavus* from *Aspergillus oryzae*. Exp Mycol 1:170–175
Klich MA, Yu J, Chang PK, Mullaney EJ, Bhatnagar D, Cleveland TE (1995) Hybridization of genes involved in aflatoxin biosynthesis to DNA of aflatoxigenic and non-aflatoxigenic aspergilli. Appl Microbiol Biotechnol 44:439–443
Koedrith P, Dubois E, Scherens B, Jacobs E, Boonchird C, Messenguy F (2008) Identification and characterization of a thermotolerant yeast strain isolated from banana leaves. Sci Asia 34:147–152

Kofli NT, Nagahisa K, Shimizu H, Shioya S (2006) Responses of different strains of *Saccharomyces cerevisiae* to osmotic stress (Tindak Balas Pelbagai Strain *Saccharomyces cerevisiae* Terhadap Tekanan Osmotik). Sains Malaysiana 35(2):9–15

Kourelis A, Kotzamanidis C, Litopoulou-Tzanetaki E, Scouras ZG, Tzanetakis N, Yiangou M (2010a) Preliminary probiotic selection of dairy and human yeast strains. J Biol Res Thessaloniki 13:93–104

Kourelis A, Kotzamanidis C, Litopoulou-Tzanetaki E, Papaconstantinou J, Tzanetakis N, Yiangou M (2010b) Immunostimulatory activity of potential probiotic yeast strains in the dorsal air pouch system and the gut mucosa. J Appl Microbiol 109:260–271

Kumeda Y, Asao T (2001) Heteroduplex panel analysis, a novel method for genetic identification of *Aspergillus* section Flavi strains. Appl Env Microbiol 67:4084–4090

Kumura H, Tanoue Y, Tsukahara M, Tanaka T, Shimazaki K (2004) Screening of dairy yeast strains for probiotic applications. J Dairy Sci 87:4050–4056

Kurtzman CP, Robnett CJ (1997) Identification of clinically important ascomycetous yeasts based on nucleotide divergence in the 50 end of the large-subunit (26 S) ribosomal DNA gene. J Clin Microbiol 35:1216–1223

Kurtzman CP, Robnett CJ (1998) Identification and phylogeny of ascomycetous yeasts from analysis of nuclear large subunit (26 S) ribosomal DNA partial sequences. Antonie Van Leeuwenhoek 73:331–371

Lee CZ, Liou GY, Yuan GF (2006a) Comparison of the aflR gene sequences of strains in *Aspergillus* section Flavi. Microbiology 152:161–170

Lee KW, Lee SK, Lee BD (2006b) *Aspergillus oryzae* as probiotic in poultry: a review. Int J Poult Sci 5:1–3

Lee YH, Tominaga M, Hayashi R, Sakamoto K, Yamada O, Akita O (2006c) *Aspergillus oryzae* strains with a large deletion of the aflatoxin biosynthetic homologous gene cluster differentiated by chromosomal breakage. Appl Microbiol Biotechnol 72:339–345

Lee SK, Kim YW, Chi SG, Joo YS, Kim HJ (2008) The Effect of *Saccharomyces boulardii* on human colon cells and inflammation in rats with trinitrobenzene sulfonic acid-induced colitis. Dig Dis Sci 54:255–263

Leuschner RGK, Bew J, Fourcassier P, Bertin G (2004) Validation of the official control method based on polymerase chain reaction (PCR) for identification of authorised probiotic yeast in animal feed. Syst Appl Microbiol 27:492–500

Li N, Zhang LM, Zhang KQ, Deng JS, Prandl R, Schoff F (2006) Effects of heat stress on yeast heat shock factor-promoter binding *in vivo*. Acta Biochimica et Biophysica Sinica 38 (5):356–362

Lopez-Daz TM, Santos JA, Prieto M, Garca-Lopez ML, Otero A (1995) Mycoflora of a traditional Spanish blue cheese. Neth Milk Dairy J 49:191–199

Lopitz-Otsoa F, Rementeria A, Elguezabal N, Garaizar J (2006) Kefir: a symbiotic yeasts-bacteria community with alleged healthy capabilities. Revista Iberoamericana de Micologia 23:67–74

Lourens A, Viljoen BC (2001) Growth and survival of a probiotic yeast in dairy products. Food Res Int 34:791–796

MacKenzie KF, Singh KK, Grown AD (1988) Water stress plating hypersensitivity of yeast: protective role of trehalose in *Saccharomyces cerevisiae*. J Gen Microbiol 134:1661–1666

MacKenzie DA, Defernez M, Dunn WB, Brown M, Fuller LJ, Seco de Herrera SRM, Gunther A, James SA, Eagles J, Philo M, Goodacre R, Roberts IN (2008) Relatedness of medically important strains of *Saccharomyces cerevisiae* as revealed by phylogenetics and metabolomics. Yeast 25:501–512

Marcus KM, Huber IT, Cramer S (1986) Influence of feeding Vitaferm during hot weather on performance of lactating cows in a large dairy herd. J Dairy Sci 69(Suppl 1):188

Marteau P, Gerhardt MF, Myara A, Bouvier E, Trivin F, Rambaud JC (1995) Metabolism of bile salts by alimentary bacteria during transit in human small bowel. Microbiol Ecol Health Dis 8:151–157

Martins CVB, Horii J, Kleiner AP (2004) Characterization of fusion products from protoplasts of yeasts and their segregates by electrophortic karyotyping and RAPD. Rev Microbial 30:1–4

Martins FS, Nardi RMD, Arantes RME, Rosa CA, Neves M, Nicoli JR (2005) Screening of yeasts as probiotic based on capacities to colonize the gastrointestinal tract and to protect against enteropathogen challenge in mice. J Gen Appl Microbiol 51:83–92

Martins FS, Rodrigues ACP, Tiago FCP, Penna FJ, Rosa CA, Arantes RME, Nardi RMD, Neves MJ, Nicoli JR (2007) *Saccharomyces cerevisiae* strain 905 reduces the translocation of *Salmonella enteric* serotype Typhimurium and stimulates the immune system in gnotobiotic and conventional mice. J Med Microbiol 56:352–359

Martins FS, Castro IM, Rosa CA, Nicoli JR, Neves MJ (2008) Effect of the trehalose levels on the screening of yeast as probiotic by *in vivo* and *in vitro* assays. Brazilian J Microbiol 39:50–55

Martins FS, Silva AA, Vieira AT, Barbosa FHF, Arantes RME, Teixeira MM, Nicoli JR (2009) Comparative study of *Bifidobacterium animalis*, *Escherichia coli*, *Lactobacillus casei* and *Saccharomyces boulardii* probiotic properties. Arch Microbiol 191:623–630

Masek T, Mikulec Z, Valpotic H, Antunac N, Mikulec N, Stojevic Z, Filipovic N, Pahovic S (2008) Influence of live yeast culture (*Saccharomyces cerevisiae*) on milk production and composition, and blood biochemistry of grazing dairy ewes during the milking period. Acta Vet Brno 77:547–554

Massot J, Descauclois J, Astoin J (1982) Protection par *Saccharomyces boulardii* de la iarrhée à *E. coli* du souriceau. Ann Pharmac Fran 40(5):445–449

McCullough MJ, Clemons KV, McCusker JH, Stevens DA (1998) Species identification and virulence attributes of *Saccharomyces boulardii* (nom. inval). J Clin Microbiol 36:2613–2617

McFarland LV (2010) Systematic review and meta-analysis of *Saccharomyces boulardii* in adult patients. World J Gastroenterol 16(18):2202–2222

McFarland LV, Surawicz CM, Greenberg RN, Fekety R, Elmer GW, Moyer KA, Melcher SA, Bowen KE, Cox JL, Noorani Z (1994) A randomized placebocontrolled trial of *Saccharomyces boulardii* in combination with standard antibiotics for *Clostridium difficile* disease. J American Med Assoc 271:1913–1918

McFarland LV, Surawicz CM, Greenberg RN, Elmer GW, Moyer KA, Melcher SA, Bowen KE, Cox JL (1995) Prevention of beta-lactam-associated diarrhea by *Saccharomyces boulardii* compared with placebo. American J Gastroenterol 90:439–448

Miller-Webster T, Hoover WH, Holt M (2002) Influence of yeast culture on ruminal microbial metabolism in continuous culture. J Dairy Sci 85:2009–2014

Miranda RLA, Mendoza MGD, Barcena-Gama JR, Gonzalez MSS, Ferrara R, Ortega CME, Cobos PMA (1996) Effect of *Saccharomyces cerevisiae* or *Aspergillus oryzae* cultures and NDF level on parameters of ruminal fermentation. Anim Feed Sci Tech 63:289–296

Molina FI, Jong SC, Huffman JL (1993) PCR amplification of the 3 V external transcribed and intergenic spacers of the ribosomal DNA repeat unit in three species of Saccharomyces. FEMS Microbiol Lett 108:259–263

Montiel D, Dickinson MJ, Lee HA, Dyer PS, Jeenes DJ, Roberts IN, James S, Fuller LJ, Matsuchima K, Archer DB (2003) Genetic differentiation of the *Aspergillus* section Flavi complex using AFLP fingerprints. Mycol Res 107:1427–1434

Morris GJ, Winters L, Coulson GE, Clarke KJ (1986) Effect of osmotic stress on the ultrastructure and viability of the yeast *Saccharomyces cerevisiae*. J Gen Microbiol 132:2023–2034

Morris GJ, Coulson GE, Clarke KJ (1988) Freezing injury in *Saccharomyces cerevisiae*: the effect of growth conditions. Cryobiology 25:471–482

Moslehi-Jenabian S, Pedersen LL, Jespersen L (2010) Beneficial effects of probiotic and food borne yeasts on human health. Nutrients 2:449–473

Murzyn A, Krasowska A, Stefanowicz P, Dziadkowiec D, Łukaszewicz M (2010) Capric Acid secreted by *S. boulardii* inhibits *C. albicans* filamentous growth, adhesion and biofilm formation. PLoS ONE 5(8):e12050. doi:10.1371/journal.pone.0012050

Neves MJ, Etchebehere L, Brandao RL, Castro IM, Lima ME, Nicoli JR (2002) Partial characterization of cholera toxin binding on membranes of *Saccharomyces boulardii*. Microecol Ther 29:185–190

Newbold CJ (1996) Probiotics for ruminants. Ann Zootech 45:329–335

Newbold CJ, Wallace RJ, McIntosh FM (1996) Mechanisms of action of the yeast *Saccharomyces cerevisiae* as a feed additive for ruminants. British J Nutr 76(2):249–261

Niault M, Thomas F, Prost J, Ansari FH, Kalfon P (1999) Fungemia due to Saccharomyces species in a patient treated with enteral *Saccharomyces boulardii*. Clin Infect Dis 28:930

Nicoli JR, Castro IM (2004) Molecular and physiological comparisons between *Saccharomyces cerevisiae* and *Saccharomyces boulardii*. Canadian J Microbiol 50:615–621

Nilson A, Peralta JMF, Miazzo RD (2004) Use of brewer's yeast (*S. cerevisiae*) to replace part of the vitamin mineral premix in finisher broiler diets. XXII Worlds Poultry Congress, Istanbul, Turkey

Nisbet DJ, Martin SA (1991) Effect of a *Saccharomyces cerevisiae* culture on lactate utilization by the ruminal bacterium *Selenomonas ruminantium*. J Anim Sci 69:4628–4633

Nivien A, Abosereh EAM, Soliman AB, Khalek AEL (2006) Mutation induction for genetic improvement of *saccharomyces boulardii* which used as a probiotic yeast. Res J Agr Biol Sci 2 (6):478–482

Ofek I, Mirelman D, Sharon N (1977) Adherence of *Escherichia coli* to human mucosal cells mediated by mannose receptors. Nature 265:623–625

Oh Y, Osato MS, Han X, Bennett G, Hong WK (2002) Folk yoghurt kills *Helicobacter pylori*. J Appl Microbiol 93:1083–1088

Omar AHM (2006) Nutritional studies on using some feed additives in poultry ration. M. Sc Thesis, Department of Animal Production, Faculty of Agriculture, Minia University, Minia Egypt

O'Sullivan DJ (2001) Screening of intestinal microflora for effective probiotic bacteria. Agri Food Chem 49(4):1751–1760

Ouwehand AC, Kirjavainen PV, Shortt C, Salminen S (1999a) Probiotics: mechanisms and established effects. Int Dairy J 9:43–52

Ouwehand AC, Kirjavainen PV, Grfnlund MM, Isolauri E, Salminen S (1999b) Adhesion of probiotic micro-organisms to intestinal mucus. Int Dairy J 9:623–630

Ouwehand AC, Tflkkf S, Kulmala J, Salminen S, Salminen E (2000) Adhesion of inactivated probiotic strains to intestinal mucosa. Lett Appl Microbiol 21:82–86

Ouwehand AC, Salminen S, Isolauri E (2002) Probiotics: an overview of beneficial effects. Antonie van Leewenhoek 82:279–289

Pardo S, Galvagno MÁ, Cerrutti P (2009) Estudios de la viabilidad y la vitalidad frente al congelado de la levadura probiótica *Saccharomyces boulardii*: efecto del preacondicionamiento fisiológico. Rev Iberoam Micol 26(2):155–160

Park DY, Namkung H, Paik IK (2001) Effect of supplementary yeast culture on the performance of laying hens. J Ani Sci Tech 43(5):639–646

Park JH, Park GH, Ryu KS (2002) Effect of feeding organic acid mixture and yeast culture on performance and egg quality of laying hens. Korea J Poult Sci 29(2):109–115

Pasha C, Kuhad RC, Rao LV (2007) Strain improvement of thermotolerant *Saccharomyces cerevisiae* VS strain for better utilization of lignocellulosic substrates. J Appl Microbiol 103(5):1480–1489

Patterson JA, Burkholder KM (2003) Application of prebiotics and probiotics in poultry production. Poul Sci 82:627–631

Periti P, Tonelli F (2001) Preclinical and clinical pharmacology of biotherapeutic agents: *Saccharomyces boulardii*. J Chemother 13:473–493

Peterson SW, Kurtzman CP (1991) Ribosomal RNA sequence divergence among sibling species of yeast. Syst Appl Microbiol 14:124–129

Piper PW (1993) Molecular events associated with acquisition of heat tolerance by the yeast *Saccharomyces cerevisiae*. FEMS Microbiol Rev 11:339–356

Piva G, Belladona S, Fusconi G, Sicbaldi F (1993) Effects of yeast on dairy cow performance, ruminal fermentation, blood components, and milk manufacturing properties. J Dairy Sci 76: 2717–2722

Pletincx M, Legein J, Vandenplas Y (1995) Fungemia with *Saccharomyces boulardii* in a 1-year-old girl with protracted diarrhea. J Pediatr Gastroenterol Nutr 21:113–115

Posteraro B, Sanguinetti M, Romano L, Torelli R, Novarese L, Fadda G (2005) Molecular tools for differentiating probiotic and clinical strains of *Saccharomyces cerevisiae*. Int J Food Microbial 103:295–304

Pothoulakis C (2009) Review article: anti-inflammatory mechanisms of action of *Saccharomyces boulardii*. Aliment Pharmacol Ther 30:826–833

Pothoulakis C, Kelly CP, Joshi MA, Gao N, O'Keane CJ, Castagliuolo I, Lamont JT (1993) *Saccharomyces boulardii* inhibits *Clostridium difficile* toxin A binding and enterotoxicity in rat ileum. Gastroenterology 104:1108–1115

Psani M, Kotzekidou P (2006) Technological characteristics of yeast strains and their potential as starter adjuncts in Greek-style black olive fermentation. World J Microbiol Biotechnol 22:1329–1336

Putnam DE, Schwab CG, Socha MT, Whitehouse NL, Kierstead NA, Garthwaite BD (1997) Effect of yeast culture in the diets of early lactation dairy cows on ruminal fermentation and passage of nitrogen fractions and amino acids to the small intestine. J Dairy Sci 80(2):374–384

Rigothier MC, Maeconis J, Gapral P (1994) Effets des levures *Saccharomyces boulardii* sur les trophozoites d'entamoeba histolytica *in vitro* et dans l'amibiase caecale du jeune rat. Parasitol Res 80:10–15

Rijnders BJ, Van Wijngaerden E, Verwaest C, Peetermans WE (2000) *Saccharomyces* fungemia complicating *Saccharomyces boulardii* treatment in a non-immunocompromised host. Inten Care Med 26:825

Rios G, Ferrando A, Serrano R (1997) Mechanisms of salt tolerance conferred by over expression of the HAL1 gene in *Saccharomyces cerevisiae*. Yeast 13:515–528

Riquelme AJ, Calvo MA, Guzman AM, Depix MS, Garcia P, Perez C, Arrese M, Labarca JA (2003) *Saccharomyces cerevisiae* fungemia after *Saccharomyces boulardii* treatment in immunocompromised patients. J Clin Gastroenterol 36:41–43

Robinson PH, Garrett JE (1999) Effect of yeast culture (*Saccharomyces cerevisiae*) on adaptation of cows to postpartum diets and on lactational performance. J Anim Sci 77:988–999

Rodrigues ACP, Nardi RM, Bambirra EA, Vieira EC, Nicoli JR (1996) Effects of *Saccharomyces boulardii* against experimental oral infection with *Salmonella typhimurium* and *Shigella flexneri* in conventional and gnotobiotic mice. J Appl Microbiol 81:251–256

Rodrigues AC, Cara DC, Fretez SH, Cunha FQ, Vieira EC, Nicoli JR, Vieira LQ (2000) *Saccharomyces boulardii* stimulates sIg A production and the phagocytic system of gnotobiotic mice. J Appl Microbiol 89:404–414

Saarela MH, Alakomi HL, Puhakka A, Matto J (2009) Effect of the fermentation pH on the storage stability of Lactobacillus rhamnosus preparations and suitability of in vitro analyses of cell physiological functions to predict it. J Appl Microbiol 106:1204–1212

Salminen SJ, von Wright AJ, Ouwehand AC, Holzapfel WH (2001) Safety assessment of probiotics and starters. In: Adams MR, Nout MJR (eds), Fermentation and Food Safety. Aspen Publishers, Gaithersburg, 239–251

Sanchez Y, Taulien J, Borkovich KA, Lindquist S (1992) Hsp104 is required for tolerance to many forms of stress. EMBO J 11:2357–2364

Sargent G, Wickens H (2004) Brewers' yeast in *C. difficile* infection: probiotic or B-group vitamins? The Pharmaceut J 273:230–231

Scanlan PD, Marchesi JR (2008) Micro-eukaryotic diversity of the human distal gut microbiota: qualitative assessment using culture-dependent and -independent analysis of faeces. The ISME J 2:1183–1193

Schingoethe DJ, Linke KN, Kalscheur KF, Hippen AR, Rennich DR, Yoon I (2004) Feed efficiency of mid-lactation dairy cows fed yeast culture during summer. J Dairy Sci 87: 4178–4181

Seguela JP, Massot J, Nesson J, Patte F (1978) Action d'un Saccharomyces lors d'une infestation expérimentale a *Candida albicans* chez le rat normal et chez le rat traité par antibiotique. Bull Soc Mycol Med 7:199–202

Serrano R (1996) Salt tolerance in plants and microorganisms: toxicity targets and defense responses. Int Rev Cytol 165:1–52

Shareef AM, Al-Dabbagh ASA (2009) Effect of probiotic (*Saccharomyces cerevisiae*) on performance of broiler Chicks. Iraqi J Vet Sci 23:23–29

Sharaf AN, Abdel N, Abosereh R, Abdalla SM, Mohamed HALA, Salim RGS (2009) Impact of Some Genetic treatments on the probiotic activities of *Saccharomyces boulardii*. Res J Cell Mol Biol 3(1):12–19

Sharma KS, Sharma KV, Kumari M, Shavani K, Mukul KS, Katoch BS (2001) Performance of laying pullets fed microbial combinations up to eighteen weeks of age/the age of sexual maturity. Indian J Ani Sci 71(6):566–569

Sharma SC (1997) A possible role of trehalose in osmotolerance and ethanol tolerance in *Saccharomyces cerevisiae*. FEMS Microbiol Lett 152(1):11–15

Shivani K, Mukul KS, Meena K, Katoch BS (2003) Biological performance of chicken fed newly isolated probiotics. Indian J Ani Sci 73(11):1271–1273

Smits GJ, Kapteyn JC, Endo HVD, Klis FM (1999) Cell wall dynamics in yeast. Curr Opin Microbiol 2:348–352

Spring P, Wenk C, Dawson KA, Newman KE (2000) The effects of dietary mannanoligosaccharides on cecal parameter and the concentrations of enteric bacteria in the ceca of Salmonella challenged broiler chicks. Poult Sci 79:205–211

Steidler L (2003) Genetically engineered probiotics. Best Prac Res Clin Gastroenterol 17:861–876

Suh SO, McHugh JV, Pollock DD, Blackwell M (2005) The beetle gut: a hyperdiverse source of novel yeasts. Mycol Res 109:261–265

Surawicz CM, Elmer GW, Speelman P, McFarland LV, Chinn J, VanBelle G (1989a) Prevention of antibiotic associated diarrhoea by *Saccharomyces boulardii*: a prospective study. Gastroenterology 96:981–988

Surawicz CM, McFarland LV, Elmer G, Chinn J (1989b) Treatment of recurrent Clostridium difficile colitis with vancomycin and *Saccharomyces boulardii*. American J Gastroenterol 84:1285–1287

Tasteyre A, Barc MC, Karjalainen T, Bourlioux P, Collignon A (2002) Inhibition of in vitro adherence of *Clostridium difficile* by *Saccharomyces boulardii*. Microbial Pathogen 32:219–225

Tuohy KM, Probert HM, Smejkal CW, Gibson GR (2003) Using probiotics and prebiotics to improve gut health. Drug Dis Today 8:692–700

Urubschurov V, Janczyk P, Pieper R, Souffrant WB (2008) Microbiota in the gastrointestinal tract of weaned piglets with emphasis on yeasts. Proc Soc Nutr Physiol 17:107

Valente P, Gouveia FC, de Lemos GA, Pimentel D, van Elsas JD, Mendonca-Hagler LC, Hagler AN (1996) PCR amplification of the rDNA internal transcribed spacer region for differentiation of Saccharomyces cultures. FEMS Microbiol Lett 137:253–256

van der Aa kuhle A, Skovgaard K, Jespersen L (2005) *In vitro* screening of probiotic properties of *Saccharomyces cerevisiae* var *boulardii* and food borne *Saccharomyces cerevisiae* strains. Int J Food Microbiol 101:29–39

Vidon N, Huchet B, Rambaud JC (1986) Influence de *Saccharomyces boulardii*, sur la sécrétion jéjunale induite chez le rat par la toxine cholérique. Gastroenterol Clin Biol 10:13–16

Wallentine MV, Johnston NP, Andrus D, Jones R, Hubex JT, Higginbotham GE (1986) The effect of feeding *Aspergillus oryme* culture-vitamin mix on the performance of lactating *dairy* cows during periods of heat stress. J Dairy Sci 69(suppl 1):189

Wang Z, Eastridge ML, Qiu X (2001a) Effects of forage neutral detergent fiber and yeast culture on performance of cows during early lactation. J Dairy Sci 84:204–212

Wang BD, Chen DC, Kuo TT (2001b) Characterization of *Saccharomyces cerevisiae* mutant with over secretion phenotype. Appl Microbiol Biotechnol 55:712–720

WHO (1995) *Saccharomyces boulardii*: a valuable adjunct in recurrent *Clostridium difficile* disease? WHO Drug Inf 9(1):15–16

Wiedmeier RD, Arambel MJ, Walters JL (1987) Effect of yeast culture and *Aspergillus oryzae* fermentation extract on ruminal characteristics and nutrient digestibility. J Dairy Sci 70: 2063–2068

Williams PEV, Tait CAG, Innes GM, Newbold CJ (1991) Effects of the inclusion of yeast culture (*Saccharomyces cerevisiae* plus growth medium) in the diet of dairy cows on milk yield and forage degradation and fermentation patterns in the rumen of steers. J Anim Sci 69:3016–3026

Wohlt JE, Finkelstein AD, Chung CH (1991) Yeast culture to improve intake, nutrient digestibility and performance by cattle during early lactation. J Dairy Sci 74:1395–1400

Wohlt JE, Corcione TT, Zajac PK (1998) Effect of yeast on feed intake and performance of cows fed diets based on silage during early lactation. J Dairy Sci 81:1345–1352

Yousefi M, Karkoodi K (2007) Effect of probiotic thepax and *Saccharomyces cerevisiae* supplementation on performance and egg quality of laying hens. Int Poult Sci 6:52–54

Yu P, Huber JT, Theurer CB, Chen KH, Nussio LG, Wu Z (1997) Effect of steam-flaked or steam rolled corn with or without *Aspergillus oryzae* in the diet on performance of dairy cows fed during hot weather. J Dairy Sci 80:3293–3297

Zhang AW, Lee BD, Lee SK, Lee KW, An GH, Song KB, Lee CH (2005) Effects of yeast (*Saccharomyces cerevisiae*) cell components on growth performance, meat quality, and ileal mucosa development of broiler chicks. Poult Sci 84:1015–1021

Zunic P, Lacotte J, Pegoix M, Buteux G, Leroy G, Mosquet B, Moulin M (1991) *Saccharomyces boulardii* fungemia. Apropos of a case. Therapie 46:498–499

Detection and Identification of Probiotic Microorganisms and Other Beneficial Organisms from the Human GI Tract

Berit Hippe, Jutta Zwielehner, Angelika Pirker, William M. Smith, and Alexander G. Haslberger

Contents

1 Introduction 58
2 Bacteriophage and Transposon Infections of Probiotic Strains 60
3 Identification of Gut Microorganisms 60
3.1 16S rRNA Sequencing 61
3.2 Alternative Targets to 16S rRNA Coding Regions 62
3.3 Quantitative PCR 62
3.4 High Resolution Melt Analysis 63
3.5 Fluorescence In Situ Hybridization 64
4 Fingerprinting Techniques 67
4.1 Denaturing DNA Gradient Gel Electrophoresis 67
4.2 Pulsed Field Gel Electrophoresis 67
4.3 Randomly Amplified Polymorphic DNA 68
4.4 Repetitive Genomic Element PCR 69
4.5 Restriction Fragment Length Polymorphism and Terminal Restriction Fragment Length Polymorphism 69
5 High-Throughput Sequencing 70
6 Microarray 74
6.1 16S rDNA Microarray for Phylogenetic Typing of Gut Communities 75
6.2 Array Tubes 77
6.3 Microarrays for Identifying the Mechanism of Action 77
7 Detection of GMO: Probiotics 78
8 Conclusions 80
References 81

B. Hippe • J. Zwielehner • A. Pirker • A.G. Haslberger (✉)
Department for Nutritional Sciences, University of Vienna, Althanstrasse 14, Vienna, Austria
e-mail: alexander.haslberger@univie.ac.at

W.M. Smith
Crop and Food Sciences, School of Land, University of Queensland, Brisbane, QLD 4072, Australia

M.-T. Liong (ed.), *Probiotics*, Microbiology Monographs 21,
DOI 10.1007/978-3-642-20838-6_3,

Abstract Most probiotics are similar to the microorganisms naturally found in the human gastrointestinal tract, and are mainly from the genera of *Lactobacillus* or *Bifidobacterium*. Conventionally, these microorganisms have been found to be fastidious, acid-tolerant, and strictly fermentative, producing lactic and acetic acids as the major end products of sugar fermentation. Exceptions from this general description have been found to occur. In addition, interaction between probiotic microorganisms or their derivatives with the gut microbiota is now a focal point of probiotics research. This requires the characterization and enumeration of all microorganisms colonizing the gut. Molecular microbiological analysis has increased the understanding of the diversity and phylogeny of beneficial strains and their functions. Modern techniques, including genotyping methods, become increasingly important for species identification and for the differentiation of probiotic strains. The precise classification and identification of probiotic strains give a strong indication of their typical habitat and origin, safety and technical applicability, and provides possibilities for monitoring and product quality. This chapter provides an overview of probiotic strain characterization, gut metagenomics, and the analytical methods (FISH, PCR, RAPD, DGGE, repPCR, PFGE, RFLP, microarray, high throughput sequencing) required for their study.

1 Introduction

Experts have debated how to define probiotics. One widely used definition, developed by the World Health Organization and the Food and Agriculture Organization of the United Nations, is that probiotics are "live microorganisms, which, when administered in adequate amounts, confer a health benefit on the host." Most probiotics are similar to microorganisms naturally found in the human gastrointestinal (GI) tract. Most often, these organisms are *Lactobacillus* or *Bifidobacterium* species. Relatively few probiotics, such as *Saccharomyces boulardii*, are yeasts (NIH, http://nccam.nih.gov/health/probiotics). Until recently, mostly only the lactic acid bacteria (LAB) were considered of relevance for food and nutrition. LABs are Gram-positive, nonspore forming, catalase-negative, aerotolerant anaerobes. They are fastidious, acid-tolerant, and strictly fermentative, producing lactic acid as the major end product of sugar fermentation (Axelsson 1998). However, exceptions from this general description do occur.

According to this view, probiotics relate to the beneficial microbiota in the gut. In contrast to the high diversity of gut microbiota, probiotic bacteria historically comprise only a few groups of this diversity. This has been discussed in a previous critical review: "Much effort has been devoted to screening bacterial isolates for properties deemed appropriate for a 'probiotic' strain, mostly characteristics that might enable the microbes to at least survive passage through the digestive tract. There must be millions of such strains to choose from, because the intestinal milieu of humans is already the home to bacteria with these properties. The concept concentrates essentially on two groups of bacteria, lactobacilli and bifidobacteria, while practically ignoring the vast array of other species that inhabit the intestinal tract of humans. It is the impact of probiotics on the composition of the intestinal microbiota, nevertheless, that forms the basis for the probiotic concept" (Tannock et al. 1999).

Further research into the mechanisms of probiotic action has extended to the ability of cell walls or particles derived from probiotic microorganisms to exert probiotic functions. These nonliving probiotic derivatives are proposed to act by receptor-mediated functions in immune or intestinal cells (Probiotics). Many of the effects obtained from viable cells of probiotics are also obtained from populations of dead cells. Heat-killed cells of *Enterococcus faecalis* stimulate the GI immune system in chickens. Likewise, dead bifidobacteria induce significant increases in TNF-α production. Administration of heat-killed *E. faecalis* to healthy dogs has also been shown to increase neutrophil phagocytes. The probiotic paradox is that both live and dead cells in probiotic products can generate beneficial biological responses. Live probiotic cells influence both the GI microbiota and the immune response whilst the components of dead cells exert an anti-inflammatory response in the GI tract (Adams 2010).

This interaction between probiotic microorganisms or their derivatives with the gut microbiota is now a focal point of probiotics research. This requires the characterization and enumeration of all microorganisms colonizing the gut. Modern molecular techniques, including genotyping methods, have become increasingly important for species identification and for the differentiation of probiotic strains. The precise classification and identification of a probiotic strain gives a strong indication of its typical habitat and origin, indicate the strain's safety and technical applicability, provides possibilities for monitoring and product quality and is a basis for patenting and commercial use. Current genotyping methods include denaturing gradient gel electrophoresis (DGGE), restriction fragment length polymorphism (RFLP), pulse field gel electrophoresis (PFGE), and others. These methods all require previously isolated pure cultures of gut microorganisms.

In gut metagenomics, methods are applied which do not require the isolation of pure cultures. It enables the analysis and characterization of species and groups of microorganisms with possibly probiotic functions, their diversity, ecology, and interaction as well as possible consequences of probiotic intervention. Clearly, methods for identification and classification of strains and metagenomic strategies follow different objectives and face specific restrictions. Compared with Sanger sequencing-based methods, new high-throughput pyrosequencing methods provide much more data on the diversity of organisms in their natural habitat, but technological biases and relative accuracy remain poorly understood (Tedersoo et al. 2010).

This chapter provides an overview (Fig. 12) of probiotic strain characterization, gut metagenomics, and the analytical methods required for their study.

2 Bacteriophage and Transposon Infections of Probiotic Strains

Bacteriophage infections of LAB pose a serious risk to the industrial production of dairy foodstuffs. Dairy fermentations are susceptible to phage infection since the starting material (mainly, raw milk) is not sterile and, in general, pasteurization processes are not adequate to deactivate viral particles. In addition, the continued use of the same starter cultures provides a constant host for phage proliferation which, consequently, can lead to slow lactic acid production or even the complete failure of fermentation, which allows the proliferation of pathogenic bacteria and can cause significant economic losses due to product rejection. Ways to improve resistances against phage infections have been described including the triggering of suicide systems (Moineau 1999; Djordjevis et al. 1997).

The available knowledge of *Lactobacillus* phages is limited when compared with that of *Lactococcus* and *Streptococcus* bacteriophages. Only a small number of *Lactobacillus* phages have been studied in detail (Alemayehu et al. 2009; Capra et al. 2006). PCR methods have been described for specific identification of strains based on phage-related sequences for probiotic *Lactobacillus rhamnosus* strains (Brandt and Alatossava 2003) and for the detection and identification of bacteriophages infecting *Lactobacillus casei*/*paracasei*, based on highly conserved regions (Binetti et al. 2008).

Complete genome sequencing of several probiotic strains provides insight into their adaptation to the gut environment and also highlights the abundance of transposable elements present in these bacteria. Transposable elements can promote genome plasticity, phenotypic changes, and contribute to bacterial adaptability. Identified phages, tempered phages, or prophages contribute to mechanisms of a horizontal gene transfer among bacteria in the human gut (Falentin et al. 2010). This genetic transfer is a serious aspect of the risk assessment required for approval of probiotic products, especially those probiotic strains that are genetically modified.

3 Identification of Gut Microorganisms

Historically, bacteria have been identified by phenotypic characterization, although more recently a number of molecular methods have been developed for the identification of a microorganism, i.e., the assignment of an unknown microorganism to a known taxon at the species or subspecies level. Depending on the objectives, a multilevel approach is recommended. The appropriate molecular methods to be used depend on the level of identification required (genus, species, or subspecies) and the target organism. The methods differ significantly in labor and personnel expenses, in the degree of discrimination, and in reproducibility within and between laboratories.

One of the first molecular methods for species differentiation is DNA–DNA hybridization (Sibley et al. 1984). The technique compares the whole genomes of two organisms. A DNA mixture is incubated so that the DNA strands can separate and reassociate. Reassociation leads to the formation of hybrid DNA double strands. Hybrid DNA can only be formed when the sequences are at least 80% complementarity. The smaller the genetic distance between the two organisms is, the more closely will the hybrids bind together and the more energy will be required to separate the two strands again. Two individuals belong to one species if the difference of the melting temperature of the DNA hybrids is less than 5% or their similarity is at least 97.5 over their whole genome (Krogius-Kurikka et al. 2009; Shneyer 2007).

The 16S rRNA gene is often used for differentiation and classification of microorganisms. The degree of similarity of the 16S rRNA genes from different individuals is representative of the variation in the whole genome of different individuals.

The first crucial step for successful identification of bacteria from feces using 16S rDNA sequence comparison is to isolate DNA from the appropriate sample. DNA from Gram-positive bacteria, including LAB and bifidobacteria (Cypionka 2010), is more difficult to extract as from Gram-negative bacteria. This is due to their thick peptidoglycan cell wall which requires additional enzymatic or mechanical lysis steps to yield useful quantities of pure DNA.

LAB colonize the small intestine at cell numbers of 10^2/g gut content. LAB that migrate through the GI tract encounter high numbers of anaerobic bacteria in the colon at cell densities of about 10^{12}–10^{13} cells/g fecal content (Fujimura et al. 2010). This considerable dilution of LAB in the fecal content renders their detection with PCR-based methods difficult and in some cases molecular methods will even fail to detect LAB in fecal samples.

The colon is a strictly anaerobic environment and is most suitable for establishment and proliferation of bifidobacteria (which comprise only 3–6% of adult GI microbiota). Their detection from fecal content will be possible in most cases, but the fact that they occur at relatively low abundance among the densely populated colon environment does pose a challenge for the molecular biologist (Fujimura et al. 2010).

Each of the methods used for the metagenomic detection and identification of beneficial microbes has its advantages and disadvantages.

3.1 16S rRNA Sequencing

DNA sequencing, developed by Frederick Sanger, has brought about a revolution in molecular biology. DNA sequence data can be compared to determine the degree of relatedness between the two sequences of DNA. This can be translated to the degree of relatedness of organisms that the DNA originates from.

During Sanger sequencing, the target sequence is amplified in a PCR reaction using a mixture of nucleotides and fluorescently labeled dideoxy-nucleotides.

Incorporation of a dideoxy-nucleotide in the newly synthesized DNA strand causes strain termination. Statistically, DNA synthesis is terminated at each possible nucleotide position in each reaction. Terminated DNA molecules of different lengths can be separated in an agarose gel according to size. The nucleotide sequence can then be determined (Loffler 2004).

It has been well documented that the different variable regions within the 16S rRNA gene are more or less suitable for phylogenetic identification of bacteria. There are nine variable regions within the 16S rRNA gene. They differ in information content, and some bacteria, such as *Escherichia coli* and *Shigella,* even do not differ at all within the 16S rRNA gene. 16S rRNA gene sequencing allows species assignment within the *Bifidobacteria*, but is not suitable for strain differentiation (Bottacini et al. 2010). A major advantage of 16S rRNA gene sequencing is that it is possible to obtain sequence data from uncultured microbes originating from environmental samples (i.e., stool samples), however 16S rRNA gene sequencing can be costly and laborious (Maukonen et al. 2008).

3.2 Alternative Targets to 16S rRNA Coding Regions

The simultaneous use of several housekeeping genes in bacterial taxonomy offers a higher resolution than 16S rRNA gene sequence data at the species level as it integrates information from different molecular markers from throughout the bacterial chromosome (Stackebrandt et al. 2002; Zeigler 2003). Common genes used as an alternative to 16S rRNA are those encoding the α subunit of the ATP synthase (*atpA*), RNA polymerase α subunit (*rpoA*) and the phenylalanine t-RNA synthase α subunit (*pheS*) can be used to differentiate the relatedness between strains of the same species. These genes are often more discriminatory among closely related species and strains than the sequence of the 16S rRNA gene (Naser et al. 2006).

Bacterial genomes often contain a significant proportion of sequences originating from prophages (Hayashi et al. 2001; Chopin et al. 2001). Such phage-related sequences have been targeted for strain-specific differentiation, termed phi-sequencing. Nucleotide primers for targeting phage-related sequences in probiotic *L. rhamnosus* strains have been successfully developed (Brandt and Alatossava 2003). The authors found their assay could discriminate well between strains of *L. rhamnosus* but described amplification of phylogenetically unrelated nontarget strains, pointing out that phi-sequencing should be preferentially used in combination with other molecular methods.

3.3 Quantitative PCR

Quantitative PCR (qPCR) is used for absolute or relative quantification of microbial DNA. The specificity of primers determines which bacteria can be identified and their accurate design is essential for sensitive and specific amplification. A lack of specificity can lead to false-positive amplification of species or strains overestimation overestimating their abundance. In probiotic research, it is important to differentiate bacteria at the strain level because only certain strains of a species possess the desired properties (Felis and Dellaglio 2007).

Bacterial enumeration using qPCR is dependent on acquiring DNA concentration in each sample used as a template by spectrophotometric measurement, e.g., nanodrop. To avoid PCR bias and to allow the linearity of measurement, the same DNA concentration for all templates should be used. In addition, impurities in the DNA suspension can be identified by A_{260}/A_{280} ratio for protein contamination and A_{260}/A_{230} ratio for salt contamination. Impurities of the DNA template can have adverse effects on the efficiency of PCR and significantly affect the accuracy of quantification.

Absolute quantification is calculated by comparing the cycle threshold value of a test sample to a standard curve generated from cycle threshold values of samples of known bacterial concentrations. Relative quantification is calculated by comparing the cycle threshold value of a test sample with the cycle threshold value of control DNA. Cycle threshold refers to the PCR cycle at which the fluorescence raises over threshold (Carey et al. 2007; Louis and Flint 2009). One commonly used florescent dye is SYBR green, which intercalates with double-stranded DNA. For reliable quantification, the concentration range of dilutions should be chosen to cover the whole concentration spectrum of the samples. If using broad specificity primers targeting phylum or class levels, a mixed standard is recommended. For comparison of different runs, the use of control sample in each run is recommended (Carey et al. 2007).

Following amplification in qPCR, the PCR products can be analyzed in melt curve analysis. PCR products are subject to an increasing temperature gradient which results in denaturation of double-stranded DNA to single-stranded DNA. As SYBR green only generates fluorescence when intertwined with double-stranded DNA, the rate of fluorescence decrease can be measured generating melt curves. Melt curve analysis can be used as a quality control, as primer dimers and unspecific PCR products might melt at a different temperature. In a special application of melt curve analysis, different isoforms of a gene involved in butyrate production were analyzed. Different peaks were shown to be representative of a different phylogenetic lineage of bacteria (Louis and Flint 2009). The 16S rRNA gene is often targeted in this technique although other house-keeping genes can be used. This method requires DNA from pure, previously characterized bacterial isolates for generation of PCR products and melt curves for comparison with unknown samples.

3.4 High Resolution Melt Analysis

High Resolution Melt (HRM) analysis is used for the detection of mutations, polymorphisms, and epigenetic differences in PCR amplicons by measuring the thermal denaturation of double-stranded DNA. In microbiology, HRM analysis is used for species- and strain characterization. It characterizes nucleic acid samples based on their disassociation (melting) behavior in a similar fashion as a basic melt analysis. HRM employs a slower increase in temperature for denaturation and acquires many times more data in a single run allowing for discrimination of highly similar sequences. Samples can be discriminated according to their sequence by a combination of length and G–C content (Wittwer 2009). Detection and differentiation of samples containing sequence variants rely on a change in the amplicon Tm and/or shape of the amplicon melting curve. Samples containing the same sequence variants are identified as groups that exhibit similar melting profiles (Radvansky et al. 2011). Even single base changes such as single nucleotide polymorphisms (SNPs) can be readily identified (Wojdacz et al. 2008). Depending on primer design and the gene targeted, bacteria can be identified and differentiated at the strain level (Ereqat et al. 2010).

3.5 Fluorescence In Situ Hybridization

By this technique it is possible to identify bacteria directly in their habitat with high-resolution microscopic techniques such as confocal laser scanning microscopy. It provides information on the number and spatial distribution of microorganisms (Amann et al. 2001). Fluorescence In Situ Hybridization (FISH) employs fluorescence-labeled DNA probes to detect or confirm genes within chromosomes or gene expression. It is a method to localize a specific DNA or RNA sequence, prove genetic changes in tissues. It can be carried out at the single-cell scale in intact cells – in situ.

The hybridization is generally described as a fusion of two complementary, exactly matching, single-stranded nucleic acid molecules. For detection of a specific DNA or RNA molecule, a complementary gene probe typically labeled with a fluorochrome is required for hybridization. Typical labels include cyanine (e.g., Cy3 and Cy5) and fluorescein molecules (Michalet et al. 2005). FISH experiments often employ several differently labeled probes of different specificity (from phylum to species) allowing for reliable identification of bacteria in complex environmental habitats as shown in Fig. 1.

Commonly used probes have a length of 15–30 nucleotides and are covalently linked at the 5′-end to a single fluorescent dye molecule (Table 1). Sequence signatures serving as suitable target site for nucleic acid probing can be conveniently and automatically identified using a probe design tool (Ludwig et al. 2004). Short probes have easier access to their target, but they might carry fewer labels and are not so selective. Not all bacterial and archaeal cells can be permeabilized by oligonucleotide probes using standard fixation protocols (Bottari et al. 2006). The accessibility of selected target sites for oligonucleotide probes can be increased by adding unlabeled oligonucleotide probes that bind adjacent to the probe target site. The aim is to unfold the nucleic acid and thus facilitate probe hybridization. These

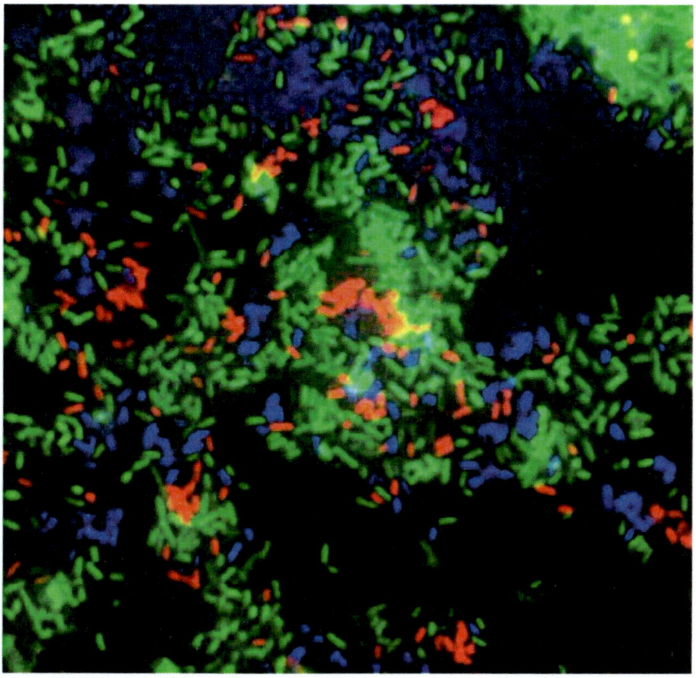

Fig. 1 Three-color fluorescent in situ hybridization image of bacteria. Three FISH probes labeled with different fluorescent dyes have been used to simultaneously identify the structure of microbial community. Each labeled single bacterial cell can be measured by Raman microspectroscopy to examine if the cell has special ecological function

Table 1 Most commonly employed fluorescent dyes to label oligonucleotides for FISH analysis (Bottari et al. 2006)

Fluorochrome	Color	Max. excitation λ (nm)	Max. emission λ (nm)
Alexa488	Green	493	517
AMCA	Blue	399	446
CY3	Red	552	565
CY5	Red	649	670
CY7	Violet	743	767
DAPI	Blue	350	456
Fluorescein	Green	494	523
Rodamine	Red	555	580
TAMRA	Red	543	575
Texas red	Red	590	615
TRITC	Red-orange	550	580

AMCA methyl cumarinic acetic acid, *CY* carbocyanine, *DAPI* 4′6-diamidino-2-phenylindole dihydrochloride, *TAMRA* tertramethyl rhodamine, *TRITC* tetramethylrodamine-isothiocyanate

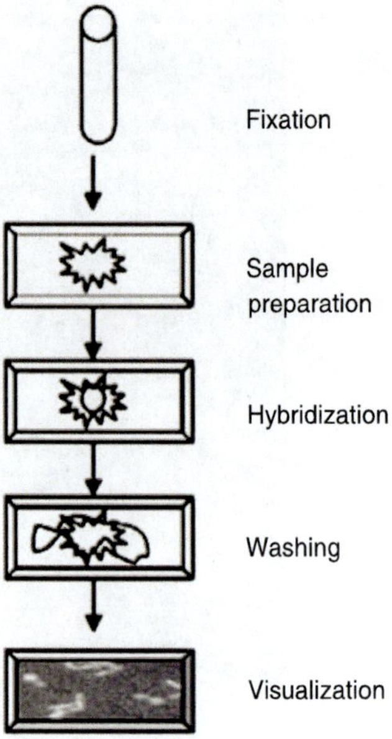

Fig. 2 Flow chart of a typical FISH. The procedure includes the following steps: (1) fixation of the specimen directly in sample tubes; (2) transfer of the sample on a glass slide and preparation of the sample, including specific pretreatment steps; (3) hybridization with the respective probes for detecting the respective target sequences; (4) washing steps to remove unbound probes; (5) mounting, visualization, and documentation of results

so-called helper probes need to be designed carefully because of their specificity to the respective probe and must have a differentiation temperature (Td) at least as high as the temperature of the probe to prevent dissociation of the helper at stringent hybridization conditions (Fuchs et al. 2000).

A typical FISH protocol (Fig. 2) includes four steps: fixation and permeabilization of the sample, hybridization, washing steps to remove unbound probe, and detection of labeled cells by microscopy or flow cytometry (Amann et al. 2001). Prior to hybridization, bacteria must be fixed and permeabilized in order to allow penetration of the fluorescent probes into the cell and to protect the RNA from degradation by endogenous RNAses (Moter and Gobel 2000). The sample is either settled on membrane filters and covered with the fixing agent (Glockner et al. 1999), or mixed with the fixing agent, incubated, sedimented by centrifugation, resuspended, transferred to glass slides and dried (Amann et al. 1990).

Probes and the stained preparations must always be stored in the dark as they are very sensitive to degradation by light. FITC (green) fades the fastest, whereas red and DAPI are very robust. When taking photos, always begin with green and use DAPI last. For a third color, mix the probes red and green to create yellow. This technique is a relatively rapid method to evaluate the presence and activity of probiotics and other gut microorganisms, directly in the sample or the process studied (Bottari et al. 2006). FISH is an in situ technique and has the potential to reveal the composition of complex microbial associations in natural systems.

4 Fingerprinting Techniques

4.1 Denaturing DNA Gradient Gel Electrophoresis

Denaturing DNA Gradient Gel Electrophoresis (DGGE) is used to determine microbial structural differences between DNA, and also to investigate broad phylogenies or specific target organisms such as pathogens or xenobiotics degraders (Sigler 2004). DGGE separates bacterial sequences by electrophoresis of PCR-amplified 16S rRNA gene fragments in a polyacrylamide gel with constantly increasing concentration of denaturants (Fischer and Lerman 1979). These denaturants are usually formamide and urea (Felske and Osborn 2005). Group-specific primers, which target groups of bacteria, such as lactobacilli (Walter et al. 2001) or bifidobacteria (Satokari et al. 2001) are available.

The oligonucleotide primers used to amplify DNA fragments for DGGE contain a GC-clamp, an approximately 40-bp region consisting mainly of guanine and cytosine. This GC-clamp prevents complete melting of the DNA fragments so that each double-stranded DNA PCR product stops migrating at a unique denaturant concentration. DNA melting depends on GC content of the fragments in contrast to agarose gel electrophoresis, which separates DNA according to length. Occasionally, some microorganisms such as *Bacteroides thetaiotaomicron*, *lactobacilli*, and some of *Fungi* produce several bands because of their multiple 16S rRNA operons. As a result the gel contains a pattern of bands and the diversity of the sample can be estimated (Sigler 2004; Blume et al. 2010). For *bifidobacteria*, the identification at the species level only is reliable (Ventura et al. 2004). The resulting bands on the DGGE gel can be compared with references amplified from known bacteria.

DGGE allows a rapid diversity assessment and comparative analysis as seen in Fig. 3. Analysis software is available for closely grouped bands at the same height to determine the degree of relatedness (Vitali et al. 2010). Further identification of fragments can be carried out by sequencing (Ventura et al. 2004).

4.2 Pulsed Field Gel Electrophoresis

Pulsed field gel electrophoresis (PFGE) enables separation of high-molecular-weight fragments which is not possible using conventional gel electrophoresis methods. During electrophoresis, the direction of the electric field changes periodically. For PFGE, rare-cutting restriction endonucleases can be used to produce a small number of large fragments from DNA. This creates an easily interpretable pattern of bands (Holzapfel et al. 2001).

PFGE is a useful tool for subtyping clinical isolates from different geographic regions and different hosts and for recognizing potentially new serovars (Galloway and Levett 2008). It is a very useful method for highlighting isolates which may represent new species or serovars (Galloway and Levett 2008). The method has

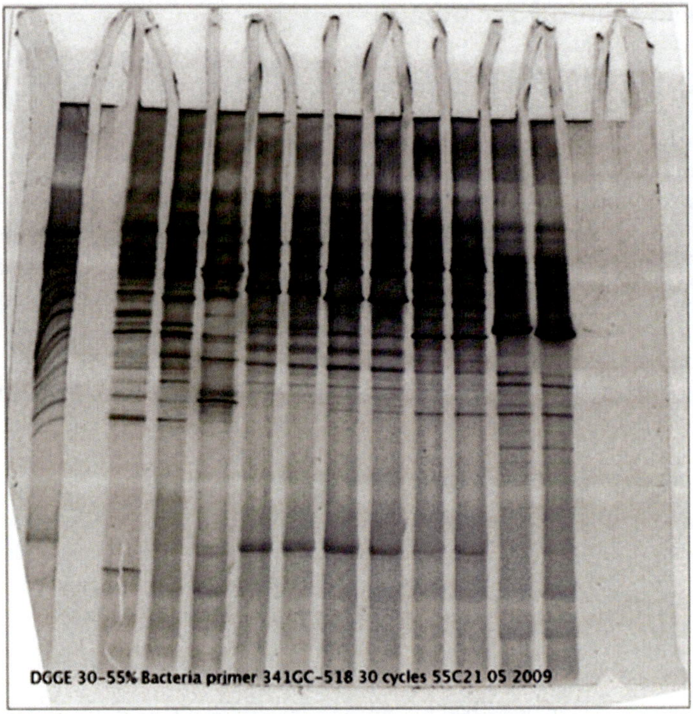

Fig. 3 PCR-DGGE band patterns of 16S rRNA coding regions of bacteria amplified with primers 341GC and 518

high reproducibility and a very good selectivity at the strain level but can be laborious and time consuming (Ventura et al. 2004; Kneifel and Domig 2009). Therefore, only a limited number of samples can be readily processed. PFGE is a frequently used method to differentiate strains of *bifidobacteria* (Wall et al. 2008; Roy et al. 1996; Ventura et al. 2004), *lactobacilli* at the strain level (Ventura and Zink 2002) but not at the genus or species level (Huys et al. 2006).

4.3 Randomly Amplified Polymorphic DNA

Randomly amplified polymorphic DNA (RAPD) is a special form of PCR which uses a short random primer at low stringency. This provides randomly sized DNA fragments (Holzapfel et al. 2001). Different species and strains can be distinguished according to the generated band patterns.

An advantage of the method is that it is simple, fast, and cheap. The disadvantage is that experiments are particularly difficult to standardize (Kneifel and Domig 2009; Ventura et al. 2004). This method makes it possible to identify lactobacilli at the strain level (Mahenthiralingam et al. 2009; Booysena et al. 2002), but comparison of genus or species levels is not possible (Huys et al. 2006). In contrast, *bifidobacteria* can be identified at species as well as at strain level (Ventura et al. 2004).

4.4 Repetitive Genomic Element PCR

All living things harbor repetitive DNA sequences in multiple sites. These sequences form the basis for several powerful tools in molecular diagnostics, medical microbiology, epidemiological analyses, and environmental microbiology (Ishii and Sadowsky 2009).

Repetitive Genomic Element PCR (RepPCR) is a simple PCR-based technique with a high discriminatory power. Furthermore repPCR is suitable for a high-throughput of strains at low cost and makes it possible to type a wide range of bacteria (Gevers et al. 2001; Olive and Bean 1999). The discriminatory power for LAB (Gevers et al. 2001) and *bifidobacteria* for example, reaches from species to strain level (Huys et al. 2006).

4.5 Restriction Fragment Length Polymorphism and Terminal Restriction Fragment Length Polymorphism

In RFLP, a PCR product is digested with one or several restriction endonucleases in order to obtain a pattern of nucleotide fragments capable of distinguishing different species or strains. Restriction enzymes cleave DNA at or around a distinct recognition site. Some restriction enzymes in nature cut DNA at random far from their recognition sites and are of little value to the molecular biologist. Other restriction enzymes recognize modified, typically methylated DNA and are exemplified by the McrBC and Mrr system of *E. coli* (BioLabs 2010).

The most common restriction enzymes used in RFLP are HhaI, HindII, and NotI that cleave DNA within their recognition sequences. Enzymes of this kind are readily available commercially. Most recognize DNA sequences that are symmetric because they bind to DNA as homodimers, but a few (e.g., BbvCI: CCTCAGC) recognize asymmetric DNA sequences because they bind as heterodimers. Some enzymes recognize continuous sequences (e.g., EcoRI: GAATTC) in which the two half-sites of the recognition sequence are adjacent, while others recognize discontinuous sequences (e.g., BglI: GCCNNNNNGGC) in which the half-sites are separated. Cleavage leaves a 3′-hydroxyl on one side of each cut and a 5′-phosphate on the other. They require only magnesium for activity and the corresponding modification enzymes require only S-adenosylmethionine. They tend to be small, with subunits in the 200–350 amino acid range (BioLabs 2010). An example for RFLP patterns is given in Fig. 4 which shows the RFLP bandpatterns of bacteria obtained from "soidon mahi" starter cultures for fermented bamboo shoot tips in

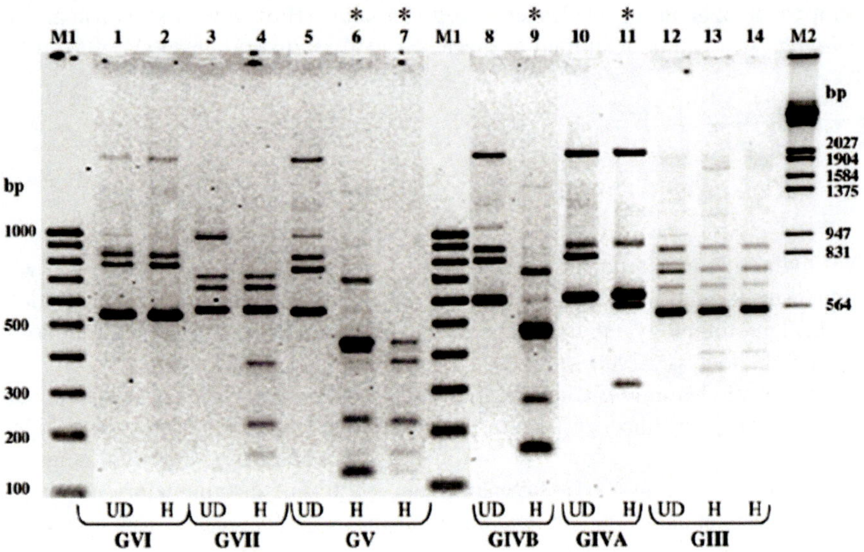

Fig. 4 ITS-PCR and HinfI-ITS-RFLP profiles of representative strains from lactic acid bacteria associated with soidon mahi starter. Lanes 1–2: *Lactobacillus plantarum* SD6B11; lanes 3–4: *Enterococcus faecium* SD3B10; lanes 5–6: *Lactobacillus brevis* SD10B10; lane 7: *L. brevis* SD10B4; lanes 8–9: *L. brevis* SD2B5; lanes 10–11: *L. brevis* SD1B8; lanes 12–13: Camobacterium sp. SD4B10; and lane 14: Carnobacterium sp. SD8B9; M1: 100 bp DNA ladder; M2: λDNA double digest (Promega). *UD* indicates undigested ITS-PCR profiles and *H* indicates HinfI digested ITS-RFLP profiles. *Asterisk* (*) indicates heterogenic HinfI-ITS-RFLP profiles of *L. brevis* isolates (Jeyaram et al. 2010)

India (Jeyaram et al. 2010). The authors found that this acidic starter was dominated by a characteristic association of *Bacillus* spp. and LAB, particularly *Lactobacillus brevis* and *Lactobacillus plantarum* (Jeyaram et al. 2010).

In terminal restriction length polymorphism analysis (T-RFLP), restriction enzymes are used on PCR products marked with a fluorescent group sitting on one primer. Thus, cleavage results in fragments of different length that can be separated using a sequencer, recognizing a fluorescent signal. Kovatcheva-Datchary et al. (2009) applied this method to analyze RNA (Fig. 5) from their RNA-stable isotope probing experiment identifying starch fermenting colonic bacteria (Kovatcheva-Datchary et al. 2009).

5 High-Throughput Sequencing

The rapid development of next-generation sequencing technologies has allowed vast numbers of partial 16S rRNA genes from uncultured bacteria to be sequenced, at a much lower cost than Sanger sequencing. In addition to bypassing previously needed cloning and/or cultivation procedures, with their associated biases,

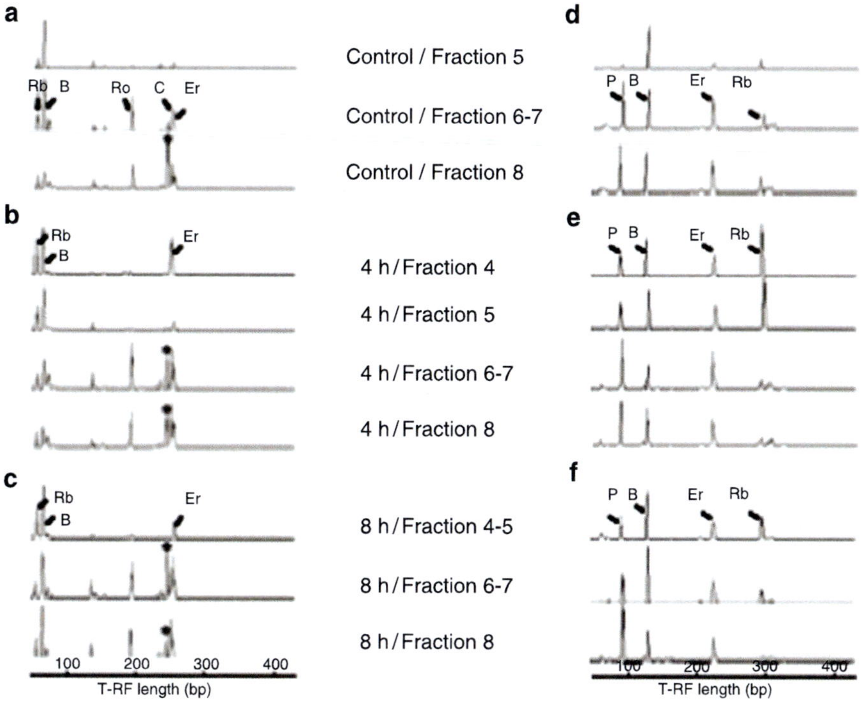

Fig. 5 T-RFLP profiles of bacterial 16S rRNA obtained from fractions of the control (**a**), 4 h gradient (**b**), and 8 h gradient (**c**), after applying AluI as restriction endonuclease; control (**d**), 4 h gradient (**e**), and 8 h gradient (**f**), after applying MspI as restriction endonuclease. Assignment of T-RF peaks to bacteria phylotypes (Rb, *Ruminococcus bromii*, B, *Bifidobacterium adolescentis*; Ro, *Ruminococcus obeum*; C, uncultured Clostridium spp.; Er, Eubacterium rectal; P, Prevotella spp.) (Kovatcheva-Datchary et al. 2009)

community structures can now be investigated at much higher resolution by revealing taxa that are much less abundant. However, this may come with lower taxonomic certainty due to the short read lengths and sometimes poorer read quality (Claesson et al. 2009). The pyrosequencing technology introduced by 454 Life Science (Margulies et al. 2005), now part of Roche, uses microscopic beads to bind individual DNA fragments. These beads with attached DNA sit in nanotiter plates and the nucleotide sequences are amplified using emulsion PCR. This process yields about 400,000 reads of 250–400bp length with an average quality score of greater than 99.5% accuracy rate (Droege and Hill 2008). Pyrosequencing has been applied to a wider range of microbial communities and variable regions of the 16S rRNA gene, such as the V6 region in microbial communities of deep-sea vents (Huber et al. 2007); V1,V2, V6, and V3 in human (Andersson et al. 2008); (Dethlefsen et al. 2008); (Turnbaugh et al. 2008) (Zhang et al. 2009) and in macaque (McKenna et al. 2008) GI tract; as well as V9 in soil-derived microbial DNA (Roesch et al. 2007).

The choice of variable regions of the 16S rRNA gene leads to considerably different reliability of sequence assignments due to differences in information content within the 16S rRNA gene as given in Table 2. Claesson et al. (2009) used sequences from human GI tract samples to investigate which region of the 16S rRNA gene would be best suited for high-throughput sequencing for this particular microbial habitat: 82.3% of sequences from the variable region V3 could be classified to genus level at a reliability of 80%, whereas only 40.4% of sequences from the V6 region could be classified to a genus at this level of reliability. The variable region V4 appeared to carry the most reliable information content, with 87.9% of sequences being assigned to genus at 80% bootstrap support (Claesson et al. 2009).

Differences in information content within the variable regions of the 16S rRNA gene lead to great differences as to which phylogenetic depth sequences can be assigned (Fig. 6). Again, reads from the variable region V4 is shown to have superior classification efficiency over the V6 region.

Table 2 Of 7,208 full-length 16 S reference sequences from the human gut, 6,054 were classified at genus level with 80% bootstrap support

Variable region	V3 (%)			V6 (%)			V4 (%)		
Bootstrap cutoff (≥)	0	50	80	0	50	80	0	50	80
Fraction of sequences classified to genus	100	92.4	82.3	100	73.5	40.4	100	97.0	87.9
Fraction of sequences correctly classified to genus	92.0	95.0	98.1	79.0	96.5	98.7	92.8	94.5	95.7

For each of the three extracted variable regions, fragments were classified at three different bootstrap thresholds, and compared with the full-length classifications (*last row*) (Claesson et al. 2009)

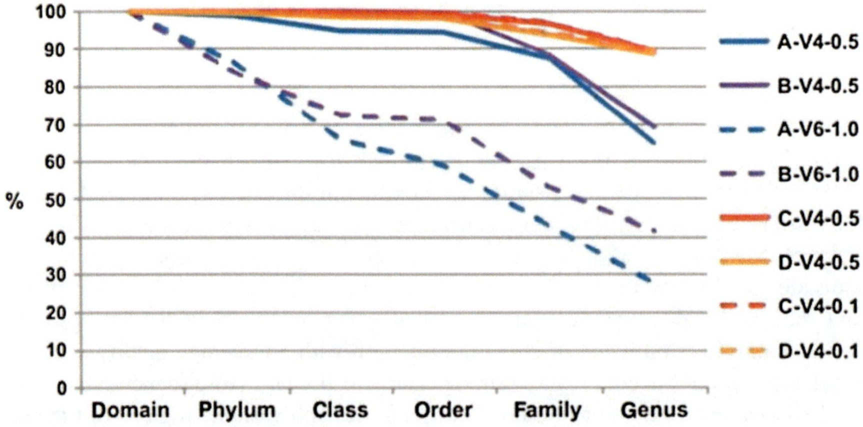

Fig. 6 Classification efficiencies at six taxonomic ranks for eight sets of sequences from four samples. The *blue* and *purple* colored *dashed lines* represent V6 amplicon reads, which have very poor classification efficiencies compared to all V4 amplicon reads, especially at the genus level. The *red* and *orange* colored *dashed lines*, representing V4-0.1 amplicon reads, show nearly identical classification efficiencies as the corresponding V4-0.5 amplicon reads (Claesson et al. 2009)

In addition to variable reliability and depth of sequence classification, the choice of variable region also influences the information on microbiota composition obtained from a high-throughput sequencing experiment (Fig. 7).

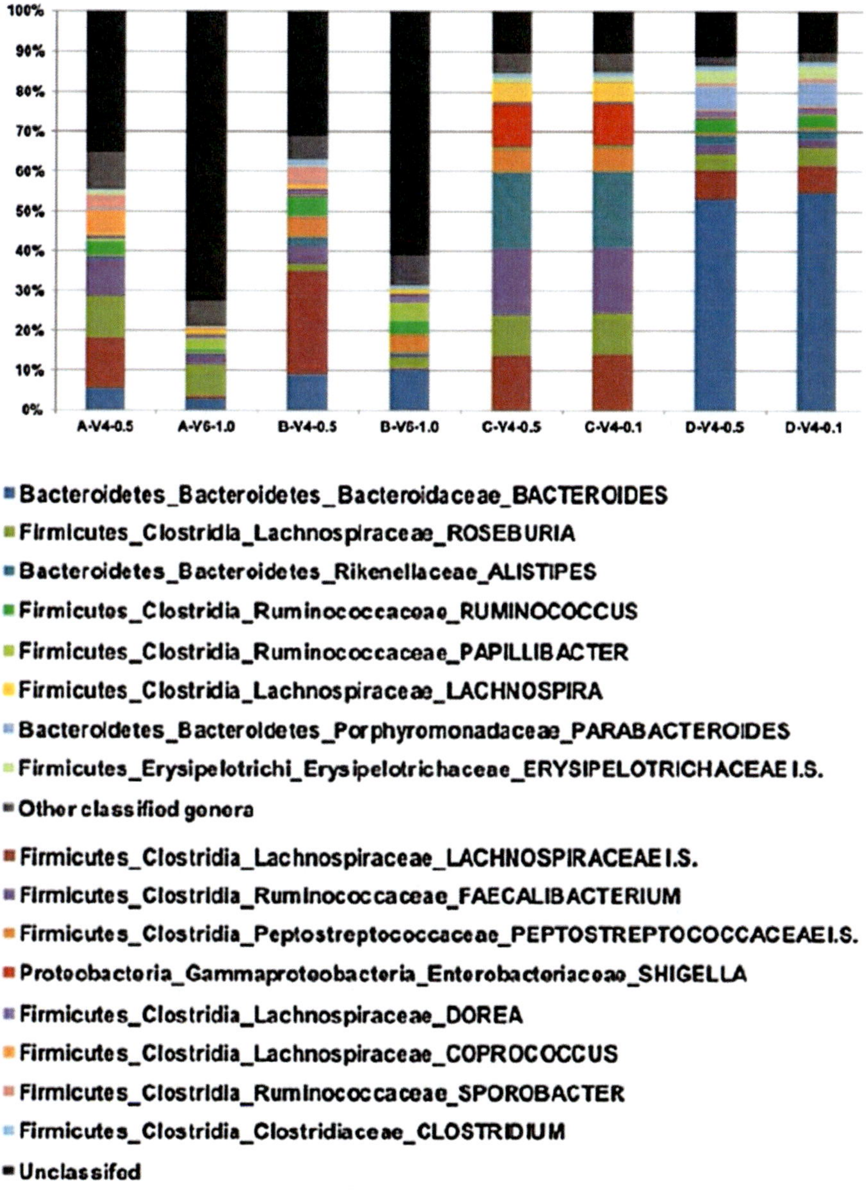

Fig. 7 Relative abundance of the 16 most abundant genera classified with at least 50% bootstrap support. Genera are labeled according to phylum_class_family_GENUS (Claesson et al. 2009)

The analysis by Claesson et al. (2009) gives a good insight into the difficulties arising with the use of high-throughput sequencing technologies. Although considered the most accurate and exhaustive technology for the characterization of microbial communities, high-throughput sequencing is not free of biases and cannot overcome the limitations in information content of the 16S rRNA gene. The application of high-throughput sequencing technologies is not limited to 16S rRNA coding regions. 454 sequencing has also been applied for sequencing an entire bacterial or viral genome in 1 day (Chen et al. 2010; Monger et al. 2010).

6 Microarray

Microarray applications allow the simultaneous comparison of the entire genome content of microorganisms by hybridization with hundreds and thousands of DNA probes. Furthermore, when targeting mRNA, microarrays can reveal which genes are differentially expressed under specific test conditions, e.g., in the food matrix or in the host. Microarrays have also been applied to characterize the strain specific properties of probiotic strains.

Although multilocus sequence typing is now regarded by many as a good standard to determine phylogenetic relationships between and within bacterial species, it does not always reflect the true genetic diversity of members of a species. Phylogenetic trees based on multilocus sequence typing may therefore differ significantly from a tree based on whole gene content (Tettelin et al. 2005).

For the construction of a microarray aiming to subtype bacterial isolates, the total number of genes potentially present in strains of a given species, the "pan-genome" (Tettelin et al. 2005; Medini et al. 2005), must be known. Thus, the development of such a microarray depends on the availability of a database of whole genome sequences of a large amount of strains. From such a database, the core genome can be defined (Fig. 8).

Based on predicted protein sequences, the *E. coli* core genome was estimated to comprise approximately 1,563 genes (Willenbrock et al.) for an infinite (or very large) number of *E. coli* genomes. Following the identification of the core genome, additional "strain-specific" genes need to be defined (Fig. 9).

For *E. coli* it was estimated that sequencing additional strains, approaching infinity, would add approximately 79 new genes. The pan-genome for *E. coli* is thus estimated to contain 9,433 different genes in total, based on the translated protein sequence homologies. The estimated pan-genome can then serve as basis for the development of a microarray application that is capable of differentiating every different strain of the same species. *E. coli* is an organism for whom exact strain differentiation is of particular interest, because this species comprises probiotic strains (e.g., *E. coli* Nissle) as well as highly pathogenic strains (e.g., O157:H7).

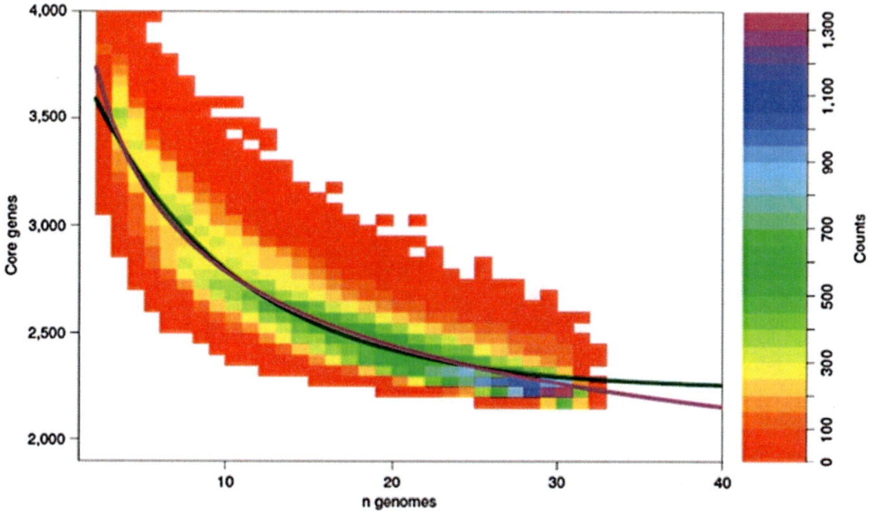

Fig. 8 Two-dimensional density plot of "core genes" for the *E. coli* pan-genome. The plot illustrates the number of *E. coli* genes for $n = 2$, 32 genomes based on a maximum of 3,200 random combinations of genomes for each n. The density colors reflect the count of combinations giving rise to a certain number of core genes; that is for $n = 3$, genome number 3 is compared to genomes 1 and 2, and the number of core genes is the number of genome 3 genes conserved in genomes 1 and 2. The *green* and *purple* lines each indicate the fit to two slightly different exponential decay functions (Tettelin et al. 2005; Willenbrock et al. 2007)

6.1 16S rDNA Microarray for Phylogenetic Typing of Gut Communities

A number of phylogenetic arrays have been constructed that permit hybridization of nucleic acids extracted from environmental samples against probes corresponding to single-stranded full or partial 16S rRNA genes (Guschin et al. 1997; Palmer et al. 2007; Harrington et al. 2008; Wilson et al. 2002). As it is technically very difficult to include more than 800,000 SSU sequences present in the databases (see http://rdp.cme.msu.edu/), microarrays with subsets of sequences specific to the ecological environment of interest are required. Recently the HIT Chip, an oligonucleotide microarray for phylogenetic profiling of human intestinal tract communities, was developed (Kovatcheva-Datchary et al. 2009). The 4,800 probes on this 16S rRNA gene tiling array consist of sets of three 18–30nt long overlapping oligonucleotides targeting the V1 and V6 region sequences from 1,140 phylotypes, respectively. Based on 98% sequence similarity, phylotypes were defined from more than 16,000 16S rRNA gene sequences identified in the human GIT. A typical work flow for phylogenetic microarray analysis is shown in Fig. 10.

With the aid of this technology it was also shown that a multispecies probiotic cocktail alleviated symptoms of irritable bowel syndrome (Kajander et al. 2008), and that starch-fermenting bacteria could be identified by using RNA stable isotope

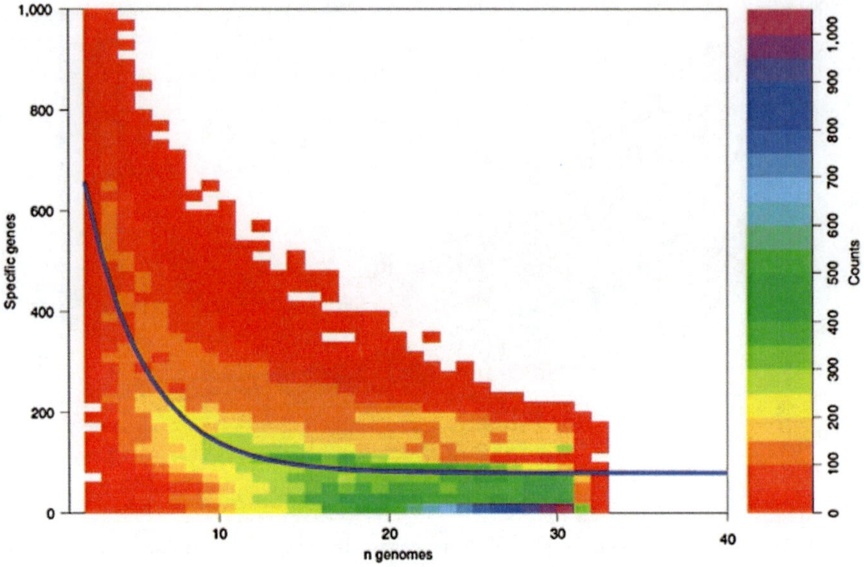

Fig. 9 Two-dimensional density plot of novel genome "specific genes" for the *E. coli* pan genome. The plot illustrates the number of novel genome specific genes for the *n*th genome when comparing $n = 2$, 32 genomes (for a maximum of 3,200 random combinations at each *n*). The density colors reflect the count of combinations giving rise to a certain number of specific genes (*y*-axis) in one genome compared to $n - 1$ other genomes; that is, for $n = 2$, genome number 2 is compared to genome number 1 and, on average, approximately 650 genes are found to be specific to strain 2 (Willenbrock et al. 2007)

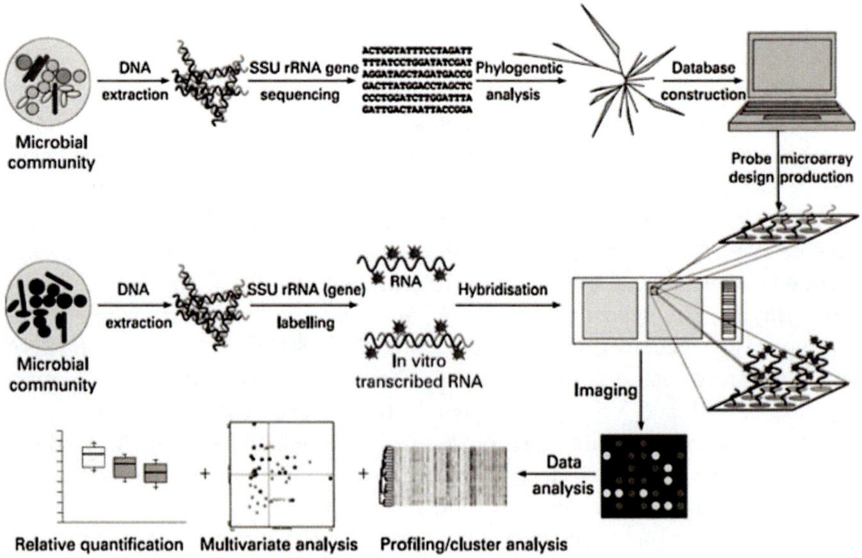

Fig. 10 Schematic representation of high-throughput analysis of human gastrointestinal (GI) tract microbiota via brute force sequencing and phylogenetic microarray analysis. *SSU rRNA* small subunit ribosomal RNA (Zoetendal et al. 2008)

probing in a human colon model with great reproducibility (Kovatcheva-Datchary et al. 2009).

6.2 Array Tubes

Array Tubes (AT) platforms consist of a custom microprobe array similar to usual glass slides microarrays, integrated into a microreaction vial. Depending on the individual assay, nucleic acid as well as protein and peptide based arrays can be manufactured. Using this method, up to 196 gene or protein sequences potentially involved in probiotic mechanisms can be analyzed simultaneously to evaluate probiotic potential of food or a bacterial strain.

6.3 Microarrays for Identifying the Mechanism of Action

An important probiotic property is the ability to adhere to the mucosa. Adherence has been linked to immune modulation and bacterial cell retention in the GI tract (Grangette et al. 2005; Hisbergues et al. 2007).

Degues et al. (2009) used DNA and mRNA microarrays to identify which properties of *Lactobacillus johnsohnii* NCC533 were responsible for the long gut persistence phenotype of this strain. NCC533 was detected for 12 days after administration to mice, whereas *L. johnsohnii* ATCC 33200, the type strain of this species, could only be detected for 5 days. For this purpose, all 1,760 open reading frames (ORFs) of NCC533 were spotted on a microarray. The DNA of the type strain ATCC 33200 was hybridized against NCC533. As a result, 233 genes differed in the genome contents of the two strains, of which about 30% were of prophage origin. To further narrow down the number of differing genes and to identify which genes are responsible for the long gut persistence, mice were monocolonized with NCC 533 and the bacterial mRNA isolated from the jejunal mouse mucosa. Fusion of the DNA- and mRNA microarray datasets revealed that only six genes, corresponding to three genetic loci, were expressed exclusively in the long-gut-persistence phenotype of NCC 533.

The first genetic locus encodes glycosyltransferase genes involved in exopolysaccharide (EPS) synthesis of the cell wall. The second encodes genes for a membrane-bound transporter complex and a cytoplasmic protein complex of a mannose-import complex. The third NCC 533-specific gene locus, which was expressed in the gut, shared significant sequence identity with IgA (immune globulin A) proteases from several pathogenic bacteria that colonize human body surfaces.

Mutants were constructed, in which these three genetic loci were deleted. It became apparent that the ability to express an IgA protease was responsible for the long gut-persistence phenotype of *L. johnsohnii* NCC 533 rather than surface characteristics of the EPS layer or sugar import ability.

A microarray was also applied to determine which genes in *Lactobacillus reuteri* ATCC 55730 are responsible for the ability to thrive under the presence of bile (Whitehead and Versalovic 2008). Bile salts have been proposed to have a wide range of cellular effects, including cell wall or membrane damage, DNA damage, protein denaturation, oxidative stress, and low intracellular pH (Begley et al. 2005). *L. reuteri* was shown to possess several multidrug resistance efflux pumps and another unknown hypothetical protein involved in bile resistance. Efflux pumps have already been shown to play important roles in the bile response of *Campylobacter jejuni*, *Listeria monozytogenes*, and *E. coli* (Lin et al. 2003; Thanassi et al. 1997; Sleator et al. 2005).

7 Detection of GMO: Probiotics

Various genetic systems have been introduced to analyze and modify LAB including plasmid vectors, selectable markers, or markers that restore an impaired function necessary for cell viability. In addition, gene-expression systems which allow the controlled expression of homologous or heterologous genes, such as those based on promoters controlled by sugar, such as the lactose operon promoter, have been introduced to LAB (Ahmed 2003).

There are no validated results for detection of already approved genetically modified microorganisms (GMM). The plasticity of the bacterial genome, the frequency of horizontal gene transfer, and the application of self-cloning complicate the detection of GMM in foods, especially when recombinant DNA is naturally present within a rich indigenous or contaminating flora. Detection usually relies on available experience on tracing of traits at the genomic and phenotypic levels. When GMMs are used for the production of food, four categories of detection can be differentiated:

1. The product is free of any DNA and impurities indicative of a GMM, for example, highly purified food ingredients or additives
2. The product contains DNA but no GMM, for example, liquid products (beer or wine) that were subjected to separation processes such as filtration or centrifugation
3. Products that contain the dead GMM with its DNA, for example, a pasteurized yoghurt or baked goods
4. Products containing the living GMM, for example, a nonpasteurized yoghurt, cheese, beer, sauerkraut, fermented sausage, etc.

Isolation of DNA is the first step in analysis for detection of the use of GMM in food by molecular methods. When the cells are not lysed, the GMMs may first be separated from the food matrix by homogenizing the food and separation of the cells by filtration and centrifugation. Detection of DNA may be hampered by the presence of inhibitors that interfere with cell lysis, degrade DNA during isolation, or interfere directly with PCR.

DNA stability in fermented foods is another concern. As long as the DNA is contained in cells, it is naturally protected against degradation. DNA released from cells into the food matrix, as occurs in the course of food processing, undergoes physical, chemical, and enzymatic degradation. Straub et al. reported that free plasmid DNA in sausages remained detectable after storage for 9 weeks (Straub et al. 1999).

For an organism-specific detection of the GMM it would need to be ensured that the recombinant DNA detected is contained in the very host strain that had been used for construction of the GMM. Detection of the recombinant DNA must be combined with identification of the host organism. This measure takes into account the possibility of horizontal transfer of recombinant DNA between different strains, species, or even less-related organisms. A combined GMM detection approach based on specific DNA probe hybridization and on diagnostic DNA in vitro amplification can be used taking into account the contaminating or indigenous microbiota present together with the GMM in foods. Living cells are required for the combined use of probes and/or PCR primers specific for the recombinant DNA as well as the microorganism. The methods available are highly sensitive, but are applicable to living GMM exclusively.

Techniques based on the detection of strain-specific DNA sequences by applying hybridization or PCR techniques with specific probes or primers are available. These unique sequences can be derived, for example, from RAPD fragments or may be obtained by the subtraction hybridization technique (Tilsala-Timisjarvi and Alatossava 1998, 2001; Matheson et al. 1997; Heller et al. 2006) (Fig. 11).

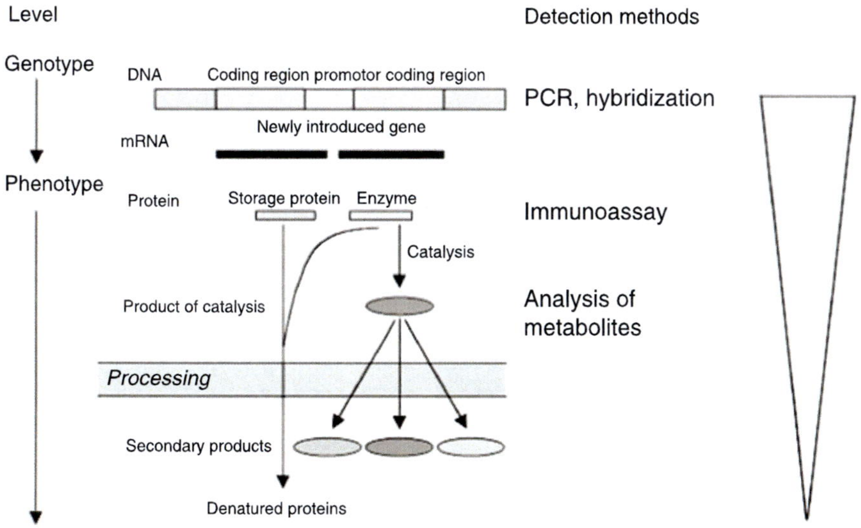

Fig. 11 The unambiguity of methods for the detection of GMO in food (Heller et al. 2006)

8 Conclusions

In conclusion, there are multiple molecular methods available for the characterization of probiotic strains as well as the analysis of metagenomic background of beneficial strains in the human gut. Analysis has increased the understanding of the diversity and phylogeny of beneficial strains and their functions tremendously. However, for many aspects a combination of both molecular and culture techniques will be necessary (Fig. 12).

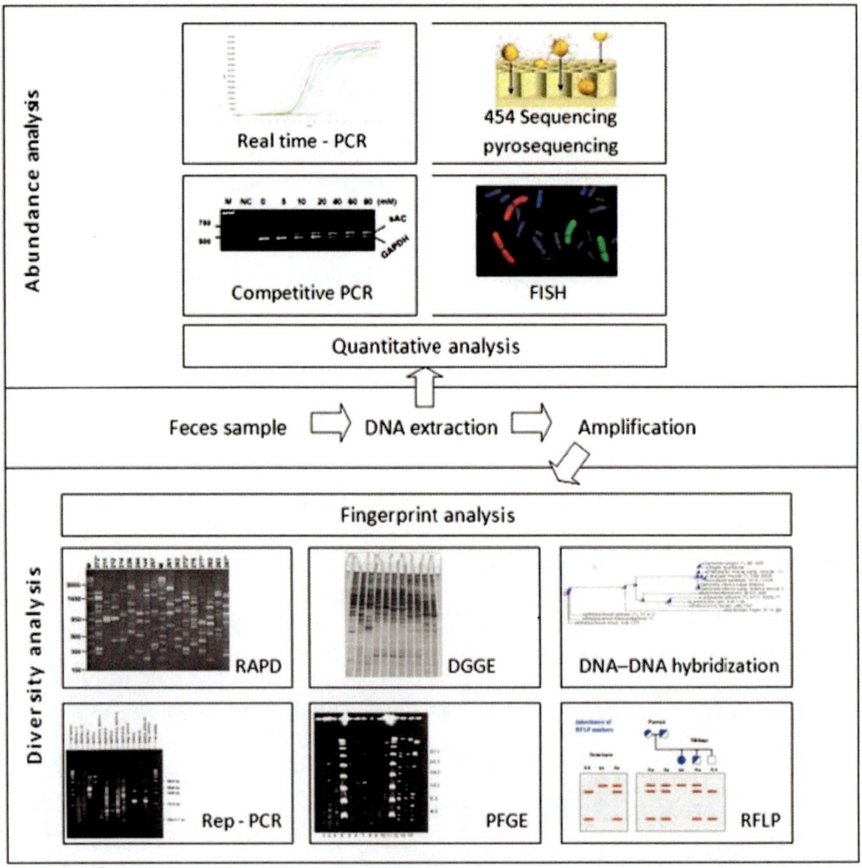

Fig. 12 Overview of methods for analyzing the gut microbiota out of feces sample (Holzapfel et al. 2001; Mahenthiralingam et al. 2009; Rothberg and Leamon 2008; Sun et al. 2004; Versalovic et al. 1994)

References

Adams C-A (2010) The probiotic paradox: live and dead cells are biological response modifiers. Nutr Res Rev 23:37–46

Ahmed F-E (2003) Genetically modified probiotics in foods. Trends Biotechnol 21:491–497

Alemayehu D, Ross R-P, O'Sullivan O, Coffey A, Stanton C, Fitzgerald G-F, Mcauliffe O (2009) Genome of a virulent bacteriophage Lb338-1 that lyses the probiotic Lactobacillus paracasei cheese strain. Gene 448:29–39

Amann R-I, Krumholz L, Stahl D-A (1990) Fluorescent-oligonucleotide probing of whole cells for determinative, phylogenetic, and environmental studies in microbiology. J Bacteriol 172:762–770

Amann R, Fuchs B-M, Behrens S (2001) The identification of microorganisms by fluorescence in situ hybridisation. Curr Opin Biotechnol 12:231–236

Andersson AF, Lindberg M, Jakobsson H, Bäckhed F, Nyrén P, Engstrand L (2008) Comparative analysis of human gut microbiota by barcoded pyrosequencing. PLoS One 3(7):e2836

Axelsson L (1998) Lactic acid bacteria: classification and physiology. Dekker, New York

Begley M, Gahan C-G, Hill C (2005) The interaction between bacteria and bile. FEMS Microbiol Rev 29:625–651

Binetti A-G, Capra M-L, Alverez M-A, Reinheimer J-A (2008) PCR method for detection and identification of Lactobacillus casei/paracasei bacteriophages in dairy products. Int J Food Microbiol 124:147–153

BIOLABS (2010) Restriction endonucleases overview [Online]. Biolabs. Available: http://www.neb.com/nebecomm/tech_reference/restriction_enzymes/overview.asp. Accessed 28 Oct 2010

Blume H-P, Brummer G-W, Horn R, Kandeler E, Kogel-Knabner I, Kretzschmar R, Stahr K, Wilke B-M (2010) Untersuchungsmethoden. In: Scheffer FW, Schachtschabel P (eds) Lehrbuch der Bodenkunde, 16th edn. Spektrum Akademischer Verlag, Berlin

Booysena C, Dicks L-M, Meijering I, Ackermann A (2002) Isolation, identification and changes in the composition of lactic acid bacteria during the malting of two different barley cultivars. Int J Food Microbiol 76:63–73

Bottacini F, Medini D, Pavesi A, Turroni F, Foroni E, Riley D, Giubellini V, Tettelin H, Van Sinderen D, Ventura M (2010) Comparative genomics of the genus *Bifidobacterium*. Microbiology 156:3243–3254

Bottari B, Ercolini D, Gatti M, Neviani E (2006) Application of FISH technology for microbiological analysis: current state and prospects. Appl Microbiol Biotechnol 73:485–494

Brandt K, Alatossava T (2003) Specific identification of certain probiotic *Lactobacillus rhamnosus* strains with PCR primers based on phage-related sequences. Int J Food Microbiol 84:189–196

Capra M-L, Quiberoni A, Reinheimer J (2006) Phages of *Lactobacillus casei/paracasei*: response to environmental factors and interaction with collection and commercial strains. J Appl Microbiol 100:334–342

Carey C-M, Kirk J-L, Ojha S, Kostrzynska M (2007) Current and future uses of real-time polymerase chain reaction and microarrays in the study of intestinal microbiota, and probiotic use and effectiveness. Can J Microbiol 53:537–550

Chen C, Kittichotirat W, Chen W, Downey J-S, Si Y, Bumgarner R (2010) Genome sequence of naturally competent Aggregatibacter actinomycetemcomitans serotype a strain D7S-1. J Bacteriol 192:2643–2644

Chopin A, Bolotin A, Sorokin A, Ehrlich S-D, Chopin M (2001) Analysis of six prophages in *Lactococcus lactis* IL1403: different genetic structure of temperate and virulent phage populations. Nucleic Acids Res 29:644–651

Claesson M-J, O'Sullivan O, Wang Q, Nikkila J, Marchesi J-R, Smidt H, De Vos W-M, Ross R-P, O'Toole P-W (2009) Comparative analysis of pyrosequencing and a phylogenetic microarray for exploring microbial community structures in the human distal intestine. PLoS One 4:e6669

Cypionka H (2010) Grundlagen der Mikrobiologie. Springer, Heidelberg

Dethlefsen L, Huse S, Sogin ML, Relman DA (2008) The pervasive effects of an antibiotic on the human gut microbiota, as revealed by deep 16S rRNA sequencing. PLoS Biol 6(11):e280

Djordjevis G-M, O'Sullivan D-J, Walker S-A, Conkling M-A, Klaenhammer T-R (1997) A triggered-suicide system designed as a defense against bacteriophages. J Bacteriol 179:6741–6748

Droege M, Hill B (2008) The genome sequencer FLX system – longer reads, more applications, straight forward bioinformatics and more complete data sets. J Biotechnol 136(1–2):3–10. Epub 21 June 2008

Ereqat S, Bar-Gal G-K, Nasereddin A, Azmi K, Qaddomi S-E, Greenblatt C-L, Spigelman M, Abdeen Z (2010) Rapid differentiation of *Mycobacterium tuberculosis* and *M. bovis* by high resolution melt curve analysis. J Clin Microbiol 48:4269–4272

Falentin H, Deutsch S-M, Jan G, Loux V, Thierry A, Parayre S, Maillard M-B, Dherbecourt J, Cousin F-J, Jardin J, Siguier P, Coulous A, Barbe V, Vacherie B, Wincker P, Bibrat J-F, Gaillardin C, Lortal S (2010) The complete genome of *Propionibacterium freudenreichii* CIRM-BIA1, a hardy actinobacterium with food and probiotic applications. PLoS One 5:e11748

Felis G-E, Dellaglio F (2007) Taxonomy of lactobacilli and bifidobacteria. Curr Issues Intest Microbiol 8:44–61

Felske A, Osborn A-M (2005) DNA fingerprinting of microbial communities. In: Osborn MA, Smith CJ (eds) Molecular microbial ecology. Taylor and Francis, New York

Fischer S-G, Lerman L-S (1979) Length-independent separation of DNA restriction fragments in two-dimensional gel electrophoresis. Cell 16:191–200

Fuchs B-M, Glockner F-O, Wulf J, Amann R (2000) Unlabeled helper oligonucleotides increase the in situ accessibility to 16 S rRNA of fluorescently labeled oligonucleotide probes. Appl Environ Microbiol 66:3603–3607

Fujimura K-E, Slusher N-A, Cabana M-D, Lynch S-V (2010) Role of the gut microbiota in defining human health. Expert Rev Anti Infect Ther 8:435–454

Galloway RL, Levett PN (2008) Evaluation of a modified pulsed-field gel electrophoresis approach for the identification of Leptospira serovars. Am J Trop Med Hyg 78(4):628–632

Gevers D, Huys G, Swings J (2001) Applicability of rep-PCR fingerprinting for identification of Lactobacillus species. FEMS Microbiol Lett 205:31–36

Glockner F-O, Fuchs B-M, Amann R (1999) Bacterioplankton compositions of lakes and oceans: a first comparison based on fluorescence in situ hybridization. Appl Environ Microbiol 65:3721–3726

Grangette C, Nutten S, Palumbo E, Morath S, Hermann C, Dewulf F-J, Pot B, Hartung T, Hols P, Mercenier A (2005) Enhanced antiinflammatory capacity of a Lactobacillus plantarum mutant synthesizing modified teichoic acids. Proc Natl Acad Sci USA 102:10321–10326

Guschin D-Y, Mobarry B-K, Proudnikov D, Stahl D-A, Rittmann B-E, Mirzabekov A-D (1997) Oligonucleotide microchips as genosensors for determinative and environmental studies in microbiology. Appl Environ Microbiol 63:2397–2402

Harrington C-R, Lucchini S, Ridgway K-P, Wegmann U, Eaton T-J, Hinton J-C, Gasson M-J, Narbad A (2008) A short-oligonucleotide microarray that allows improved detection of gastrointestinal tract microbial communities. BMC Microbiol 8:195

Hayashi T, Makino K, Ohnishi M, Kurokawa K, Ishii K, Yokoyama K, Han C-G, Ohtsubo E, Nakayama K, Murata T, Tanaka M, Tobe T, Iida T, Takami H, Honda T, Sasakawa C, Ogasawara N, Yasunaga T, Kuhara S, Shiba T, Hattori M, Shinagawa H (2001) Complete genome sequence of enterohemorrhagic *Escherichia coli* O157:H7 and genomic comparison with a laboratory strain K-12. DNA Res 8:11–22

Heller K-J, Hammes W-P, Hertel C, Bauer T (2006) Methods for detection of genetically modified microorganisms used in food fermentation processes. Wiley On line library

Hisbergues M, Magi M, Rigaux P, Steuve J, Garcia L, Goudercourt D, Pot B, Pestel J, Jacquet A (2007) In vivo and in vitro immunomodulation of Der p 1 allergen-specific response by *Lactobacillus plantarum* bacteria. Clin Expr Allerg 37:1286–1295

Holzapfel W-H, Harberer P, Geisen R, Bjorkroth J, Schillinger U (2001) Taxonomy and important features of probiotic microorganisms in food and nutrition. Am J Clin Nutr 73:365S–373S

Huse SM, Huber JA, Morrison HG, Sogin ML, Welch DM (2007) Accuracy and quality of massively parallel DNA pyrosequencing. Genome Biol 8(7):R143

Huys G, Vancanneyt M, D'haene K, Vankerckhoven V, Goosens H, Swings J (2006) Accuracy of species identity of commercial bacterial cultures intended for probiotic or nutritional use. Res Microbiol 157:803–810

Ishii S, Sadowsky MJ (2009) Applications of the rep-PCR DNA fingerprinting technique to study microbial diversity, ecology and evolution. Environ Microbiol 11(4):733–740

Jeyaram K, Romi W, Singh T-A, Devi A-R, Devi S-S (2010) Bacterial species associated with traditional starter cultures used for fermented bamboo shoot production in Manipur state of India. Int J Food Microbiol 143:1–8

Kajander K, Myllyluoma E, Rajilic-Stojanovic M, Kyronpalo S, Rasmussen M, Jarvenpaa S, Zoetendal E-G, De Vos W-M, Vapaatalo H, Korpela R (2008) Clinical trial: multispecies probiotic supplementation alleviates the symptoms of irritable bowel syndrome and stabilizes intestinal microbiota. Aliment Pharmacol Ther 27:48–57

Kneifel W, Domig K-J (2009) Taxonomie von Milchsäurebakterienmit probiotischer Kapazität. In: Bischoff SC (ed) Probiotika, Präbiotika und Synbiotika. Thieme Verlag, Stuttgart

Kovatcheva-Datchary P, Egert M, Maathuis A, Rajilic-Stojanovic M, De Graaf A, Smidt H, De Vos W-M, Venema K (2009) Linking phylogenetic identities of bacteria to starch fermentation in an in vitro model of the large intestine by RNA-based stable isotope probing. Environ Microbiol 11:914–926

Krogius-Kurikka L, Lyra A, Malinen E, Aarnikunnas J, Tuimala J, Paulin L, Makivuokko H, Kajander K, Palva A (2009) Microbial community analysis reveals high level phylogenetic alterations in the overall gastrointestinal microbiota of diarrhoea-predominant irritable bowel syndrome sufferers. BMC Gastroenterol 9:95

Lin J, Sahin O, Michel L-O, Zhang Q (2003) Critical role of multidrug efflux pump CmeABC in bile resistance and in vivo colonization of Campylobacter jejuni. Infect Immun 71:4250–4259

Loffler G (2004) Basiswissen Biochemie mit Pathobiochemie. Springer, Heidelberg

Louis P, Flint H-J (2009) Diversity, metabolism and microbial ecology of butyrate-producing bacteria from the human large intestine. FEMS Microbiol Lett 294:1–8

Ludwig W, Strunk O, Westram R, Richter L, Meier H, Yadhukumar, Buchner A, Lai T, Steppi S, Jobb G, Förster W, Brettske I, Gerber S, Ginhart AW, Gross O, Grumann S, Hermann S, Jost R, König A, Liss T, Lüssmann R, May M, Nonhoff B, Reichel B, Strehlow R, Stamatakis A, Stuckmann N, Vilbig A, Lenke M, Ludwig T, Bode A, Schleifer KH (2004) ARB: a software environment for sequence data. Nucleic Acids Res 32(4):1363–1371

Mahenthiralingam E, Marchbank A, Drevinek P, Garaiova I, Plummer S (2009) Use of colony-based bacterial strain typing for tracking the fate of Lactobacillus strains during human consumption. BMC Microbiol 9:251

Margulies M, Egholm M, Altman WE, Attiya S, Bader JS, Bemben LA, Berka J, Braverman MS, Chen YJ, Chen Z, Dewell SB, Du L, Fierro JM, Gomes XV, Godwin BC, He W, Helgesen S, Ho CH, Irzyk GP, Jando SC, Alenquer ML, Jarvie TP, Jirage KB, Kim JB, Knight JR, Lanza JR, Leamon JH, Lefkowitz SM, Lei M, Li J, Lohman KL, Lu H, Makhijani VB, McDade KE, McKenna MP, Myers EW, Nickerson E, Nobile JR, Plant R, Puc BP, Ronan MT, Roth GT, Sarkis GJ, Simons JF, Simpson JW, Srinivasan M, Tartaro KR, Tomasz A, Vogt KA, Volkmer GA, Wang SH, Wang Y, Weiner MP, Yu P, Begley RF, Rothberg JM (2005) Genome sequencing in microfabricated high-density picolitre reactors. Nature 437(7057):376–380. Epub 31 July 2005. Erratum in: Nature (2006);441(7089):120

Matheson V-G, Munakata-Marr J, Hopkins G-D, Mccarty P-L, Tiedje J-M, Forney L-J (1997) A novel means to develop strain-specific DNA probes for detecting bacteria in the environment. Appl Environ Microbiol 63:2863–2869

Maukonen J, Matto J, Suihko M-L, Saarela M (2008) Intra-individual diversity and similarity of salivary and faecal microbiota. J Med Microbiol 57:1560–1568

McKenna P, Hoffmann C, Minkah N, Aye PP, Lackner A, Liu Z, Lozupone CA, Hamady M, Knight R, Bushman FD (2008) The macaque gut microbiome in health, lentiviral infection, and chronic enterocolitis. PLoS Pathog 4(2):e20

Medini D, Donati C, Tettelin H, Masignani V, Rappuoli R (2005) The microbial pan-genome. Curr Opin Genet Dev 15:589–594

Michalet X, Pinaud F-F, Bentolila L-A, Tsay J-M, Doose S, Li J-J, Sundaresan G, Wu A-M, Gambhir S-S, Weiss S (2005) Quantum dots for live cells, in vivo imaging, and diagnostics. Science 307:538–544

Moineau S (1999) Applications of phage resistance in lactic acid bacteria. Antonie Van Leeuwenhoek 76:377–382

Monger W-A, Alicai T, Ndunguru J, Kinyua Z-M, Potts M, Reeder R-H, Miano D-W, Adams I-P, Boonham N, Glover R-H, Smith J (2010) The complete genome sequence of the Tanzanian strain of Cassava brown streak virus and comparison with the Ugandan strain sequence. Arch Virol 155:429–433

Moter A, Gobel U-B (2000) Fluorescence in situ hybridization (FISH) for direct visualization of microorganisms. J Microbiol Methods 41:85–112

Naser N-M, Hagen K-E, Vancanneyt M, Cleenwerck I, Swings J, Tompkins P-A (2006) Lactobacillus suntoryeus Cachat and Priest 2005 is a later synonym of *Lactobacillus helveticus* (Orla-Jensen 1919) Bergey et al. 1925 (Approved Lists 1980). Int J Syst Evol Microbiol 56:355–360

Olive D-M, Bean P (1999) Principles and applications of methods for DNA-based typing of microbial organisms. J Clin Microbiol 37:1661–1669

Palmer C, Bik E-M, Digiulio D-B, Relman D-A, Brown P-O (2007) Development of the human infant intestinal microbiota. PLoS Biol 5:e177

Probiotics, DOA http://www.functionalingredientsmag.com/article/Science-Now/probiotics-dead-or-alive-.aspx. http://www.functionalingredientsmag.com/article/Science-Now/probiotics-dead-or-alive-.aspx [Online]

Radvansky J, Bazsalovicsova E, Kralova-Hromadova I, Minarik G, Kadasi L (2011) Development of high-resolution melting (HRM) analysis for population studies of Fascioloides magna (Trematoda: Fasciolidae), the giant liver fluke of ruminants. Parasitol Res 108:201–209

RFLP: http://www.ncbi.nlm.nih.gov/projects/genome/probe/doc/TechRFLP.shtml

Roesch LF, Fulthorpe RR, Riva A, Casella G, Hadwin AK, Kent AD, Daroub SH, Camargo FA, Farmerie WG, Triplett EW (2007) Pyrosequencing enumerates and contrasts soil microbial diversity. ISME J 1(4):283–290. Epub 5 July 2007

Rothberg J-M, Leamon J-H (2008) The development and impact of 454 sequencing. Nat Biotechnol 26:1117–1124

Roy D, Ward P, Champagne G (1996) Differentiation of bifidobacteria by use of pulsed-field gel electrophoresis and polymerase chain reaction. Int J Food Microbiol 29:11–29

Satokari R-M, Vaughan E-E, Akkermans A-D, Saarela M, De Vos W-M (2001) Bifidobacterial diversity in human feces detected by genus-specific PCR and denaturing gradient gel electrophoresis. Appl Environ Microbiol 67:504–513

Shneyer V-S (2007) On the species-specificity of DNA: fifty years later. Biochemistry (Mosc) 72:1377–1384

Sibley P-E, Harper M-E, Peeling W-B, Griffiths K (1984) Growth hormone and prostatic tumours: localization using a monoclonal human growth hormone antibody. J Endocrinol 103:311–315

Sigler V (2004) Denaturing gradient gel electrophoresis (DGGE) [Online]. Laboratory for Microbial Ecology, Department of Earth, Ecological and Environmental Sciences, University of Toledo. Accessed 24 Aug 2010

Sleator R-D, Wemekamp-Kamphuis H-H, Gahan G-C, Abee T, Hill C (2005) A PrfA-regulated bile exclusion system (BilE) is a novel virulence factor in *Listeria monocytogenes*. Mol Microbiol 55:1183–1195

Stackebrandt E, Frederiksen W, Garrity GM, Grimont PA, Kämpfer P, Maiden MC, Nesme X, Rosselló-Mora R, Swings J, Trüper HG, Vauterin L, Ward AC, Whitman WB (2002) Report of the ad hoc committee for the re-evaluation of the species definition in bacteriology. Int J Syst Evol Microbiol 52(Pt 3):1043–1047

Straub J-A, Hertel C, Hammes W-P (1999) A 23 S rDNA-targeted polymerase chain reaction-based system for detection of Staphylococcus aureus in meat starter cultures and dairy products. J Food Prot 62:1150–1156

Sun X-C, Cui M, Bonanno J-A (2004) [HCO3-]-regulated expression and activity of soluble adenylyl cyclase in corneal endothelial and Calu-3 cells. BMC Physiol 4:8

Tannock G-W, Tilsala-Timisjarvi A, Rodtong S, Ng J, Munro K, Alatossava T (1999) Identification of Lactobacillus isolates from the gastrointestinal tract, silage, and yoghurt by 16 S-23 S rRNA gene intergenic spacer region sequence comparisons. Appl Environ Microbiol 65:4264–4267

Tedersoo L, Nilsson R-H, Abarenkov K, Jairus T, Sadam A, Saar I, Bahram M, Bechem E, Chuyong G, Koljalg U (2010) 454 Pyrosequencing and Sanger sequencing of tropical mycorrhizal fungi provide similar results but reveal substantial methodological biases. New Phytol 188:291–301

Tettelin H, Masignani V, Cieslewicz M-J, Donati C, Medini D, Ward N-L, Angiuoli S-V, Brabtree J, Jomes A-L, Durkin A-S, Deboy R-T, Davidsen T-M, Mora M, Scarselli M, Margarit Y-R, Peterson J-D, Hauser C-R, Sundaram J-P, Nelson W-C, Madupu R, Brinkac L-M, Dodson R-J, Rosovits M-J, Sullivan S-A, Daugherty S-C, Haft D-H, Selengut J, Gwinn M-L, Zhaou L, Zafar N, Khouri H, Radune D, Dimitriv G, Watkins K, O'Conner K-J, Smith S, Utterback T-R, White O, Rubens C-E, Grandi G, Madoff L-C, Kasper D-L, Telford J-L, Wessels M-R, rapuoli R, Fraser C-M (2005) Genome analysis of multiple pathogenic isolates of Streptococcus agalactiae: implications for the microbial "pan-genome". Proc Natl Acad Sci USA 102:13950–13955

Thanassi D-G, Cheng L-W, Nikaido H (1997) Active efflux of bile salts by *Escherichia coli*. J Bacteriol 179:2512–2518

Tilsala-Timisjarvi A, Alatossava T (1998) Strain-specific identification of probiotic *Lactobacillus rhamnosus* with randomly amplified polymorphic DNA-derived PCR primers. Appl Environ Microbiol 64:4816–4819

Tilsala-Timisjarvi A, Alatossava T (2001) Characterization of the 16 S-23 S and 23 S-5 S rRNA intergenic spacer regions of dairy propionibacteria and their identification with species-specific primers by PCR. Int J Food Microbiol 68:45–52

Turnbaugh PJ, Hamady M, Yatsunenko T, Cantarel BL, Duncan A, Ley RE, Sogin ML, Jones WJ, Roe BA, Affourtit JP, Egholm M, Henrissat B, Heath AC, Knight R, Gordon JI (2009) A core gut microbiome in obese and lean twins. Nature 457(7228):480–484. Epub 30 Nov 2008

Ventura M, Zink R (2002) Specific identification and molecular typing analysis of *Lactobacillus johnsonii* by using PCR-based methods and pulsed-field gel electrophoresis. FEMS Microbiol Lett 217:141–154

Ventura M, Van Sinderen D, Fitzgerald G-F, Zink R (2004) Insights into the taxonomy, genetics and physiology of bifidobacteria. Antonie Van Leeuwenhoek 86:205–223

Versalovic J, Schneider M, De Bruijn F-J, Lupski J-R (1994) Genomic fngerprinting of bacteria using repetitive sequence-based polymerase chain reaction. Methods Mol Cell Biol 5:25–40

Vitali B, Ndagijimana M, Cruciani F, Carnevali P, Candela M, Guerzoni M-E, Brigidi P (2010) Impact of a symbiotic food on the gut microbial ecology and metabolic profiles. BMC Microbiol 10:4

Wall R, Hussey S-G, Ryan C-A, O'Neill M, Fitzgerald G, Stanton C, Ross P (2008) Presence of two Lactobacillus and Bifidobacterium probiotic strains in the neonatal ileum. ISME J 2:83–91

Walter J, Hertel C, Tannock G-W, Lis C-M, Munro K, Hammes W-P (2001) Detection of Lactobacillus, Pediococcus, Leuconostoc, and Weissella species in human feces by using group-specific PCR primers and denaturing gradient gel electrophoresis. Appl Environ Microbiol 67:2578–2585

Willenbrock H, Hallin P-F, Wassenaar T-M, Ussery D-W (2007) Characterization of probiotic *Escherichia coli* isolates with a novel pan-genome microarray. Genome Biol 8:R267

Wilson K-H, Wilson W-J, Radosevich J-L, Desantis T-Z, Viswanathan V-S, Kuczmarski T-A, Andersen G-L (2002) High-density microarray of small-subunit ribosomal DNA probes. Appl Environ Microbiol 68:2535–2541

Wittwer C-T (2009) High-resolution DNA melting analysis: advancements and limitations. Hum Mutat 30:857–859

Wojdacz T-K, Dobrovic A, Algar E-M (2008) Rapid detection of methylation change at H19 in human imprinting disorders using methylation-sensitive high-resolution melting. Hum Mutat 29:1255–1260

Zeigler DR (2003) Gene sequences useful for predicting relatedness of whole genomes in bacteria. Int J Syst Evol Microbiol 53(Pt 6):1893–1900

Zhang H, DiBaise JK, Zuccolo A, Kudrna D, Braidotti M, Yu Y, Parameswaran P, Crowell MD, Wing R, Rittmann BE, Krajmalnik-Brown R (2009) Human gut microbiota in obesity and after gastric bypass. Proc Natl Acad Sci USA 106(7):2365–2370. Epub 21 Jan 2009

Zoetendal E-G, Rajilic-Stojanovic M, De Vos W-M (2008) High-throughput diversity and functionality analysis of the gastrointestinal tract microbiota. Gut 57:1605–1615

Maintenance and Protection of Probiotics

Benjamin D. Huckle and Zhibing Zhang

Contents

1 Maintenance of Probiotics ... 88
1.1 Cell Viability Problems ... 88
1.2 Stomach Acid Susceptibility ... 89
1.3 The Prebiotic Concept ... 90
1.4 Medium Manipulation ... 90
1.5 Freeze-Dried Probiotic Bacteria ... 91
1.6 Encapsulation ... 91
2 Protective Compression Coating ... 95
2.1 Compression Coating Materials ... 95
2.2 Compression Pressure ... 96
2.3 Cell Stability During Storage ... 97
3 Controlled Release ... 98
3.1 Gel-Forming Polymers ... 98
3.2 pH Sensitive Materials ... 99
3.3 Disintegrants ... 100
3.4 Compression Coated Cell Release and Viability Profile ... 100
4 Conclusions ... 105
References ... 105

Abstract Beneficial effects of probiotic bacteria depend on the viability of cells once delivered to the intestines. Cells tend to lose their viability with time during storage and cannot survive passage through acidic gastric fluids, and hence cannot adhere to the colon and exert their beneficial effects on the host, unless they are conferred some degree of protection. This chapter debates various protection techniques, with a particular focus on encapsulation and compression coating. Enteric coating materials are discussed as suitable for compression coating, while also improving cell storage stability and ensuring cell survival during exposure to

B.D. Huckle • Z. Zhang (✉)
Fermentation Improvement Group, GlaxoSmithKline, Worthing, UK
e-mail: z.zhang@bham.ac.uk

M.-T. Liong (ed.), *Probiotics*, Microbiology Monographs 21,
DOI 10.1007/978-3-642-20838-6_4, © Springer-Verlag Berlin Heidelberg 2011

harsh acidic gastric fluids. Techniques for controlled release of cells to the colon, including the use of hydrophobic disintegrants such as pectin, are also described.

1 Maintenance of Probiotics

Due to peristalsis and the antimicrobial effects of gastric acid, the stomach and duodenum contain relatively few bacteria. The jejunum contains a very small number of aerobic species, while the terminal ileum represents a transition zone of microflora prior to the colon. The colon contains a dense population of anaerobes, and thus in order for probiotic bacteria to exert their beneficial effects on the host they must adhere to and colonise this site, a process that can be promoted with targeted delivery.

A minimum number of probiotic bacteria must reach the targeted site alive and in a viable state, in order to exert their probiotic effects. Minimum therapeutic doses, as outlined in Table 1, vary significantly, making general dose recommendations for unspecified products difficult. Where the dosage is not met the efficacy of the product will not be fully seen, creating a problem for manufacturers hoping to maintain and develop further consumer confidence towards their products.

1.1 Cell Viability Problems

To be effective, probiotic bacteria must be capable of being prepared in a viable manner on an industrial scale and must remain viable during storage and use. They must also be able to survive the intestinal ecosystem of the host, to then ultimately be capable of exerting their beneficial effects. However, there are many factors which affect cell viability, such as oxygen, moisture, heat and acid.

Most probiotic products have a short shelf life, even when stored at low temperature (Lee and Salminen 1995). In addition, a number of probiotic products have received criticism regarding their composition and labelling, concerning the number or type of bacteria stated on the labels (Shortt 1998). Several studies have indicated that the quantity of viable bacteria in some commercially available products is actually below the desired level (Kailasapathy and Chin 2000; Shortt 1998). This poses a problem for both consumers and manufacturers alike, since the beneficial effects of the bacteria can only be seen if a sufficient number of viable cells are

Table 1 Suggested minimum therapeutic doses (viable cell numbers)

Minimum dose	References
10^8–10^9	Kurmann and Rasic (1991)
10^6–10^9	Lee and Salminen (1995)
$>10^9$	Fuller and Gibson (1998)
10^9–10^{12}	Douglas and Sanders (2008)

present at the time of consumption. It is also thought that some cells may be viable-but-non-culturable (VBNC) (Huckle et al. 2008), and as such will remain undetected when analysed with traditional techniques such as dilution plating.

1.2 Stomach Acid Susceptibility

In order to reach the intestine and exert their beneficial effects, probiotics must first overcome a number of physical and chemical barriers in the gastrointestinal (GI) tract. These barriers include exposure to stomach acid, bile salts and enzymes. Many strains of *L. acidophilus* and *Bifidobacterium* sp. intrinsically lack the ability to survive such stomach acidity. It has been reported that three *Lactobacillus* strains tested showed an intense loss of cell viability at a pH of below 3 (Conway et al. 1987). It has also been shown that the number of viable *L. acidophilus* BG2FO4 rapidly declined at pH 2, but did not decrease significantly at pH 4 (Hood and Zottola 1988). Similar results were obtained by Charteris et al. (1998) where it was shown that 14 out of 15 *Lactobacillus* and *Bifidobacterium* sp. strains lost more than 90% of their reproductive viability when exposed to simulated gastric juice at pH 2, and by Chan and Zhang (2005) where the loss of viability of *L. acidophilus* was pH dependent (Fig. 1).

Due to such damaging effects, probiotics even if they are still viable in stomach may be in a sub-lethally injured state upon reaching the colon and their chances of survival may be compromised, along with their ability to colonise the intestine and have any advantageous effect on the host.

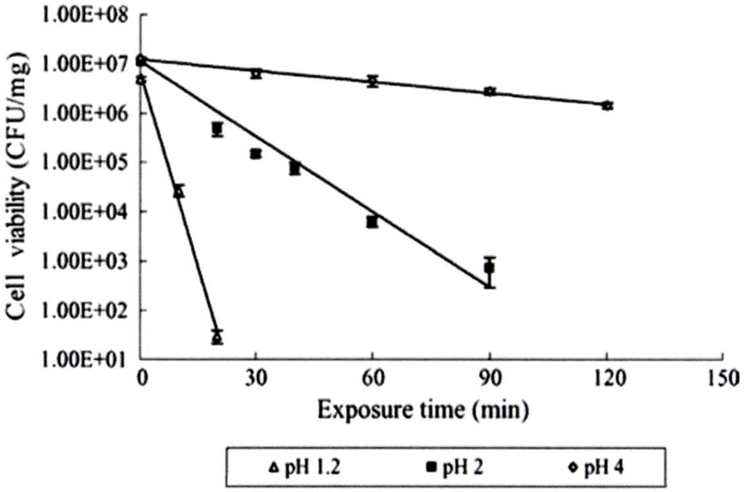

Fig. 1 Loss of viability of *L. acidophilus* during exposure to simulated gastric fluid (Chan and Zhang 2005). Adapted from Chan and Zhang, Bioencapsulation by compression coating of probiotic bacteria for their protection in an acidic medium, Process Biochemistry. Copyright Elsevier

1.3 The Prebiotic Concept

A prebiotic was first defined as a "non-digestible food ingredient that beneficially affects the host by selectively stimulating the growth and/or activity of one or a limited number of bacteria in the colon, and thus improves host health" (Gibson and Roberfroid 1995). In order for a food ingredient to be classified as a prebiotic it must first resist host digestion, adsorption and absorption; then be fermented by the microflora colonising the GI tract and lastly it must selectively stimulate the growth and/or activity of one or a limited number of bacteria within the GI tract.

Many carbohydrates have been reported to possess prebiotic activity, including fructooligosaccharides, galactooligosaccharides, isomaltooligosaccharides, inulin and lactulose (Gibson 2004). These forms of dietary fibre are readily fermentable by specific *Bifidobacteria* sp. and *Lactobacilli* sp., increasing their cell number and viability. Where a pre- and probiotic are used simultaneously, the combination is termed a synbiotic (Schrezenmeir and de Vrese 2001). Recently, several efforts have been made to develop functional foods or preparations containing so-called synbiotic products (Homayouni et al. 2008; Ooi et al. 2010), whereby a twofold positive effect on the gut microflora can be expected.

1.4 Medium Manipulation

Probiotic cells can be enhanced via the supplementation of certain compounds such as fructose, magnesium, manganese and bakers yeast to the initial fermentation medium. This acts to stimulate the growth of lactic acid bacteria resulting in higher reproductively viable cell counts, shorter generation times and improved sugar utilisation abilities (Kailasapathy and Chin 2000).

Product enrichment of probiotic yoghurt with skimmed milk and whey protein concentrate has proved effective in increasing the buffering capacity of such yoghurt (Supriadi et al. 1994). Ascorbic acid can be used to improve the stability of *L. acidophilus* during storage due to its oxygen scavenging capabilities (Dave and Shah 1997). Charteris et al. (1998) demonstrated that the addition of milk proteins could generally increase the survival of lactic acid bacteria, when exposed to simulated gastric juice. This finding is supported by Kos et al. (2000) who found that the addition of casein, skimmed milk, mucin or whey protein concentrate could significantly enhance the survival of *L. acidophilus* M92 when exposed to a simulated gastric juice of pH 2. The protective effect of such additives is most likely due to their buffering capability.

Improved viability can also be achieved by exposing cells to a sub-lethal stress, such as acid (Sanz 2007), oxygen or heat (Kosin and Rakshit 2010), which will prime the inherent adaptive stress response of cells, and enhance their survival in otherwise lethal conditions during industrial processing and gastric transit.

Manipulating the manufacturing conditions of yoghurt in this way can increase the storage stability of lactic acid bacteria. This can be achieved in a number of ways (Kailasapathy and Chin 2000), some of which include: terminating a fermentation at a higher pH, reducing the storage temperature to below 3–4°C, heat shocking the yoghurt for 5 min at 58°C and lowering the incubation temperature to 37°C.

1.5 Freeze-Dried Probiotic Bacteria

Probiotics are most commonly used as foods in the form of fermented milk products or yoghurt (Lauland 1994). As a consequence they require refrigeration and their shelf life is limited to only a few weeks, causing logistical problems for manufacturers and inconvenience for consumers. This also makes probiotics very unsuitable for travellers who rely on probiotic bacteria to prevent travellers' diarrhoea. As a solution to this problem, probiotics have been developed into freeze-dried solid forms that have a longer shelf life and do not need refrigeration (Otero et al. 2007), thus making them more convenient for both manufacturers and consumers alike. The freeze-drying process involves a solvent, typically water, being removed from a frozen solution via sublimation.

However, freeze-drying can lead to cell injury and decreased viability due to the exposure of cells to the attenuating effects of freezing and dehydration. Loss of viability during freeze-drying is due to temperature changes, phase changes and dehydration, resulting in damage to cell membranes, cell walls, ribosomes and DNA (Castro et al. 1997). Despite this, freeze-drying is still gentler than other techniques such as spray drying, permitting higher probiotic survival rates. Probiotic viability can be enhanced during freeze-drying by selecting specific strains, which exhibit greater survival rates than others. In addition, several factors have been identified as critical to cell survival, including initial cell mass, growth conditions, the composition of growth and drying media, and rehydration conditions (Carvalho et al. 2004).

Freeze-dried probiotics can be divided into two categories, those that are non-encapsulated and those that are encapsulated.

1.6 Encapsulation

Probiotic bacteria can be freeze-dried to form a powder, giving them improved storage stability. Freeze-dried lactic acid bacteria showed good survival after 12 months of storage at 4°C, in comparison with those stored at 20°C for the same period of time (Champagne et al. 1996). However, despite giving superior stability the buffering capacity of powder is very limited, meaning that only a very small number of bacteria survive when passing through the stomach (Lauland 1994). Freeze-dried lactic acid bacteria are also very sensitive to humidity (Lauland

1994), so much so that even short exposure to unfavourable conditions may cause a rapid loss of viable cells. The storage stability of the cells may be jeopardised if they are exposed to atmospheric conditions, or if they are misused by consumers, e.g. by scooping the powders with a wet spoon.

One technique that permits the stabilisation of freeze-dried cells during storage and when exposed to an acidic medium is called encapsulation. Encapsulation of probiotic bacteria has been proven successful by several studies, using various materials and methods. These methods are outlined in the following sections.

1.6.1 Hard Capsules

The majority of probiotic products commercially available today are available in formulations as hard capsules (Lauland 1994). This is because dried cells can easily be placed into capsules, and their stability during storage is improved (Zárate and Nader-Macias 2006). It was demonstrated by Lal et al. (1978) that freeze-dried cell powders placed in gelatine capsules could retain more activity than those simply stored in the powder form. However, the capsules still require an enteric-film coating to enhance the resistance of the cells against gastric acid (Lauland 1994). The application of enteric-film coatings to capsules also holds other advantages (Cole 1995; Porter and Bruno 1990), such as masking any unpleasant taste or odour, making the product easier to handle, improving the aesthetic qualities of the product and the ability to modify the cell release profile of the product.

1.6.2 Film Coating

A cell encapsulation method based on film coating was developed by Kim et al. (1988) with the aim of improving the storage stability and acid tolerance of cells. This involved mixing freeze-dried cells homogeneously with some other excipients using a wet granulation technique. The mixture was then fed into an extruder with the extruded filaments subsequently placed in a spheroniser to form rounded particles of approximately 1.2 mm diameter. The spheres containing cells were then dried and spray-coated in a fluidised bed coater.

When *L. plantarum* cells were coated with carboxymethylcellulose (CMC) their storage stability was significantly improved (Kim et al. 1988). It was also shown that cell survival remained unaltered when exposed to an acidic medium of pH 2 for up to 6 h, after initially being coated with polyvinyl acetate phthalate (PVAP) (Kim et al. 1988). Furthermore, Yokota et al. (1990) utilised the same coating technique for capsules containing freeze-dried lactic acid bacteria and obtained results showing improved cell stability during storage, and when exposed to an acidic medium. Recently, Stummer et al. (2010) have coated three bacterial species, *Enterococcus faecium*, *Bifidobacterium bifidum* and *Lactobacillus reuteri*, with formulations comprising different concentrations of shellac and additives using a fluidised-bed, and found the coating protected the microorganisms during storage, against acidic pH and provided the best release profile in simulated intestinal fluid (SIF), but *E. faecium* and *B. bifidum* were more resistant to manufacturing processing than *L. reuteri*.

However, film coating involves significant cost and can also severely damage cells due to the liquid–solid phase transition that is involved. Additionally, if organic solvents (e.g. alcohols, ketones, esters and chlorinated hydrocarbons) are used to aid the film drying, then the technique may not be appropriate for probiotic products since the solvents are either non-food grade or can be toxic to cells (Matsumoto et al. 2004).

1.6.3 Hot-Melt Coating

Conventional hot melt coating involves using molten materials, such as solid fats, waxes and certain polymers, to provide a shell with excellent water barrier properties. Hot melted coating material is sprayed onto the core material, which is then solidified using cold air. This technique is low cost, but is poorly adapted for use with cells due to the high temperatures involved. A similar technique, more appropriate for use with cells, is detailed below.

A process called 'Continuous Multi-Purpose Melt Technology' (CMT) was developed by Appelgren and Eskilson (1990) to encapsulate cells. In this process, a solid or viscous material is heated until an appropriate level of viscosity is reached, at which point the fluid is then pumped into a specially designed mixer where, in a narrow chamber, a turbine wheel revolves at high velocity. The speed and pressure of this turbine wheel causes the fluid to instantly disintegrate into a fine mist, which then envelops freeze-dried cells that are simultaneously fed into the turbine chamber via a separate inlet. After spending only a fraction of a second in the turbine chamber the processed material is then discharged from the mixer as a coated product, which is then cooled. Over 10% of freeze-dried *L. acidophilus* cells coated using this technique survived after being exposed to simulated gastric juice of pH 2 for 1 h, whereas viability of the same strain was only 0.1% when other methods were used (Appelgren and Eskilson 1990). A commercial product based on hot-melt coating was introduced in Sweden during 1988 and was termed Docidus®.

1.6.4 Matrix Cell Entrapment

Calcium alginate is the most commonly used matrix for cell entrapment because it is mild for cells, easy to produce and low cost (Sheu and Marshall 1993). Furthermore, alginate is non-toxic and widely accepted in foods. An additional advantage of this technique is that immobilisation is reversible, i.e. alginate gels can be solubilised by sequestering calcium ions, causing entrapped cells to be released.

Immobilisation can be carried out using either an emulsion or extrusion technique. The emulsion technique typically involves 2–3% (m/v) of sodium alginate mixed with a cell culture which is then homogenised with a large volume of vegetable oil, to form a water-in-oil emulsion. Once the emulsion is formed, the alginate is then insolubilised using calcium chloride to form gel beads within the oil phase, which are then later harvested by filtration (Lacroix et al. 1990).

The extrusion method is the oldest and most common approach to matrix cell entrapment (Favaro-Trindade and Grosso 2000; Champagne et al. 1992; Kearney et al. 1990). The extrusion technique typically involves 1–2% (w/v) of sodium alginate mixed homogeneously with a cell culture, which is then forced through a nozzle via a peristaltic pump and added drop wise to an agitated solution of calcium chloride. This forms calcium alginate beads, which are then kept for some time to permit gel strengthening. The storage stability of freeze-dried *Lactococcus lactis* encapsulated in a calcium alginate matrix was found to be significantly higher than that of non-encapsulated freeze-dried cells (Champagne et al. 1992). It was generally shown that the immobilised cells required a longer period of time to obtain a 1-log cycle drop in cell survival. The storage stability of cells in yoghurt when in an encapsulated state is also greater than when they are merely in a free state (Sultana et al. 2000). Unfortunately, the cells encapsulated in this way did not show any improvement in acid tolerance when compared to free cells (Sultana et al. 2000). This is not surprising since the calcium alginate matrix is porous, which does not stop acid penetration from the suspending liquid to its interior. It was shown by Favaro-Trindade and Grosso (2000) that the survival of both immobilised cells and free cells was unaffected upon exposure to a simulated gastric juice of pH 2, but the survival of both types of cells decreased by approximately 6 log cycles when exposed to a simulated gastric juice of pH 1. However, matrix cell entrapment is very difficult and costly to perform on a large scale (Picot and Lacroix 2004).

A promising method to entrap probiotic cells in matrix is based on using emulsification followed by gelation. Recently, a new method for encapsulating probiotic cells *Lactobacillus paracasei* ssp. *paracasei* F19 and *Bifidobacterium lactis* Bb12 (Heidebach et al. 2009) was developed based on a transglutaminase-catalysed gelation of casein suspensions containing the cells. It has been shown that the cells were protected from damage at pH levels similar to those in the human stomach. This method should be relatively easy to scale up.

1.6.5 Compression Coating

Major advances in compression coating were made at the start of the 1950s as machines for compressing a coating around a tablet core appeared on the market. Such techniques were accepted enthusiastically in the 1960s, although until very recently have only been used to coat traditional drugs (Gunsel and Dusel 1989). It is thought that the method of compression coating is suitable for large-scale use with probiotics, as the tableting process has already been well established and is widely used throughout the pharmaceutical industry.

The advantages and disadvantages of compression coating are discussed in detail elsewhere (Gunsel and Dusel 1989; Burlinson 1968; Cole 1995). Briefly, the advantages include the ability to coat an active ingredient that is moisture or heat sensitive, to separate incompatible materials by placing one in the core and the other in the coating and to provide a modified release profile of the active ingredient. Disadvantages include possible difficulties in binding the core and coating material together, the possibility of the core expanding and splitting the coating material and also the amount of coating material required could be at least twice the weight of the core when compared to film coating where the weight increase is negligible. Finally, the compression coating process is relatively slow with a production rate of around 500–1,500 coated tablets per min, whereas 10,000 uncoated tablets could be produced per min with industrial machinery which can then be coated via film coating.

2 Protective Compression Coating

The compression coating technique used for the encapsulation of lyophilised probiotic cells has been proven successful in this area (Huckle et al. 2008). In the following section compression coating will be described in greater detail, with a focus on how the technique can be optimised for use with probiotic cells.

2.1 Compression Coating Materials

Sodium alginate is an ideal coating material, as it is natural, safe and widely accepted in the food, drink and pharmaceutical industries (Gacesa 1998). However, pure sodium alginate is poorly compressible due to its elastic properties (Takeuchi et al. 1998), and thus cannot form a rigid tablet. Therefore, to obtain a more rigid tablet using sodium alginate powders it is necessary to incorporate a binder. In work carried out by Chan (2002), hydroxypropyl cellulose (HPC) was used as a binder, and the cell pellets obtained showed a much greater rigidity than if no binder was used. In a similar study, it was found that a high content of hydroxypropylmethylcellulose acetate succinate (HPMCAS) as well as a medium to high compaction force was necessary to achieve gastric juice resistance, or alternatively, low amounts of HPMCAS with the addition of sodium alginate were necessary to provide the same gastric resistance (Stadler and Viernstein 2003).

Compression coating of *L. acidophilus* was carried out by Olvera (2001, Personal Communication) with various enteric coating materials, namely; Sureteric Hydroxypropyl Methylcellulose Phtalate (HPMCP), Eudragit S100, Eudragit L100, Eudragit L100-55, Opadry AMB and Acryl Eza. Not every enteric coating material could be compressed to form a tablet, a fact that highlights the need for characterisation of the mechanical behaviour of the powder as well as assessment of the mechanical strength of the compacts. Eudragit L100-55 and Sureteric were found to be the two most suitable materials. Eudragit L100-55 has been proven to be a particularly suitable material for the compression coating technique, as it provides the cell tablet with a high degree of protection when exposed to simulated gastric fluid (SGF) while also dissolving readily in SIF, and showing relatively little loss in cell viability (Huckle et al. 2008). Eudragit is also widely accepted in the pharmaceutical and food industries.

Calcium phosphate is a directly compressible filler which is commonly used and produced from phosphoric acid and slaked lime. Although mainly used in vitamin and mineral supplements because of its high calcium and phosphorous content, the use of calcium phosphate is increasing in pharmaceutical preparations due to its low cost and desirable flow and compression characteristics (Wade and Weller 1994). Tablets made with calcium phosphate are difficult to eject from dies, and so the addition of a lubricant is necessary. Alkaline lubricants, such as magnesium stearate, are ideal for use with calcium phosphate as they have virtually no effect on its binding properties (Wade and Weller 1994). Due to the hydrophilic nature of calcium phosphate, it is rapidly penetrated by water (Lerk et al. 1976). Despite this rapid water penetration, calcium phosphate is relatively insoluble, and so when using calcium phosphate it may be necessary to include a disintegrant with an active mechanism for swelling, for example pectin, to induce the disintegration forces necessary for tablet break up (Huckle et al. 2008).

2.2 *Compression Pressure*

Chan (2002) found that cell reproductive viability was inversely related to the compression pressure applied to the cell containing powder. As compression pressure was applied to the cell powder the powder density increased, this was due to decreases in the inter- and intra-particulate pore spaces (Chan and Zhang 2002). In the initial stages of compaction, the powders (mixture of cells and excipients, e.g. skimmed milk) were probably rearranged to fill large void spaces (Paronen and Iikka 1996). This is further demonstrated by a sharp increase in displacement as compaction pressure is increased. Further increases in pressure led to a consolidation of the powders, and also fragmentation of the skimmed milk particles as these were larger than the cells. No significant loss in cell reproductive viability was reported up to a pressure of 30 MPa, although such pressure still caused some damage to cell walls (Chan and Zhang 2002). A gradual increase in loss of reproductive viability is associated with a pressure increase from 30 MPa to 90 MPa, this may be because cells are under increased mechanical stress and some weaker cells are unable to tolerate the compaction. Stronger cells tolerated such compression pressure, but increasingly more cells were damaged at their wall. When cell containing powders were exposed to pressures greater than 90 MPa, damage occurred to both the cell wall and membranes (Chan and Zhang 2002). Cell reproductive viability was found to decrease almost linearly as the compression pressure increased. Such a linear relationship was also observed by Blair et al. (1991) while studying the reproductive viability of

Staphylococcus aureus during tablet compression. At pressures exceeding 90 MPa, the movement of skimmed milk particles and bacteria is likely to become very limited.

Pressure should be as low as possible in order to minimise any cell loss. However, the tablets require a certain degree of rigidity in order to survive further handling. Yanagita et al. (1978) suggested that conventional tableting processes employ a pressure of around 30–50 MPa for compaction. In work carried out by Chan and Zhang (2002), a compression pressure of 60 MPa was used to produce cell pellets for further compression coating. Compression coating of bacterial cells in this way requires the cells to survive two separate compression processes, the first to form the cell containing core, and the second to coat the core with coating material. It was found by Chan and Zhang (2002) that the reduction in cell reproductive viability due to compression coating, for the formulation and processing conditions investigated, was insignificant. This may be because the pressure used for the compression coating was the same as that used to compact the cell powders into the initial core. Thus, as the bacteria were able to survive the forces from the first compression, they were able to withstand the second compression because there was no additional force to cause any movement in the cell pellet. Recently, Yap et al. (2006a, b) investigated the compaction behaviour of several industrial excipients, Eudragit® L100-55, Eudragit® L100, Eudragit® S100, Advantose™ 100, Barcroft™ CS90 calcium carbonate and Starlac™, and identified Eudragit® L100-55 can be compressed into a rigid table at relatively low pressure.

2.3 Cell Stability During Storage

The stability of freeze-dried cells decreases during storage, and higher survival rates are recorded at lower storage temperatures. In addition, subsequent storage conditions, such as temperature, atmosphere, light exposure and humidity, are important for the recovery of freeze-dried cells. It has been demonstrated (Castro et al. 1996) that changes in the lipid profile occur during storage of freeze-dried concentrates of *L. bulgaricus*, thus suggesting membrane lipid oxidation. In order to prevent or reduce such oxidation and increase storage survival, cell powders should ideally be stored under vacuum or controlled water activity.

Storage results obtained by Chan and Zhang (2002) showed that the isolation of bacteria from the environment via compression coating did result in greater stability. The stability of the compression coated bacteria, during storage at ambient temperature, was improved by approximately ten times. This is thought to be because the coating material may act as a barrier to prevent the movement of water into or out of the bacterial cell, to protect the bacterium from light and perhaps to even slow down thermal death. It was additionally noted that the compression pressure used to form the cell pellets also affected the long-term storage stability of bacteria, although the environmental factors were still dominant. This may be because when higher compression pressures are used, the cells sustain more damage to the cell wall, thus making them more susceptible to lethal injury during storage.

3 Controlled Release

Cells must be protected from gastric juice and acidity while passing through the stomach, and then released upon entering the colon. It is important for cells to be readily released when entering the colon to permit colonisation before being excreted. The most popular controlled release techniques based on compression coating are discussed in the following sections.

3.1 Gel-Forming Polymers

In recent years, there has been a renewed interest in using compression coating to coat drugs with a gel-forming polymer for controlled release purposes. Colon-specific delivery systems of this nature have been developed for compression coating drugs with pectin (Hodges et al. 2009) or guar gum (Sinha et al 2005). When exposed to a dissolution fluid, these materials form a viscous gel layer around a drug containing core. Drug release does not occur before sufficient erosion of the coat is achieved. The polymers were also rapidly degraded in the presence of simulated enzymes, for example pectinolytic enzymes which degrade pectin and are localised in the colonic region of the GI tract, therefore significantly increasing drug release and hence targeting release to the colon. The coating was also found to protect the core from a dissolution medium for a period of time due to the formation of a hydrogel layer, which delayed the permeation of the dissolution fluid to the core. This was demonstrated when sodium alginate was used to coat the soluble drug theophylline, and a delayed release profile of the drug during dissolution was observed (Kaneko et al. 1998).

In a different study, Turkoglu et al. (1999) proposed a pectin-HPMC compression coat system for colonic drug delivery. Turkoglu and Ugurlu (2002) undertook a further study, to develop and evaluate a colonic drug delivery system in vitro, based on a compression coated tablet containing 5-ASA as the core. A pectin HPMC mixture was used as the coating layer, based on the GI transit time concept. High molecular weight HPMC was used as it increases the mechanical strength of the tablet wall around the core during its transportation in the GI tract and acts to partially modify the high solubility of pectin. It was thus hypothesised that with the positive contribution of HPMC, pectin would be a good candidate for designing a colonic delivery system. Results showed that the pectin:HPMC ratio was a significant factor in controlling the degree of coating material erosion during dissolution (Turkoglu and Ugurlu 2002). The higher the pectin:HPMC ratio, the higher the rate of dissolution. Although probiotic bacteria such as *L. acidophilus* are different from water soluble or non-soluble drug molecules, similar gel-forming polymers may also be used for controlled delivery of the cells to colon, which is described below.

3.2 pH Sensitive Materials

The use of pH sensitive coatings is one of the most widely used techniques for the specific delivery of oral drugs to the colon (Colinet et al. 2010) and involves coating drug tablets or pellets with a pH sensitive material. The pH sensitive material is insoluble or almost impermeable when exposed to dissolution fluids of low pH, but can be readily dissolved in those fluids with a pH ranging from 5 to 7 (Khan et al. 2000). It has been shown via the use of a radiotelemetric capsule that the pH in the stomach and small intestine ranges from 1 to 2.5 and 6.6 to 7.5, respectively (Evans et al. 1986, 1988), while at the inlet of the colon and in the distal colon the pH is 6.4 and 7, respectively. Materials with varying dissolution pH can thus be used to target drug release specifically to the colon.

This can be done in one of two ways. The first approach involves using materials that dissolve at pH 6 or above. Such materials begin to dissolve as they enter the small intestine, and so a significant coating thickness is required to ensure the drug is delivered to the colon. This approach is also heavily dependent on the GI transit time that varies between individuals (Chan et al. 2001), for example if the transit time increases, then only a small amount of drug will be delivered to the colon. Another problem encountered with this technique is that after the ingestion of food the stomach pH rises to around 6.2, before slowly returning to the basal level (Russell et al. 1993), and thus some of the coating material may be dissolved while in the stomach.

The second approach involves using materials that dissolve at pH 7 or above (Myers et al. 1987). The use of such materials reduces any drug release in the upper GI tract, although there is a possibility that little or no drug will be released in the colon if the coating is too thick. Methylacrylic acid-methylmethacrylate copolymers (e.g. Eudragit L100-55) are made up of around 30% methacrylic acid units, which tend to dissolve at a pH of over 7 (Chan et al. 2001). This makes it a suitable coating material for colonic drug/probiotic delivery, with an example tablet shown in Fig. 2.

pH sensitive coatings can also be used for specific delivery to other parts of the gastric intestinal tract, with the choice of material being dictated by the pH conditions of the site for delivery.

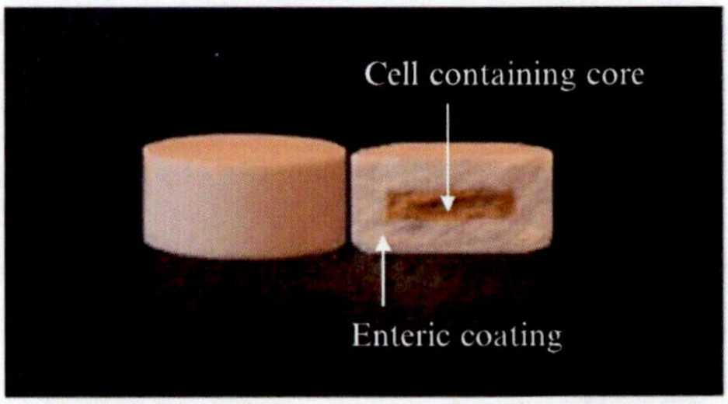

Fig. 2 Compression coated tablet before and after diametral compression

3.3 Disintegrants

Relying solely on pH dependent release can be imprecise as Eudragit L100-55, although soluble in the colon, may not fully release probiotic cells prior to excretion. Alternatively, after the ingestion of food, pH in the stomach can rise to around 6.2 before gradually returning to the basal level, which could dissolve some of the coating material before the colon is reached. In addition, the pH and GI transit time can vary between individuals. Therefore, a polysaccharide tablet disintegrant such as pectin could be used to speed up release of cells in the colon. The colonic flora does not appear to present many modifications and remains qualitatively similar from one individual to another, and hence pectinolytic enzymes will always be present between individuals to degrade pectin. Pectin is ideal as it is widely accepted by the food industry. The overall distribution of hydrophilic and hydrophobic groups on a pectin molecule determines its solubility or tendency to gel. High methyl ester (methoxylation degree >50%) pectins are less soluble than low methyl ester (methoxylation degree <50%) pectins. Therefore, the pectin excipient used should be a low methoxy biocompatible pectin with 15% esterification (GENU Pectin Slendid TYPE 100), thus making it more hydrophilic and increasing its swelling characteristics upon hydration, resulting in tablet disintegration. A calcium phosphate excipient can also be added for nutritional purposes.

3.4 Compression Coated Cell Release and Viability Profile

Dissolution studies were conducted by Huckle et al. (2008) using lyophilised *L. acidophilus* initially as loose cell powder, then as a tablet and lastly as a compression coated tablet. Cells were exposed sequentially to SGF of pH 2 for 2 h, followed by SIF of pH 6.8 for the remainder of the study. Reproductive viability (CFU/ml) was measured using traditional techniques, and total cell counts (TCC/ml), dead cell counts (DCC/ml) and viable cell counts (VCC/ml) based on TO-PRO-3 positive staining were made using a Microcyte (Aber Instruments) dual parameter flow cytometer.

3.4.1 Cell Powder

Figure 3 shows the effect of a dissolution study on the viability of lyophilised *L. acidophilus* cell powder. The CFU/ml drops vary rapidly when exposed to SGF, however this does increased slightly when subsequently exposed to SIF, although the counts still remain very low. The TCC/ml remains constant for the duration of the study. The VCC/ml is similar to the TCC/ml when exposed to SGF and then decreases upon exposure to SIF. This then remains relatively constant at a level significantly higher than the CFU/ml for the remainder of the experiment, thus demonstrating that a significant number of cells were intact, possibly metabolically active, but not capable of division at that time. Such cells are thought of as VBNC or dormant.

3.4.2 Cell Tablet

If *L. acidophilus* lyophilised cell powder is compressed into a tablet, and then subjected to a dissolution study, cell viability is as shown in Fig. 4 (Huckle 2006). CFU/ml drops sharply, as seen in Fig. 3 with loose cell powder, although the starting CFU/ml count at 0 h was significantly less with a tablet compared to

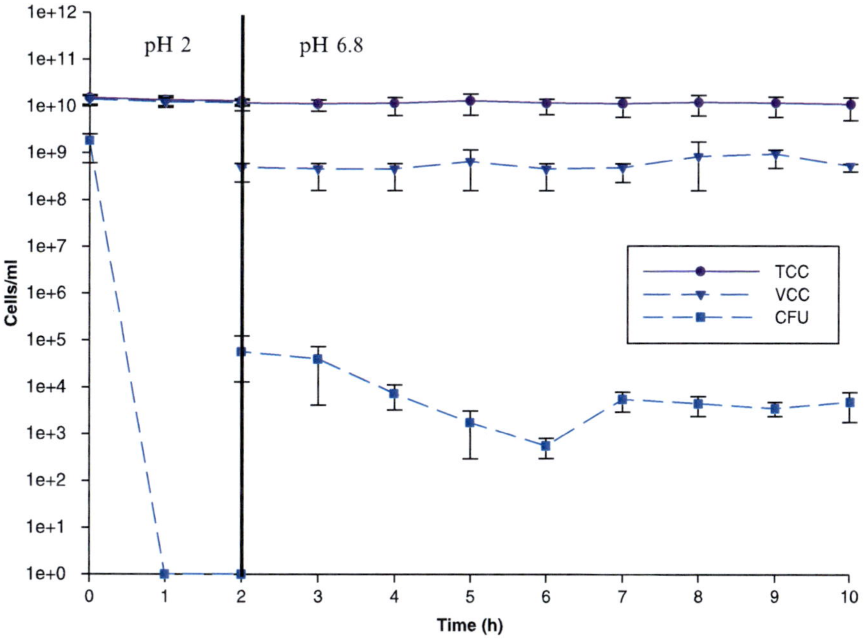

Fig. 3 Total cell count/ml (TCC) and viable cell count/ml (VCC) as measured by flow cytometry, and colony forming units/ml (CFU) as measured by dilution plating, during the dissolution study of lyophilised *L. acidophilus* cell powder when exposed to SGF (pH 2) for 2 h and subsequently SIF (pH 6.8). Error bars represent the range from the mean of triplicate studies (Huckle 2006)

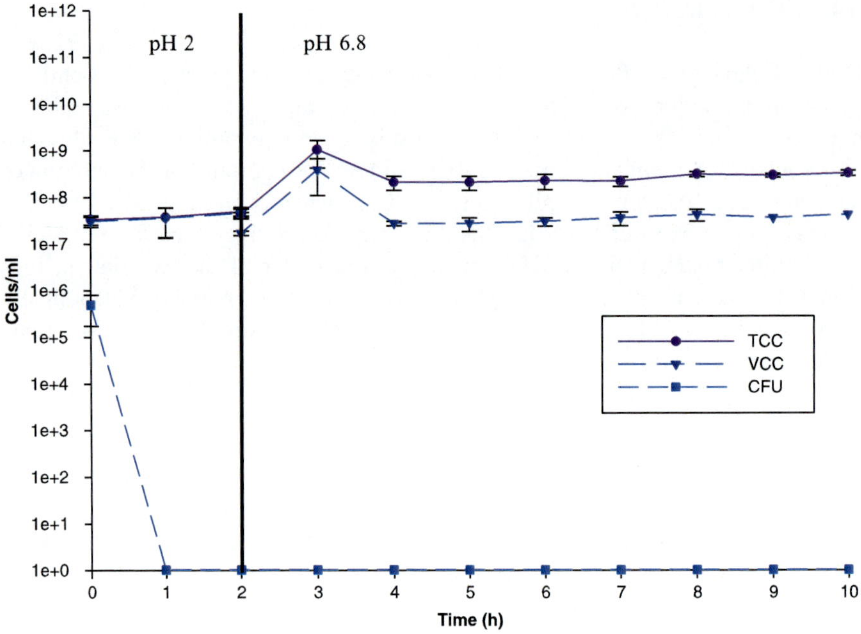

Fig. 4 Total cell count/ml (TCC) and viable cell count/ml (VCC) as measured by flow cytometry, and colony forming units/ml (CFU) as measured by dilution plating, during the dissolution study of a *L. acidophilus* core tablet when exposed to SGF (pH 2) for 2 h and subsequently SIF (pH 6.8). Error bars represent the range from the mean of triplicate studies (Huckle 2006)

loose cell powder, most likely because cells within the tablet were yet to be released. As with loose cell powder, the CFU/ml drops to 0 after 1 h. The TCC/ml and VCC/ml are both very similar and do not change significantly over the first 2 h of the study during exposure to SGF. Upon exposure to SIF the TCC/ml rises and peaks at 3 h, a result thought likely to be due to the pH change and the effect this may have on the hydration and swelling characteristics of the tablet. The TCC/ml then drops after 4 h to a point mid-way between that seen at 2 and 3 h, this fall could result from cell lysis, with cells perhaps unable to repair any sub-lethal injuries caused by compression after the additional stress imposed from the pH change. The hypothesis for cell lysis is further supported by the lack of any drop in TCC/ml for cell powder (Fig. 3), suggesting that only those cells sub-lethally injured during compression are susceptible. After 4 h the TCC/ml remains relatively constant for the remainder of the study, increasing slightly but not by a significant amount. The VCC/ml drops slightly upon exposure to SIF, and then follows the trend of the TCC/ml but at a reduced level. After 10 h the final TCC/ml is somewhat lower for a tablet (3.24×10^8/ml) than for loose cell powder (1.11×10^{10}/ml), this is thought to be due to a number of cells in the tablet being destroyed through the compaction process and also later by cell lysis after the pH change, but more significantly due to a lack of complete cell release due to the tablet not fully dissolving.

3.4.3 Compression Coated Tablet

Huckle et al. (2008) also coated a tablet of lyophilised *L. acidophilus* cell powder with Eudragit L100-55 by compression. Figure 5 shows the results of a dissolution study on this compression coated tablet. The CFU/ml count remains at 0 when exposed to SGF, but when exposed to SIF it gradually increases, reaching a maximum of around 5.4×10^9/ml at a dissolution time of 10 h. The TCC/ml remains relatively constant during exposure to SGF, and then increases significantly as the coating material dissolves when exposed to SIF, to a level similar to that found with cell powder after 10 h. The VCC/ml closely matches the TCC/ml during exposure to SGF, and is significantly higher than the number of CFU/ml, indicating that the majority of cells are viable although not reproductively so when exposed to SGF. When exposed to SIF the CFU/ml increases to closely match the VCC/ml after 5 h dissolution, showing that the majority of viable cells are now also reproductively viable. The rapid rise in reproductive viability highlights the effectiveness of the coating formulation in protecting cells during exposure to SGF, suggesting that cells are able to recover far more readily from any sub-lethal injury caused during compression when protected from the effects of pH stress.

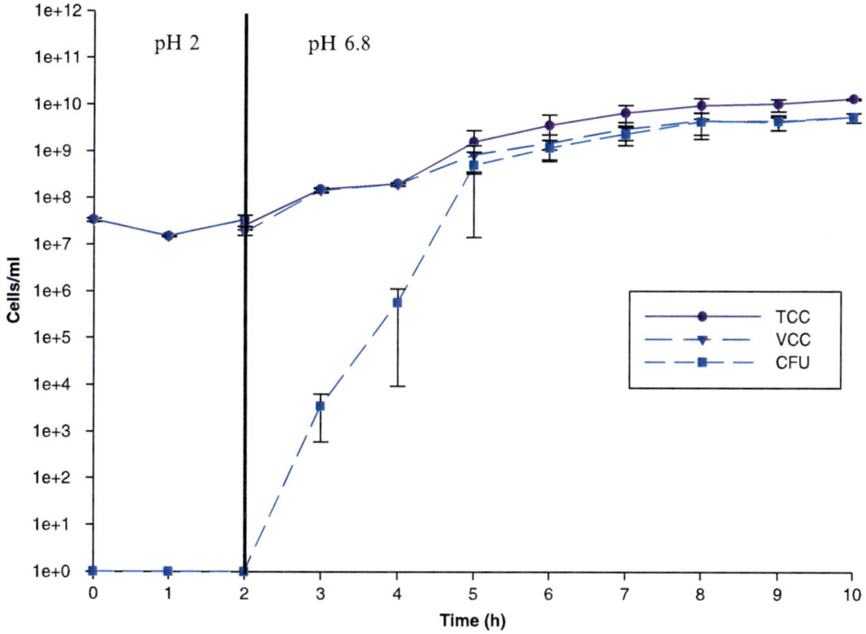

Fig. 5 Total cell count/ml (TCC) and viable cell count/ml (VCC) as measured by flow cytometry, and colony forming units/ml (CFU) as measured by dilution plating, during the dissolution study of a compression coated *L. acidophilus* tablet when exposed to SGF (pH 2) for 2 h and subsequently SIF (pH 6.8). Error bars represent the range from the mean of triplicate studies (Huckle 2006)

The TCC/ml after 5 h was 1.6×10^9/ml and after 10 h was around 1.3×10^{10}/ml, thus in both cases the minimum therapeutic dose of 1×10^9 cells was met. However, despite the minimum therapeutic dose being met, only 12% of cells were released after 5 h, this could potentially result in cells not being released adequately in time to adhere to the colon prior to excretion.

3.4.4 Compression Coated Tablet with Disintegrant

The release profile of a compression coated tablet can be modified with a suitable disintegrant, as demonstrated by Huckle et al. (2008). Pectin was included in the cell tablet formulation at a 1:1 ratio with cell powder, again with a coating of Eudragit L100-55.

If the results shown in Fig. 6 are compared with those of Fig. 5, it can clearly be seen that the addition of pectin to the tablet formulation is beneficial with regard to increasing cell release. Once the Eudragit coating begins to dissolve in intestinal fluid, the pectin in the core is seen to become hydrated and rapidly swell, resulting in faster tablet disintegration. At 3 h the number of CFU was 1.5×10^8/ml when formulated with pectin, whereas without pectin the number of CFU was much less

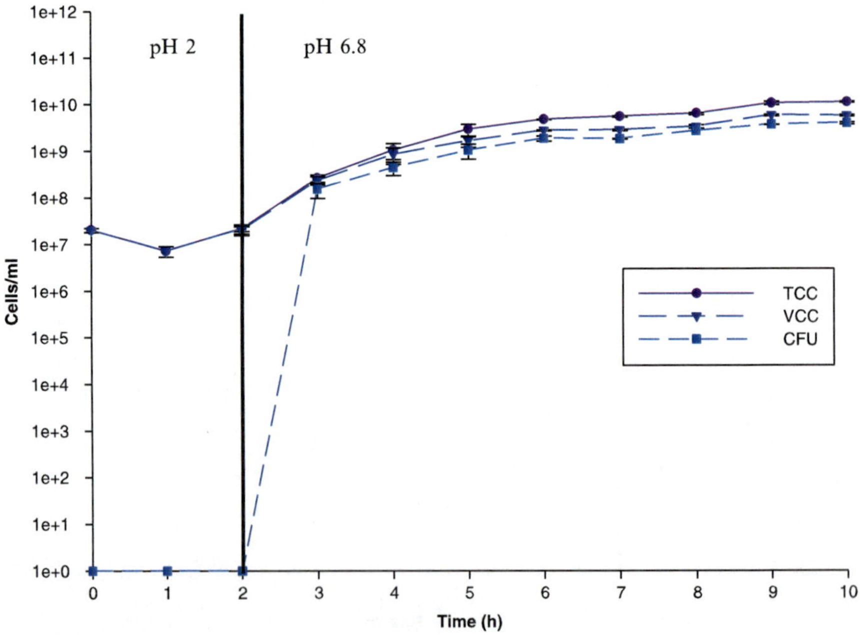

Fig. 6 Total cell count/ml (TCC) and viable cell count/ml (VCC) as measured by flow cytometry, and colony forming units/ml (CFU) as measured by dilution plating, during the dissolution study of a compression coated *L. acidophilus* tablet including a core of pectin, when exposed to SGF (pH 2) for 2 h and subsequently SIF (pH 6.8). Error bars represent the range from the mean of triplicate studies (Huckle 2006)

at 3.2 × 10^3/ml, thus significantly more reproductively active cells are released when pectin is present. In both cases, the number of CFU/ml is significantly less than the number of VCC/ml during exposure to SGF, but when exposed to SIF the number of CFU/ml recovers to a level similar to the VCC/ml after only 1 h when pectin is included, compared with a period of around 5 h when no pectin is included. Therefore, the majority of cells reach a reproductively viable state faster when using pectin as a release agent.

4 Conclusions

Unprotected probiotic bacteria can lose viability during storage due to various detrimental factors in the environment including moisture, oxygen and heat, cannot survive passage through the GI tract in sufficient quantity to meet the minimum therapeutic dose of 1 × 10^9 cells, and thus cannot confer any significant beneficial effect on the host. Various encapsulation techniques have been developed to protect probiotic cells, including compression coating which is relatively cheap and easy to implement. Using this technique cells can be compressed into a core and then further coated using an enteric coating formulation such as Eudragit L100-55, providing protection during storage and exposure to SGF and allowing recovery when exposed to SIF, in which the coating can dissolve. The use of a hydrophobic disintegrant (e.g. pectin) in the core of a compression coated formulation, significantly increases the release rate of cells to the colon, improving targeted release. When coated in this manner, reproductive viability successfully meets the minimum therapeutic dose of 1 × 10^9 cells for the investigated bacteria *L. acidophilus*, specifically at the site of delivery, the colon based on in vitro studies. Compression coating may be further exploited to realise stabilisation and controlled release of other probiotic strains, but more research is required in order to develop new formulations which can be used to prepare coated tables with desirable properties.

References

Appelgren C, Eskilson C (1990) A novel method for the granulation and coating of pharmacologically active substances. Drug Dev Ind Pharm 16:2345–2351

Blair TC, Buckton G, Bloomfield SF (1991) On the mechanism of kill of microbial contaminants during tablet compression. Int J Pharm 72:111–115

Burlinson H (1968) The coating of tablets. In: Tablets and Tabletting. William Heinemann Medical Books, London, pp 48–54

Carvalho AS, Silva J, Ho P, Teixeira P, Malcata FX, Gibbs P (2004) Relevant factors for the preparation of freeze-dried lactic acid bacteria. Int Dairy J 14:835–847

Castro HP, Teixeira PM, Kirby R (1996) Changes in the membrane of Lactobacillus bulgaricus during storage following freeze-drying. Biotechnol Lett 18:99–104

Castro HP, Teixeira PM, Kirby R (1997) Evidence of membrane damage in Lactobacillus bulgaricus following freeze-drying. J Appl Microbiol 82:87–94

Champagne CP, Morin N, Couture R, Gagnon C, Jelen P, Lacroix C (1992) The potential of immobilised cell technology to produce freeze-dried, phage-protected cultures of Lactococcus lactis. Food Res Int 25:419–427

Champagne CP, Mondou F, Raymond Y, Roy D (1996) Effect of polymers and storage temperature on the stability of freeze-dried lactic acid bacteria. Food Res Int 29:555–562

Chan ES (2002) Stabilisation of probiotic bacteria. PhD Thesis, University of Birmingham, Birmingham, UK

Chan ES, Zhang Z (2002) Encapsulation of probiotic bacteria Lactobacillus acidophilus by direct compression. Trans IChemE 80:78–82

Chan ES, Zhang Z (2005) Bioencapsulation by compression coating of probiotic bacteria for their protection in an acidic medium. Process Biochem 40:3346–3351

Chan WA, Boswell CD, Zhang Z (2001) Comparison of the release profiles of a water soluble drug carried by Eudragit-coated capsules in different in vitro dissolution liquids. Powder Technol 119:26–32

Charteris WP, Kelly PM, Morelli L, Collins JL (1998) Development and application of an in vitro methodology to determine the transit tolerance of potentially probiotic Lactobacillus and Bifidobacterium species in the upper human gastrointestinal tract. J Appl Microbiol 84:759–768

Cole GC (1995) Introduction and overview of pharmaceutical coating. In: Cole GC (ed) Pharmaceutical coating. Taylor and Francis, London, pp 1–5

Colinet I, Dulong V, Mocanu G, Picton L, Le Cerf D (2010) Effect of chitosan coating on the swelling and controlled release of a poorly water-soluble drug from an amphiphilic and pH-sensitive hydrogel. Int J Biol Macromol 47:120–125

Conway PL, Gorbach SL, Goldin BR (1987) Survival of lactic acid bacteria in the human stomach and adhesion to intestinal cells. J Dairy Sci 70:1–12

Dave RI, Shah NP (1997) Effectiveness of ascorbic acid as an oxygen scavenger in improving viability of probiotic bacteria in yoghurts made with commercial starter cultures. Int Dairy J 7:435–443

Douglas LC, Sanders ME (2008) Probiotics and prebiotics in dietetics practice. J Am Diet Assoc 108:510–521

Evans DF, Bramley R, Pye G, Dyson TJ, Hardcastle JD (1986) Gastrointestinal pH profiles in man. Gastroenterology 90:1410

Evans DF, Pye G, Bramley R, Clark AG, Dyson TJ, Hardcastle JD (1988) Measurement of gastrointestinal pH profiles in normal ambulant human subjects. Gut 29:1035–1041

Favaro-Trindade CS, Grosso CRF (2000) The effect of the immobilisation of Lactobacillus acidophilus and Bifidobacterium lactis in alginate on their tolerance to gastrointestinal secretions. Milchwissenschaft 55:496–499

Fuller R, Gibson GR (1998) Probiotics and prebiotics: microflora management for improved gut health. Clin Microbiol Infect 4:477–480

Gacesa P (1998) Alginates. Carbohydr Polym 8:161–182

Gibson GR (2004) Fibre and effects on probiotics (the prebiotic concept). Clin Nutr Suppl 1:25–31

Gibson GR, Roberfroid MB (1995) Dietary modulation of the human colonic microbiota: introducing the concept of prebiotics. J Nutr 125:1401–1412

Gunsel WC, Dusel RG (1989) Compression-coated and layer tablets. In: Herbert A et al (eds) Pharmaceutical dosage forms: tablets, vol 1, 2nd edn. Dekker, New York, pp 196–245

Heidebach T, Först P, Kulozik U (2009) Transglutaminase-induced caseinate gelation for the microencapsulation of probiotic cells. Int Dairy J 19:77–84

Hodges LA, Connolly SM, Band J, O'Mahony B, Ugurlu T, Turkoglu M, Wilson CG, Stevens HNE (2009) Scintigraphic evaluation of colon targeting pectin–HPMC tablets in healthy volunteers. Int J Pharm 370:144–150

Homayouni A, Azizi A, Ehsani MR, Yarmand MS, Razavi SH (2008) Effect of microencapsulation and resistant starch on the probiotic survival and sensory properties of synbiotic ice cream. Food Chem 111:50–55

Hood SK, Zottola EA (1988) Effect of low pH on the ability of Lactobacillus acidophilus to survive and adhere to human intestinal cells. J Food Sci 53:1514–1516
Huckle BD (2006) Compaction of probiotic cells with targeted delivery and analysis of cell physiological state using flow cytometry. PhD Thesis, University of Birmingham, UK
Huckle BD, Hewitt CJ, Jüsten P, Yan Y, Zhang Z (2008) Determination of the viability and release profile of compression coated probiotic cells using fluorescent staining and flow cytometry. Minerva Biotec 18:11–16 (Special Edition on Bioencapsulation)
Kailasapathy K, Chin J (2000) Survival and therapeutic potential of probiotic organisms with reference to Lactobacillus acidophilus and Bifidobacterium spp. Immunol Cell Biol 78:80–88
Kaneko K, Kanada K, Miyagi M, Saito N, Ozeki T, Yuasa H, Kanaya Y (1998) Formation of water-insoluble gel in dry-coated tablets for the controlled release of theophylline. Chem Pharm Bull 46:728–729
Kearney L, Upton M, McLoughlin A (1990) Enhancing the viability of Lactobacillus plantarum inoculum by immobilising the cells in calcium-alginate beads incorporating cryoprotectants. Appl Environ Microbiol 56:3112–3116
Khan MZI, Stedul HP, Kurjakovic N (2000) A pH-dependent colon-targeted oral drug delivery system using methacrylic acid copolymers: II. Manipulation of drug release using Eudragit (R) L100 and Eudragit S 100 combination. Drug Dev Ind Pharm 26:549–554
Kim HS, Kamara BJ, Good IC, Enders GL Jr (1988) Method for the preparation of stable microencapsulated lactic acid bacteria. J Ind Microbiol 3:253–257
Kos B, Suskovic J, Goreta J, Matosic S (2000) Effect of protectors on the viability of Lactobacillus acidophilus M92 in simulated gastrointestinal conditions. Food Technol Biotech 38:121–127
Kosin B, Rakshit SK (2010) Induction of heat tolerance in autochthonous and allochthonous thermotolerant probiotics for application to white shrimp feed. Aquaculture 306:302–309
Kurmann JA, Rasic JL (1991) The health potential of products containing Bifidobacteria. In: Robinson RK (ed) Therapeutic properties of fermented milks. Elsevier, London, pp 117–158
Lacroix C, Paquin C, Arnaud JP (1990) Batch fermentation with entrapped growing cells of Lactobacillus casei. Optimisation of the rheological properties of the entrapment. Appl Microbiol Biotechnol 32:403–408
Lal M, Tiwari MP, Sinha RN, Ranganathan B (1978) Activity of the freeze-dried Streptococcus lactis sub sp Diacetilactic DRC in gelatin capsules and strip packed tablets during storage. Egypt J Dairy Sci 6:33–37
Lauland S (1994) Commercial aspects of formulation, production and marketing of probiotic products. In: Gibson SAW (ed) Human health: the contribution of microorganisms. Springer, London, pp 159–173
Lee YK, Salminen S (1995) The coming of age of probiotics. Trends Food Sci Tech 6:241–245
Lerk CF, Schoonen AJ, Fell JT (1976) Contact angles and wetting of pharmaceutical powders. J Pharm Sci 65:843–847
Matsumoto M, Mochiduki K, Kondo K (2004) Toxicity of ionic liquids and organic solvents to lactic acid-producing bacteria. J Biosci Bioeng 98:344–347
Myers B, Evans DNW, Rhodes J, Evans BK, Hugher BR, Lee MG, Richens A, Richards D (1987) Metabolism and urinary excretion of 5-aminosalicylic acid in healthy volunteers when given intravenously or released from absorption at different sites of gastrointestinal tract. Gut 28:196–200
Ooi L-G, Bhat R, Rosma A, Yuen K-H, Liong M-T (2010) A synbiotic containing Lactobacillus acidophilus CHO-220 and inulin improves irregularity of red blood cells. J Dairy Sci 93:4535–4544
Otero MC, Espeche MC, Nader-Macías ME (2007) Optimization of the freeze-drying media and survival throughout storage of freeze-dried Lactobacillus gasseri and Lactobacillus delbrueckii subsp. delbrueckii for veterinarian probiotic applications. Proc Biochem 42:1406–1411
Paronen P, Iikka J (1996) Porosity-pressure functions. In: Alderborn G, Nystrom C (eds) Pharmaceutical powder compaction technology. Dekker, New York, USA, pp 55–97

Picot A, Lacroix C (2004) Encapsulation of bifidobacteria in whey protein-based microcapsules and survival in simulated gastrointestinal conditions and in yoghurt. Int Dairy J 14:505–515
Porter SC, Bruno CH (1990) Coating of pharmaceutical solid-dosage form. In: Lieberman HA (ed) Pharmaceutical dosage forms: tablets, vol 3. Dekker, New York
Russell TL, Berardi RR, Barnett JL, Dermentzoglou LC, Jarvenpaa KM, Schmaltz SP, Dressman JB (1993) Upper gastrointestinal pH in seventy nine healthy, elderly, North American men and women. Pharm Res 10:187–196
Sanz Y (2007) Ecological and functional implications of the acid-adaptation ability of Bifidobacterium: a way of selecting improved probiotic strains. Int Dairy J 17:1284–1289
Schrezenmeir J, de Vrese M (2001) Probiotics, prebiotics and synbiotics-approaching a definition. Am J Clin Nutr 73:361S–364S
Sheu TY, Marshall RT (1993) Microentrapment of Lactobacilli in calcium alginate gels. J Food Sci 54:557–561
Shortt C (1998) Living it up for dinner. Chem Ind London 8:300–303
Sinha VR, Mittal BR, Kumria R (2005) In vivo evaluation of time and site of disintegration of polysaccharide tablet prepared for colon-specific drug delivery. Int J Pharm 289:79–85
Stadler M, Viernstein H (2003) Optimisation of a formulation containing viable lactic acid bacteria. Int J Pharm 256:117–122
Stummer S, Salar-Behzadi S, Unger FM, Oelzant S, Penning M, Viernstein H (2010) Application of shellac for the development of probiotic formulations. Food Res Int 43:1312–1320
Sultana K, Godward G, Reynolds N, Arumugaswamy R, Peiris P, Kailasapathy K (2000) Encapsulation of probiotic bacteria with alginate-starch and evaluation of survival in simulated gastrointestinal conditions and in yoghurt. Int J Food Microbiol 62:47–55
Supriadi D, Kailasapathy K, Hourigan JA (1994) Effect of partial replacement of skim milk powder with whey protein concentrate on buffering capacity of yoghurt. Proceedings of the XXIV international dairy congress, Melbourne
Takeuchi H, Yasuji T, Hino T, Yamamoto H, Kawashima Y (1998) Spray-dried composite particles of lactose and sodium alginate for direct tabletting and controlled releasing. Int J Pharm 174:91–100
Turkoglu M, Ugurlu T (2002) In vitro evaluation of pectin-HPMC compression coated 5-aminosalicylic acid tablets for colonic delivery. Eur J Pharm Biopharm 53:65–73
Turkoglu M, Takka S, Baran H, Sakr A (1999) Pectin hydroxymethyl cellulose drug delivery system for colon targeting. Pharm Ind 61:662–665
Wade A, Weller PJ (1994) Handbook of pharmaceutical excipients, 2nd edn. The pharmaceutical, London, UK
Yanagita T, Miki T, Sakai T, Horikoshi I (1978) Microbiological studies on drugs and their raw materials. I. Experiments on the reduction of microbial contaminants in tablets during processing. Chem Pharm Bull 26:185–190
Yap SF, Adams MJ, Seville JPK, Zhang Z (2006a) Single and bulk compression of pharmaceutical excipients: evaluation of mechanical properties. Powder Technol 185:1–10
Yap SF, Adams MJ, Seville JPK, Zhang Z (2006b) Understanding the mechanical properties of single micro-particles and their compaction behaviour. China Particuol 4:35–40
Yokota T, Sato T, Uemitsu N, Kitabatake K (1990) Development of enteric coated tablets containing lactic acid bacteria. International Bifidobacterium conference, 12–13 Sept 1990, Keidanren Kaikan, Tokyo, Japan
Zárate G, Nader-Macias ME (2006) Viability and biological properties of probiotic vaginal lactobacilli after lyophilization and refrigerated storage into gelatin capsules. Proc Biochem 41:1779–1785

Genetic Engineering of Probiotic Microorganisms

T.C. Bamunuarachchige, H.A.M. Wickramasinghe, D.M.J.C. Dissanayaka, and N.A.D. Wickramarathna

Contents

1 Genomics of Probiotic Microorganisms ... 110
  1.1 Structural Genomics of Probiotic Bacteria ... 110
  1.2 Functional Genomics of Probiotic Bacteria ... 111
  1.3 Comparative Genomics of Probiotic Bacteria ... 118
2 Genetic Modifications in Probiotic Bacteria ... 119
  2.1 Molecular Alterations of Probiotic Bacteria for Enhanced Digestive Tract Survival ... 119
3 Genetically Modified Lactic Acid Bacteria as Mucosal Delivery Vectors ... 122
  3.1 Advantages of LAB as Mucosal Delivery Vectors ... 122
  3.2 Route of Delivery of LAB and Immunogenic Response ... 123
  3.3 Development of Expression Systems for LAB ... 123
  3.4 GM LAB Strain Models as Live Vaccine ... 124
  3.5 Use of LAB to Deliver Single-Chain Variable Fragment Antibody Production ... 127
  3.6 LAB as DNA Vaccine Delivery Agents ... 127
4 Mutant Probiotics, Issues, and Ethics ... 129
  4.1 Type of Mutation Events in Probiotics ... 129
  4.2 Issues of Mutant Probiotics and Possible Solutions ... 129
  4.3 Ethics Concerning Mutant Probiotics ... 133
5 Conclusion ... 134
References ... 134

T.C. Bamunuarachchige (✉)
Faculty of Applied Sciences, Department of Biological Sciences, Rajarata University, Mihintale, Sri Lanka
e-mail: chathush@rjt.ac.lk

H.A.M. Wickramasinghe
Faculty of Agriculture, Department of Agricultural Biology, University of Peradeniya, Peradeniya, Sri Lanka

D.M.J.C. Dissanayaka
Teaching hospital, Anuradhapura, Sri Lanka

N.A.D. Wickramarathna
University of Alberta, Edmonton, AB, Canada

M.-T. Liong (ed.), *Probiotics*, Microbiology Monographs 21,
DOI 10.1007/978-3-642-20838-6_5,

Abstract With the advent of the era of genomics and proteomics, the molecular mechanisms of beneficial characteristics of probiotics are gradually being elucidated. These studies while paving the way for concrete evidence of the beneficial effects of probiotics have also lead to the idea of improved probiotics through genetic modification. Genetically Modified (GM) probiotics are mainly concerned with the improved survival and persistence of probiotics within the human and animal gut, tolerance of packing and storage conditions of food and the delivery of therapeutics by live probiotics. With the identification of genetic elements that confer tolerance to increased osmolarity, bile salt and reduced water activity, the improvement of probiotics for survival has become a reality. With the expression of various molecules such as antigens, enzymes, and molecules of immunological importance within probiotic microbes, the use of GM probiotics in the field of therapeutics looks promising. However, the safety issues of GM probiotics and the consumer perceptions need to be addressed properly in order to reap their full benefits.

1 Genomics of Probiotic Microorganisms

The term "probiogenomics" was first coined by Ventura et al. (2009) to explain the genome-scale analysis of probiotic mechanisms and potentials of health-promoting microorganisms colonized in their host's gut. *Lactobacillus* and *Bifidobacterium* species are such probiotic microorganisms typically used for genomic-scale analysis. Lactobacilli are phylogenetically a diverse group of bacteria with more than hundred species. On the other hand, bifidobacteria comprises of about 30 species.

Genomic-scale analysis is usually started with sequencing of genomic DNA of organisms. After sequencing the free living bacteria, *Haemophilus influenza* in 1995 (Fleischmann et al. 1995) and with the rapid improvements in sequencing technology, many probiotic microorganisms have been sequenced. Lactobacilli and bifidobacteria received a significant attention in such genome sequencing projects. Though there are several *Lactococcus* and *Streptococcus* species used for genomic sequencing (Table 1), the available information is comparatively low.

Bifidobacterium longum NCCC2705 isolated from stools of human infants was the first probiotic bacteria sequenced (Schell et al. 2002). After the establishment of the lactic acid bacterium genome consortium (LABGC) in the United States, many industrially important strains have been sequenced. Thus there are about 89 such genomic sequences available to-date (NCBI n.d.).

1.1 Structural Genomics of Probiotic Bacteria

About 67 lactobacilli genome sequences are currently available (Table 2). The size of the genome of most of these is around 2 Mb and it varies from 1.28 Mb (*Lactobacillus iners*) to 3.32 Mb (*Lactobacillus plantarum*). The G + C content

Table 1 Genome sequence information of other probiotic bacteria; *Lactococcus* and *Streptococcus* species

			Genome		Genes			
Species	Strain	Accession no:	size (Mb)	% GC	Total	Structural RNAs:	Protein coding	Pseudogenes
L. lactis subsp. *cremoris*	*MG1363*	AM406671	–	35	2,597	81	2,434	82
L. lactis subsp. *cremoris*	*SK11*	CP000425	–	35	2,610	82	2,384	144
L. lactis subsp. *lactis*	*Il1403*	–	–	35	2,425	79	2,321	1
L. lactis subsp. *lactis*	*KF147*	CP001834	2.6	34	2,624	87	2,444	93
S. thermophilus	*LMD-9*	CP000419	–	39	2,002	87	1,709	206
S. thermophilus	*LMG 18311*	CP000023	1.8	39	1,974	85	1,889	None
S. thermophilus	*CNRZ1066*	CP000024	1.8	39	2,000	85	1,915	None

Source: NCBI (http://www.ncbi.nlm.nih.gov/), October 10 2010

of the genomes of lactobacilli is usually low and it ranges from 32 to 51%. Number of genes identified from a lactobacilli genome varies from 3,137 in *L. plantarum* subsp. *plantarum* to 1,258 in *L. iners*. Interestingly, there are many examples for presence of pseudogenes in their genomes.

Based on the published data at National Center for Biotechnology Information (NCBI), 21 genome sequences are currently available for bifidobacteria (Table 3). The size of the genome ranges from 1.9 to 2.8 Mb with 1,603 to 2,588 genes in the genome. The G + C content is usually higher than 55%.

1.2 Functional Genomics of Probiotic Bacteria

Almost all probiotic lactibaciili meet the demand of energy through fermentation of carbohydrates by Embden–Meyerhof pathway (EMP) or phosphoketolase pathway (PKP). Apart from EMP or PKPs, some lactobacilli species such as *L. salvarius* possesses all genes responsible for the production of major enzymes for glucogenesis. This mechanism is active during glucose starvation. Few lactobacilli species are capable of synthesizing some vitamins, amino acids, and also degradation of polypeptides. *L. reuteri* CRL1098 is capable of synthesizing cobolamin (vitamin B-12) (Morita et al. 2008). Kleerebezem et al. (2003) proved that *L. plantarum* synthesizes all amino acids except branched amino acids. *L. acidophilus* and *L. juhnsonii* contain cell-wall-bound proteinases that may involve in protein degradation (Pridmore et al. 2004). Additionally, most probiotic bacteria have genes for carbohydrate and amino acid transportation (Denou et al. 2008).

Table 2 Basic structural features of lactobacilli

						Genes			
No.	Species	Strain	Accession no:	Genome size (Mb)	% G + C	Total	Structural RNAs	Protein coding	Pseudogenes
1	*L. acidophilus*	*NCFM*	CP000033	2.0	34	1,938	74	1,864	None
2	*L. acidophilus*	*ATCC 4796*	ACHN00000000	–	34	2,084	64	2,020	None
3	*L. amylolyticus*	*DSM 11664*	ADNY00000000	1.54	38	1,746	62	1,684	None
4	*L. antri*	*DSM 16041*	ACLL00000000	2.24	51	2,286	62	2,224	None
5	*L. buchneri*	*ATCC 11577*	ACGH00000000	–	39	3,065	63	3,002	None
6	*L. brevis*	*ATCC 367*	CP000416	2.3	46	2,314	82	2,185	49
7	*L. brevis* subsp. *gravesensis*	*ATCC 27305*	ACGG00000000	–	39	3,106	65	3,041	None
8	*L. casei*	*ATCC 334*	CP000423	2.9	46	2,906	76	2,748	82
9	*L. casei*	*Zhang*	CP001084	2.86	46	2,906	75	2,804	27
10	*L. casei*	*BL23*	FM177140	3.08	46	3,090	75	3,015	None
11	*L. coleohominis*	*101-4-CHN*	ACOH00000000	–	41	1,709	57	1,652	None
12	*L. crispatus*	*MV-1A-US*	ACOG00000000	–	35	2,213	62	2,151	2
13	*L. crispatus*	*125-2-CHN*	ACPV00000000	–	35	2,139	57	2,082	1
14	*L. crispatus*	*JV-V01*	ACKR00000000	–	36	2,278	69	2,209	None
15	*L. crispatus*	*ST1*	FN692037	2.04	36	2,100	76	2,024	None
16	*L. crispatus*	*214-1*	ADGR00000000	2.07	36	2,221	58	2,163	1
17	*L. delbrueckii* subsp. *bulgaricus*	*PB2003/044-T3-4*	AEAT00000000	1.98	49	1,981	72	1,909	None
18	*L. delbrueckii* subsp. *bulgaricus*	*ATCC BAA-365*	CP000412	1.86	49	2,034	127	1,715	192
19	*L. delbrueckii* subsp. *bulgaricus*	*ATCC 11842*	CR954253	1.87	49	2,184	122	1,529	533
20	*L. fermentum*	*28-3-CHN*	ACQG00000000	1.96	51	1,933	53	1,880	None
21	*L. fermentum*	*ATCC 14931*	ACGI00000000	–	52	1,929	63	1,866	None
22	*L. fermentum*	*IFO 3956*	AP008937	2.1	51	1,912	69	1,843	None
23	*L. gasseri*	*202-4*	ACOZ00000000	–	34	1,821	48	1,773	1

24	*L. gasseri*	*JV-V03*	ACGO00000000	2.01	34	2,036	59	1,977	None
25	*L. gasseri*	*MV-22*	ABWH00000000	–	34	1,956	39	1,917	None
26	*L. gasseri*	*ATCC 33323*	CP000413	1.9	35	1,898	98	1,755	48
27	*L. gasseri*	*224-1*	ADFT00000000	2.01	35	2,354	102	2,252	1
28	*L. helveticus*	*DSM 20075*	ACLM00000000	–	36	2,129	51	2,078	None
29	*L. helveticus*	*DPC 4571*	CP000517	2.1	37	1,838	73	1,610	155
30	*L. hilgardii*	*ATCC 8290*	ACGP00000000	–	39	2,854	63	2,791	None
31	*L. iners*	*AB-1*	ADHG00000000	1.29	32	1,258	49	1,209	None
32	*L. iners*	*DSM 13335*	ACLN00000000	–	32	1,264	50	1,214	None
33	*L. jensenii*	*208-1*	ADEX00000000	3.33	34	3,264	82	3,182	None
34	*L. jensenii*	*JV-V16*	ACGQ00000000	1.6	34	1,518	68	1,450	None
35	*L. jensenii*	*SJ-7A-US*	ACQD00000000	1.68	33	1,679	49	1,630	None
36	*L. jensenii*	*115-3-CHN*	ACQN00000000	1.61	33	1,520	50	1,470	None
37	*L. jensenii*	*27-2-CHN*	ACOF00000000	–	33	1,526	50	1,476	None
38	*L. jensenii*	*1153*	ABWG00000000	–	34	1,626	48	1,578	1
39	*L. jensenii*	*269-3*	ACOY00000000	–	34	1,624	49	1,575	None
40	*L. johnsonii*	*FI9785*	FN298497	1.76	34	1,780	70	1,710	None
41	*L. johnsonii*	*ATCC 33200*	ACGR00000000	–	34	1,896	58	1,838	None
42	*L. johnsonii*	*NCC 533*	AE017198	2.0	34	1,918	97	1,821	None
43	*Lactobacillus jensenii*	*1153*	ABWG00000000		48	1,626	34	1,578	1
44	*L. plantarum*	*WCFS1*	AL935263	3.32	44	3,135	86	3,007	42
45	*L. plantarum*	*JDM1*	CP001617	3.2	44	3,029	78	2,948	3
46	*L. plantarum* subsp. *plantarum*	*ATCC 14917*	ACGZ00000000	3.19	44	3,223	69	3,154	None
47	*L. plantarum* subsp. *plantarum*	*ST-III*	CP002222	3.25	44	3,137	79	2,996	62
48	*L. paracasei* subsp. *paracasei*	*8700:2*	ABQV00000000	–	46	3,070	49	3,021	1
49	*L. paracasei* subsp. *paracasei*	*ATCC 25302*	ACGY00000000	–	46	3,103	61	3,042	None

(continued)

Table 2 (continued)

No.	Species	Strain	Accession no:	Genome size (Mb)	% G + C	Genes: Total	Genes: Structural RNAs	Genes: Protein coding	Genes: Pseudogenes
50	*L. reuteri*	*100-23*	AAPZ00000000	–	38	2,269	88	2,181	None
51	*L. reuteri*	*SD2112*	ACGW00000000	2.26	38	2,362	100	2,262	None
52	*L. reuteri*	*MM2-3*	ACLB00000000	–	38	2,105	60	2,045	None
53	*L. reuteri*	*CF48-3A*	ACHG00000000	–	38	2,223	59	2,164	None
54	*L. reuteri*	*MM4-1A*	ACGX00000000	–	38	2,082	59	2,023	None
55	*L. reuteri*	*JCM 1112*	AP007281	2.04	38	1,901	81	1,820	None
56	*L. reuteri*	*DSM 20016*	CP000705	2.0	38	2,027	88	1,900	39
57	*L. rhamnosus*	*HN001*	ABWJ00000000	–	46	2,811	53	2,758	None
58	*L. rhamnosus*	*LMS2-1*	ACIZ00000000	–	46	3,209	54	3,155	None
59	*L. rhamnosus*	*GG*	FM179322	3.0	46	2,985	72	2,913	None
60	*L. rhamnosus*	*Lc 705*	FM179323	2.97	46	2,954	76	2,878	None
61	*L. ruminis*	*ATCC 25644*	ACGS00000000	–	43	2,332	68	2,264	None
62	*L. sakei* subsp. *sakei*	*23 K*	CR936503	1.9	41	1,963	84	1,879	None
63	*L. salivarius*	*ACS-116-V-Col5a*	AEBA00000000	2.04	32	2,184	63	2,121	1
64	*L. salivarius*	*ATCC 11741*	ACGT00000000	–	32	2,046	70	1,976	None
65	*L. salivarius*	*UCC118*	CP000233	1.83	32	1,864	99	1,717	49
66	*L. ultunensis*	*DSM 16047*	ACGU00000000	–	35	2,271	61	2,210	None
67	*L. vaginalis*	*ATCC 49540*	ACGV00000000	–	40	1,937	67	1,870	None

Source: NCBI (http://www.ncbi.nlm.nih.gov/), October 2010

Table 3 Structural genomic features of bifidobacteria genomes

No.	Species	Strain	Accession no:	Genome size (Mb)	% GC	Genes: Total	Genes: Structural RNAs	Genes: Protein coding	Genes: Pseudo genes
1	*B. adolescentis*	*L2-32*	AAXD00000000	2.4	59	2,499	71	2,428	None
2	*B. adolescentis*	*ATCC 15703*	AP009256	2.1	59	1,702	70	1,632	None
3	*B. angulatum*	*DSM 20098*	ABYS00000000	2.0	59	1,811	63	1,748	None
4	*B. animalis* subsp. *lactis*	*AD011*	CP001213	1.9	60	1,603	59	1,527	17
5	*B. animalis* subsp. *lactis*	*DSM 10140*	CP001606	1.9	60	1,629	63	1,566	None
6	*B. animalis* subsp. *lactis*	*Bl-04*	CP001515	1.9	60	1,631	64	1,567	None
7	*B. animalis* subsp. *lactis*	*HN019*	ABOT00000000	1.9	60	1,632	54	1,578	None
8	*B. bifidum*	*NCIMB 41171*	ABQP00000000	2.2	62	1,888	55	1,833	None
9	*B. breve*	*DSM 20213*	ACCG00000000	2.3	58	2,323	60	2,263	None
10	*B. catenulatum*	*DSM 16992*	ABXY00000000	2.1	56	2,011	61	1,950	None
11	*B. dentium*	*ATCC 27679*	AEEQ00000000	2.6	58	2,402	66	2,336	None
12	*B. dentium*	*Bd1*	CP001750	2.6	58	2,197	68	2,129	None
13	*B. dentium*	*ATCC 27678*	ABIX00000000	2.6	58	2,500	70	2,430	None
14	*B. gallicum*	*DSM 20093*	ABXB00000000	2.0	57	1,980	62	1,918	None
15	*B. longum* subsp. *longum*	*JDM301*	CP002010	2.5	59	2,035	64	1,958	13
16	*B. longum*	*DJO10A*	CP000605	2.4	60	2,061	73	1,989	None
17	*B. longum*	*NCC2705*	AE014295	2.3	60	1,798	70	1,727	None
18	*B. longum* subsp. *infantis*	*ATCC 55813*	ACHI00000000	2.4	60	2,171	62	2,109	None
19	*B. longum* subsp. *infantis*	*CCUG 52486*	ABQQ00000000	2.5	59	2,296	56	2,240	None
20	*B. longum* subsp. *infantis*	*ATCC 15697*	CP001095	2.8	59	2,588	94	2,416	80
21	*B. pseudocatenulatum*	*DSM 20438*	ABXX00000000	2.3	56	2,220	69	2,151	None

Source: NCBI (http://www.ncbi.nlm.nih.gov/), October 2010

1.2.1 Carbohydrate Degradation

Utilization of carbohydrates which cannot be digested and utilized by the host is important for the stable colonization of probiotic bacteria. Thus most probiotic bacteria utilize plant carbohydrates which cannot be digested by humans. Resistant starches such as pectin, ruffinose, pullulan, and oligosaccharides such as galacto-oligosaccharides (GOS), fructose-oligosaccharides (FOS) are best sources of carbohydrates for probiotic microorganisms. Additionally, some bacteria especially bifidobacterial strains have the ability to digest host-derived compounds such as musin (Klijn et al. 2005).

The carbohydrate uptake, transport, and utilization capability of a probiotic bacterium is considered as a competitive advantage for its host adaptation. More than 10% of bifidobacterial genome usually encodes for such activities (Schell et al. 2002). Different probiotic bacteria show capabilities of utilization of different sources of carbohydrates. *L. plantarum* adapts to many different environments and thus it encodes a large number of enzymes for utilization of many different carbohydrates (Kleerebezem et al. 2003).

Most carbohydrate-digesting genes are present as gene clusters or as operons. The *L. acidophilus msm* operon is an example for FOS metabolism (Barrangou et al. 2003). It encodes proteins for ATP-dependent binding cassette (ABC) transporters and a cytoplasmic β-fructofuranosidase for hydrolysis of FOS.

1.2.2 Bile Tolerance

There are several mechanisms involved in bile tolerance in probiotic bacteria. Expression of bile salt hydrolases (BSH), genes for cell envelop biosynthesis, bile exporters, acid and oxidative stress responsers have shown to be associated with bile tolerance (Begley et al. 2006). Cyclopropane fatty acid synthase and putative phosphotidyl glycerophosphatase are some examples identified from *L. acidophilus* NCFM and *L. reuteri* ATCC55730, respectively, as genes activated during acid/bile stress (Wall et al. 2007). Furthermore, a putative esterase gene, lr1516 (Whitehead et al. 2008), a putative cell wall modifying enzyme, cdp-A (Altermann et al. 2004) and dlt operon (Bron et al. 2006) have been identified as genes involved in maintaining cell envelop integrity during bile stress.

The bile stress directly affects the structure and function of major macro-molecules such as DNA and protein as acid/bile stress damages such molecules. Therefore, genes associated with protein denaturation and DNA damage are also up-regulated with bile stress while genes involved in substrate transport and metabolism are down-regulated.

In *L. acidophilus*, a 7 kbp operon with eight structural genes (two CRS, a transporter, a putative oxidoreductase, and four unknown proteins) have shown to be up-regulated during bile stress (Pfeiler et al. 2007). Further analysis with mutants showed that those eight proteins are associated with bile sensitivity and tolerance.

1.2.3 Stress Responses

The maintenance of intracellular pH is very important for lactic-acid-producing bacteria and they posses several mechanisms for it. The *L. plantarum* genome encodes an F_1F_0-ATPase for F_1F_0-ATPase proton pumps, ten sodium-proton antipracters and three paralogous alkaline-shock proteins (Kleerebezem et al. 2003).

Pieterse et al. (2005) demonstrated that several cell wall surface proteins are expressed with the presence of lactic acid and thus it has been proposed that these proteins may be responsible for acid tolerance properties of cells in *L. plantarum*. Association of acid tolerance with expression of proteins responsible for cell wall biosynthesis (Wall et al. 2007), amino acid decarboxylation (Ascarte-Peril et al. 2004) have also been demonstrated. Therefore, it could be suggested that there should be several mechanisms for acid tolerance in lactic acid bacteria.

1.2.4 Bacteriocin Biosynthesis

Probiotic bacteria reduce or prevent the growth of pathogenic bacteria in the gut of their host by several ways such as reduction of the pH and making the gut environment unfavorable for bacterial growth, secreting antimicrobial peptides, inhibiting bacterial adherence to gut epithelial cells and preventing bacterial translocations and production of antimicrobial compounds such as bacteriocins.

According to Flynn et al. (2002), the first bacteriocin to be characterized up to molecular level was ABP-118, a class IIb bacteriocin with two polypeptides. Class II bacteriocins are heat-stable, minimally modified bacteriocins produced by gram-positive bacteria. Generally, class II bacteriocins increase the cell membrane permeability and thereby activate leakages of molecules from target organisms. Lactacin B is also a class II bacteriocin and its production and processing in *L. acidophilus* are encoded by 12 putative genes located in a 9.5 kb region of the genome (Altermann et al. 2005). Birri et al. (2010) found a new bacteriocin produced by *Enterococcus avium* and named it as avicin A. Molecular analysis revealed that genes responsible for the production of avicin A are organized into four operon-like structures. They include structural genes for avicin A and B, their immunity genes to protect the bacterium from bacteriocin, bacteriocin transporters, and some regulatory genes.

1.2.5 Other Host Adaptation Mechanisms

Effective probiotic bacteria should show mechanisms for adaptation to their host GIT. Such mechanisms would be the presence of adherence factors, integration to host mucus, and host immunomodulation. Production of extracellular proteins is important for the host adaptation of probiotic bacteria, especially for adherence and communication. There are about 211 proteins identified from *L. plantarum* as putative extracellular proteins of which, transporters, regulators, and enzymes are predominant. *L. johnsonii* also produces about 117 such proteins (Boekhorst et al. 2004). Though the external polysaccharide layer has not been detected around *L. acidophilus* NCFM, a gene cluster with 14 genes for EPS production has been identified. Further studies have implied that this cluster shows high synteny to reported EPS clusters of *L. gasseri* and *L. johnsonii* (Altermann et al. 2005).

1.3 Comparative Genomics of Probiotic Bacteria

Because of the availability of many published genome sequence information on probiotic bacteria, scientists have a great opportunity to analyze genomic information in comparative means. This provides further knowledge about genetic stability and diversity among species and strains of a single species. Most such comparative genomic analysis proved that the essential housekeeping gene functions are highly conserved while showing diversity in genes responsible for adaptation to specific environments. Thus based on the changes in the ecological adaptations, especially to nutrient-rich environments, losses or gains of genes and thereby changes in the size of the genome have been observed.

Boekhorst et al. (2004) compared the genomes of *L. plantarum* and *L. johnsonii* and found a great diversity in terms of structural and functional properties. *L. plantarum* consists of a larger genome (3.3 Mb) and a large number of different proteins compared to those of *L. johnsonni*. This could mainly be due to the fact that *L. plantarum* is a versatile bacterium adapted to many different environments ranging from plants to human GIT, whereas *L. johnsonni* commonly colonizes the human GIT (Kleerebezem et al. 2003).

Boekhorst et al. (2004) demonstrated that *L. plantarum* encodes 268 proteins for metabolism and transport of amino acids but only 125 proteins were found from *L. johnsonii* for the same. Because of the presence of a large number of genes in *L. plantarum*, it has the capability of biosynthesis of all amino acids except leucine, isoleucine, and valine while *L. johnsonii* has no ability to synthesis most amino acids.

Molenaar et al. (2005) explored the *L. plantarum* genome diversity using microarray and compared the genomes of 20 strains of the species. Based on their findings, genes involved in biosynthesis and degradation of structural compounds such as proteins, lipid, and DNA are highly conserved. However, a high variation was observed with the genes involved in sugar catabolism and transport. They identified two highly variable regions of the genome and based on these, they differentiated the tested 20 strains into two clusters. Those two clusters were exactly the same as the two subdivisions that proposed earlier for the species through conventional methods.

2 Genetic Modifications in Probiotic Bacteria

As the importance of probiotic bacteria has become clear in fields such as food industry, medicine and cosmetics, the quest for improvement has begun. The major avenues of genetic modifications in probiotic bacteria include strategies for enhanced survival within the human GI tract, strategies for delivery of antigens and strategies for better survival during food processing and storage.

2.1 Molecular Alterations of Probiotic Bacteria for Enhanced Digestive Tract Survival

The therapeutic potential of probiotic bacteria, whether as natural organisms or modified organisms delivering an important protein or an antigen, depends on how well they are adapted to the conditions in the gastrointestinal tract for survival and persistence. While many probiotic bacteria such as bifidobacteria have evolved to tolerate the stresses of the gastrointestinal transit, significant variation exists among different strains. For instance, strains such as *Lactococcus lactis* show very low persistence. Thus, the methods that could improve the physiological robustness of probiotic cultures inside the GI tract are a clinical imperative (Sleator and Hill 2006).

2.1.1 Environmental Challenges for Survival Within the Gastrointestinal Tract

During the gastric passage, bacteria encounter several challenging conditions such as low pH of the stomach, low oxygen content, elevated osmolarity and bile in the upper small intestine. Out of these, bile in particular, provides a key challenge to bacteria that survive and transit the stomach and enter the small intestine. Bile acids, which are synthesized from cholesterol, play a vital role in the physiochemical defenses of the host through direct degradation of bacterial membranes (Hofmann 1999). Hence, selection of probiotics for human use is hugely influenced by their ability to tolerate bile. These challenges can be met either by understanding and improving systems that aid tolerance of harsh conditions or by introducing them from other sources including pathogens.

2.1.2 Systems Found in LAB That Contribute to Their Survival Within the GI Tract

With the ever-increasing knowledge on genomes of lactic acid bacteria, the mechanisms and the systems that are responsible for their survival inside the gastrointestinal tract have been elucidated (Sánchez et al. 2005). Out of these, bile tolerant and osmotolerant systems have been studied extensively. The importance of the understanding of these mechanisms and systems is that they can be introduced, manipulated, or combined to improve the survival of LAB in the GI tract.

Bile Salt Hydrolase Activity

The ability to hydrolyze bile salts has been given prominence in probiotic strain selection in most instances. To date, the BSH activity has been identified in different probiotic strains such as *Lactobacillus* sp. (Lundeen and Savage 1992) and *Bifidobacteium* sp. (Kim et al. 2004) but not in bacteria such as *Lactococcus lactis*, isolated from bile salt-absent environments (Moser and Savage 2001). BSHs catalyze deconjugation of bile salts by hydrolyzing amide bonds as they liberate the taurine/glucine moiety from the steroid core. This deconjugation has been linked to several possible functions of *bsh* including nutritional advantage, alteration of membrane characteristics by facilitating the incorporation of cholesterol or bile into membranes (Taranto et al. 2003) and bile detoxification (Grill et al. 2000). Some strains possess more than one BSH homolog which are not identical to each other but to a homolog of another strain (Mc Auliffe et al. 2005). This is a clear indication that the *bsh* genes are horizontally transferred and these homologs must have been acquired from different sources. In fact it is strongly believed that the pathogen *Listeria monocytogenes* must have acquired the *bsh* gene from other intestinal microorganisms (Begley et al. 2006). The identification and characterization of different *bsh* genes will no doubt lead to genetic engineering applications such as overexpression of the genes, and combination of heterologous genes, for improved survival in the GI tract.

Osmotolerant Systems in LAB

Several systems that cater to osmotolerance in LAB have been characterized. These include the osmotically inducible high-affinity glycine betain uptake system BusA (*opuA*) of *L. lactis* (Obis et al. 1999) and quaternary ammonium compound transporter QacT of *Lactobacillus plantarum* (Glaasker et al. 1998).

2.1.3 Pathobiotechnology as a Tool for Improved Survival of the Probiotic Bacteria

The ability of some of the pathogens to evolve strategies to combat the harsh conditions within the GI tract (Sleator and Hill 2005) has provided a source to mine for these systems. In fact the term pathobiotechnology which has been coined by Sleator and Hill (2006) describes the exploitation of pathogenic stress-survival strategies in the design of more versatile probiotic cultures (Sleator and Hill 2006). Out of these pathogens *L. monocytogenes*, a gram-positive intracellular foodborne pathogen, has shown a remarkable ability to withstand the extreme conditions of the gastric passage (Gahan and Hill 2005). In fact *L. monocytogenes* has even been isolated from gallbladder indicating its ability to tolerate very high concentration of bile salt (Hardy et al. 2004).

The BilE System to Enhance the Bile Tolerance

Sleator et al. (2005) have identified a novel bile-resistant system designated as the BilE. BilE functions by excluding bile from the cell and it has shown to facilitate improved gastrointestinal transit in infected mouse models. Watsen et al. (2008) have used this BilE system to enhance the bile tolerance and improve the survival of *Bifidobacterium* and *Lactococcus*. A 2.9 kb PCR fragment has been isolated from *L. monocytogenes* EGD-e using specific primer PCR. This fragment corresponds to the *bileE* region with its own promoter. This region cloned in pNZ80048 vector has been subsequently transferred to *Bifidobacterium breve* strain UCC2003 and *Lactococcus lactis* strain NZ9000. BilE harboring pNZ8008 has shown considerable stability in both the strains. When challenged with 1% w/v porcine bile, the engineered *L. lactis* has shown a 2.5-log enhanced resistance to bile over the 20-min course of the kill curve while the engineered *B. brevis* has shown 2.5-log enhanced survival at 15- and 20-min postexposure levels.

The BetL System for Osmotolerance

Some bacterial species have the ability to tolerate environmental stress, which can be attributed, at least to an extent, to the accumulation of protective compounds termed compatible solutes (Sleator and Hill 2001). The preferred compatible solute for the majority of bacteria is the trimethylammonium compound, glycine betaine. *Listeria monocytogenes* contains a secondary glycine betaine transporter (BetL) which is highly characterized at both transcriptional (Sleator et al. 2000) and posttranslational levels (Sleator et al. 2003b). Although similar transport systems have been identified in some lactic acid bacteria (Glaasker et al. 1998), it has been shown that the betain uptake through BetL transporter of *L. monocytogenes* improves osmotolerance significantly (Sleator et al. 2003a). Hence, the use of *L. monocytogenes* BetL transporter to improve osmotolerance and thereby improve the persistence of LAB is a promising approach. In addition to the improved osmotolerance, BetL is also capable of improving barotolerance and providing protection during spray and freeze drying, attributes that help keep LAB viable during production and storage. Sheehan et al. (2006) introduced the *bet*L gene under the transcriptional control of *nis*A into *Lactobacillus salivarius* along with the helper plasmid encoding the nisin RK two-component system (NICE). According to their results, nisin induction of *bet*L in genetically engineered

Lactobacillus salivarius resulted in a significantly higher growth rate and a higher final optical density when grown at 7% NaCl than those of the control, which failed to grow under similar conditions. This significant increase in osmotolerance suggests that BetL is not only expressed but also functional in the *Lactobacillus* background. A similar research with *Bifidobacterium breve* UCC2003 has also demonstrated the ability of BetL to improve the survival in simulated gastric juice at pH 2.5.

3 Genetically Modified Lactic Acid Bacteria as Mucosal Delivery Vectors

Mucosal surfaces which act as the windows to the outside world through which nutrients are taken in and end products of metabolism are expulsed, also attracts many pathogens leading to a number of infectious diseases. Hence mucosal immunity has become a focal point in the field of medicine, opening avenues for mucosal vaccines. In addition to the stimulation of local responses, the mucosal immunization can give rise to a strong mucosal IgA antibody production in distant mucosal effector sites (Hvalbye et al. 1999). Mucosal immunization is also known to stimulate systemic immune responses and induce cytotoxic T lymphocytes. Mucosal vaccines are easier to administer and are also known to have low production costs than systemic vaccines. There are a number of ways of introducing the immunogenic antigens to mucosal surfaces. Out of these, the use of attenuated pathogens such as attenuated *Salmonella* spp. have required years of research to achieve the balance between maximum immunogenecity and minimum side effects in experimental models and humans (Tacket and Levine 2007). So the ideal alternative would be to generate an efficient immunogenic response with minimum or no side effects. Since LAB are considered safe they offer an exiting approach as delivery systems to the mucosal surfaces.

3.1 Advantages of LAB as Mucosal Delivery Vectors

In addition to being considered safe, works on LAB have identified a number of traits that make them ideal vectors for mucosal delivery. Due to their long association with the GI tract, LAB can survive passage through the stomach acid and effects of bile. One of the biggest advantages of LAB is their ability to elicit antigen-specific secretory IgA responses at mucosal surfaces while they themselves induce low levels of immune response following mucosal administration. It has been proven that LAB can also be engineered to express multiple proteins and other molecules (Gilbert et al. 2000) which make them ideally suited to deliver multiple antigens expressed in the same strain.

3.2 Route of Delivery of LAB and Immunogenic Response

LAB vectors could be delivered via a number of routes such as intranasal, gastrointestinal and vaginal. Out of these the most studied routes are the intranasal and gastrointestinal. While some studies have shown a higher immune response with the gastrointestinal route (Ramasamy et al. 2006) others have shown a higher immune response after delivery via the intranasal route (Cortes-Perez et al. 2007). However, Cortes-Perez et al. (2007) showed that intranasal administration of recombinant LAB strains (delivering HPV 16 E7 antigen) produced a significant amount of anti-E7 IgG in the serum of mice, compared to the gastrointestinal administration. The variability of immunogenic response is also affected by a number of other factors such as the LAB strain, choice of the antigen, cellular presentation, and scheme of immunization.

3.3 Development of Expression Systems for LAB

One of the major requirements of a good expression system for mucosal delivery of antigens is a strong promoter, especially when the amount of antigen expressed would affect the immune response. Arguably, it is the strong constitutive promoters that are most welcome for better expression of the antigens when they are delivered to the mucosal surfaces. In addition to the natural constitutive promoters found in LAB such as amylase A (*amy*A), an exciting approach of use of synthetic promoters obtained from consensus sequences of *L. lactis* promoters has also been attempted (Jensen and Hammer 1998). However, some antigens such as the gp50 protein of the Aujeszky disease virus (ADV) seemingly require regulatable promoters for stable expression (Kleerebezem et al. 1997). These inducible promoters also help investigate the effect of quantity of antigen delivered on the magnitude and duration of the immune response.

The antigen expressing genetically modified LAB (GM LAB) could present the antigen in the cytoplasm, secrete it or keep it anchored to the cell wall (Bermúdez-Humarán et al. 2004). The location of expression directly affects the immunogenic response and the survival and stability of the foreign antigen. Pouwels et al. (1998) developed a set of high-efficiency *Lactobacillus* expression vectors based on expression signals and replication elements derived from *Lactobacillus* DNA sequences. The expression regions of these vector constructs consists of Pamy or Pldh as promoters, signal sequences such as *amy*A and *prt*P for membrane transportation of the antigen, a multiple cloning site for insertion of the foreign antigen, *gus*A as the marker gene, the anchor sequence of the proteinase gene of *L. lactis* and transcription terminator of *cbh* gene of *L. planatarum* These constructs have been used successfully to express antigens such as tetanus toxin fragment C (TTFC), urease A, and B subunits of *Helicobacter pylori*. Pouwels et al. (1998) have also considered the bacterial surface layer (S-layer) proteins as means of producing the antigens on the surface of *Lactobacillus* species. The natural function of the S-layer proteins is adherence to the proteins of the mammalian extracellular matrix (Toba et al. 1995).

3.4 GM LAB Strain Models as Live Vaccine

There are three major groups which are heavily studied for recombinant live vaccine work: *Lactococcus lactis*, *Streptococcus gordini*, and *Lactobacilli*. In fact the LABVAC European Network is evaluating these three hosts as models for potential vaccine vehicles. Currently, there are a number of molecules that are being cloned in LAB and studied in murine systems as potential future therapeutics.

3.4.1 The TTFC (Tetanus Toxin Fragment) Model

TTFC is a 47 kDa nontoxic polypeptide with a ganglioside binding domain. The protection offered by the use of this antigen relies on the induction of a systemic immune response. GM *L. lactis*, a noncolonizing LAB expressing TTFC has been heavily studied as a model (Chamberlain et al. 1997). In mice *L. lactis* is involved in only passive transit as its persistence is less than 24 h; hence, TTFC has been either expressed intracellularly or anchored to the surface-associated proteins. The expression systems studied have been either induced or constitutive. While the induced expression systems such as the lactoccocal T7 system (derived from pLET vectors) have shown high intracellular expression levels of TTFC, the constitutive expression systems such as pTREX have produced at levels of 1–3% of total cell protein (Chamberlain et al. 1997). However, the high levels of intracellular expression of TTFC have also been observed to kill the host cells, stressing the importance of the constitutive expression systems which produce the antigen at low levels. In murine models, both the pLET and the pTREX systems have shown to be successful in the immune response even when introduced through the oral route (Robinson et al. 1997). Other LAB such as *Lactobacillus* spp. and *S. gordoni* have also shown promise with the TTFC antigen in terms of immuno-response in murine models (Hanniffy et al. 2004). The strains may differ in their efficacy to evoke immune responses but the use of different dosages and methodologies make it difficult to compare them. In a study that compared alenine recemose cell wall mutants, *L. lactis* and *L. plantarum* as delivery vectors of TTFC, it was concluded that *L. plantarum* is more immunogenic than *L. lactis* (Grangette et al. 2004).

3.4.2 Delivery of the Peptides of the Trefoil Factor Family

Secretory trefoil family factor peptides (TFF1, 2 and 3) are well known for their protective and healing effects after mucosal damage, and are of interest as potential therapeutic agents in inflammatory bowel disease (IBD) (Taupin and

Podolsky 2003). When the Trefoil Factor Families (TFFs) are administered orally, they adhere to mucus of the small intestine and get absorbed at the caecum. Hence, the use of GM LAB to deliver these peptides to the mucosa is an exciting prospect. Intragastric administration of TFF secreting *L. lactis* has shown to effectively prevent and heal acute DSS (dextran sulfate sodium)-induced colitis and spontaneous colitis in the murine model of IBD (Vandenbroucke et al. 2004). The ability of the GM *L. lactis* to make basolateral contact with the colonocytes probably enables the accumulation of the TFFs out of reach of the mucins allowing them to make contact with the TFF receptors (Thim and Mortz 2000). Ptgs2, a known TFF target gene, which contributes to the healing and downregulation of the inflammatory responses in the gastrointestinal tract (Ehrlich et al. 1998), was found to be strongly induced in the intestine of mice treated with GM *L. lactis* secreting TFF. The prophylactic effect of TFF producing *L. lactis* has found to be substantially abrogated by inhibition of Ptgs2 with meloxicam, clearly indicating that Ptgs induction is a critical event in the success of TFF application.

3.4.3 Prevention of HIV-1 Transmission

Several approaches have been carried out to handle HIV through GM LAB. An in vitro study on *Lb. jensenii*, expressing two-domain CD4 proteins, has shown to be successful in inhibiting HIV-1 entry into HeLa cells in vitro. The first step in viral entry, binding of gp120 to the extracellular domains of CD4 of the immune cells, can be blocked by the high-affinity interaction of the microbiocidal cyanovirin-N (CV-N) protein with the mannose structures present on gp120 (Boyd et al. 1997). *S. gordonii* was the first human commensal LAB strain that was used as a host for CV-N (Giomarelli et al. 2002). In this study CV-N, expressed as a fusion protein using the M-protein secretion and cell wall anchoring domains, was able to capture HIV-1 in a concentration-dependent manner.

3.4.4 *S. hyricus* Lipase Expression in the Cytoplasm of *L. lactis*

Drouault et al. (2002) expressed *Staphilococcus hyicus* lipase gene in the cytoplasm of *L. lactis* and demonstrated that the use of this strain in pigs with ligated pancreatic duct (to create pancreatic inefficiency) increased fat absorption. Dead cells of *L. lactis*, preloaded with lipase, has demonstrated a higher fat absorption compared to the expression strains. This approach has the potential to improve the pancreatic inefficiency in humans.

3.4.5 Pneumococcal Surface Protein A

Intranasal application of GM *L. lactis* expressing the pneumococcal surface protein A (PspA) has shown better protection to respiratory challenge of pneumococci than intranasally applied PspA or PspA applied with alum (Hanniffy et al. 2007). It is believed that this superior performance is due to a shift toward TH1 response. Oliveira et al. (2006) reported that *L. lactis* expressing PspA in mice did not produce detectable levels of anti-PspA-IgA in bronchoalveolar lavage (BAL) washes, despite measurable levels of anti-PspA antibodies in serum samples. However, Cortes-Perez et al. (2007) who have observed a similar phenomenon working with HPV-16 E 7 attribute this to the early production of mucosal IgA by B cells residing in organized nasal-associated lymphoid tissues. So if the anti-PspA-IgA quantification is attempted after a considerable time lapse of the immunization, the IgA found in BAL washes are deemed to be low.

3.4.6 Cytokine Secreting LAB

The immune system depends on relatively small simple messenger proteins such as cytokines, IFNs, growth factors, and chemokines for long systemic immunoresponse. Although cytokines are considered as simple structures that do not depend on glycosylation or other secondary modifications, many of them depend on correct disulphide bridge formation (Zhang et al. 1997). However, the biological activity of the recombinant cytokines has been shown both in vitro and in vivo, clearly indicating the ability of *Lactococci* to carry out correct disulphide bridge formation of cytokones (Steidler et al. 2000). *L. lactis* have been designed to secrete cytokines with a view to redirect the immune system (Steidler et al. 1998; Schotte et al. 2000). This ability of lactococci and LAB to express cytokines has lead to their use in therapy such as coadministration with other antigens to improve the immune response and as a method of suppressing IBD.

3.4.7 Coexpression of Antigens and Cytokines

A strong adjuvant effect has been observed with live recombinant cultures of lactococci secreting IL-2 or IL-6 together with the TTFC antigen (Steidler et al. 1998). According to the results of this study, anti-TTFC serum responses were 10–15 folds higher in mice that were also immunized with IL-2 or IL-6. In addition, bacteria expressing TTFC in combination with IL-6 were also capable of eliciting serum anti-TFFC IgA responses. Since this effect was completely nullified with the killed bacteria, it is believed that the live bacteria secreting the cytokines are essential to the mechanism. Cortes-Perez et al. (2007) showed that significant regression of the established HPV-16-induced tumors (challenged with TC-1 tumor cell line) of mice takes place when the E7 antigen is coadministrated with IL-12 via lactococci whereas the treatment with E7 expressing lactococci was unable to give such protection.

3.4.8 IL-10 for Inflammatory Bowel Disease

IBD is most likely be due to breakdown of immune tolerance toward intestinal microflora. The cytokine IL-10 is a powerful anti-inflammatory agent and a central factor in induction and maintenance of immune tolerance (Thompson and Powrie 2004). However, administration of IL-10 by injection is known to induce side effects (Tilg et al. 2002) especially, the induction of pro-inflammatory IFN-γ. Hence, long-term high dose application of IL-10 becomes pointless. Even the oral application is impossible due to extremely acid-sensitive nature of IL-10 (Schotte et al. 2000). Sleator et al. (2000) have produced a genetically engineered *L. lactis* that secretes IL-10 and they have observed that daily oral dosage of this strain cures chronic murine DSS colitis. The healing capability was observed to be on par with systemically given anti-inflammatory drugs such as dexamethasone and anti-IL-12. The mouse experimental colitis model mimics the human IBD; hence, the use of IL-10 secreting *L. lactis* has great potential as an inexpensive, easily administered therapeutic.

3.5 Use of LAB to Deliver Single-Chain Variable Fragment Antibody Production

Single-chain variable fragment antibody (scFv) are antibodies which are synthesized without the Fc region. This arrangement is helpful when the antibodies are expected to penetrate into the targets such as cancer cells. The normal antibodies are too bulky for this purpose. An scFv with structural surface similarity to a broad-spectrum killer toxin of *Pichia anomala* has been produced (Polonelli and Morace 1987). *S. gordonii* strains engineered to express the surface-attached scFv or secretory scFv were capable of clearing *Candia albicans* from experimental mice (Fig. 1a). The secreted scFv showed a therapeutic efficiency similar to the conventional antifungal antibiotic fluconazole. However, scFv can also be used to neutralize antigens by binding to them. Hence, surface-expressed scFv producing *L. lactis* or LAB can be used to prevent the action of toxin by neutralizing them. scFv could also be used to clear bacteria from a surface, interfering with their binding or not allowing access. For example, an scFv that binds to a major adhesion molecule of *Streptococcus mutants* (a causative agent of dental caries) has the ability to protect the enamel from being colonized by agglutination and clearance of the pathogen (Fig. 1b).

3.6 LAB as DNA Vaccine Delivery Agents

DNA vaccines help produce the antigens in the host cells. Hence, this gives the advantage of potent immuno-response in addition to the production of antibodies.

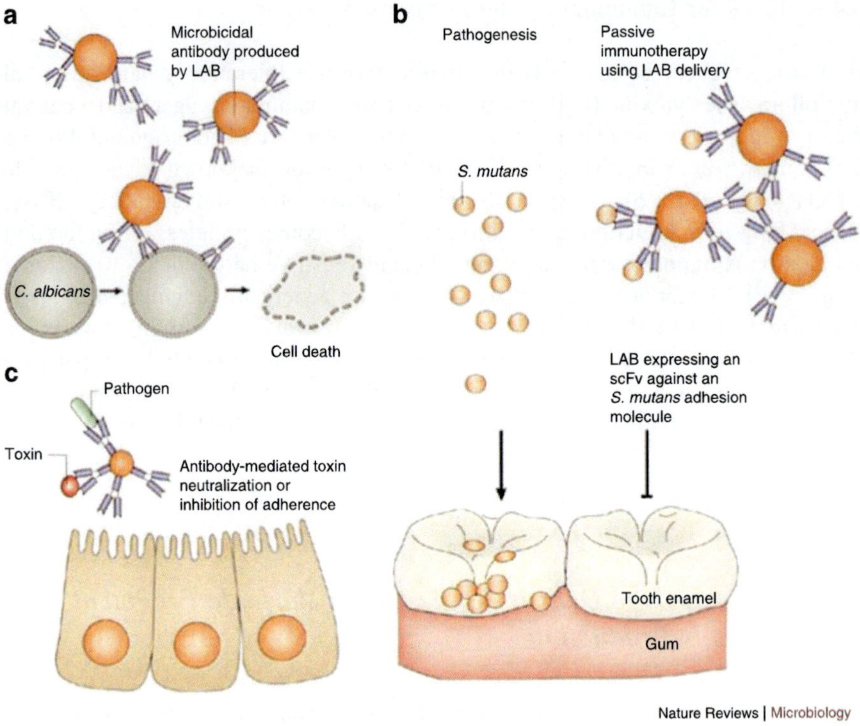

Fig. 1 Anti-infective strategies based on mucosal delivery of single-chain variable fragment antibodies. (**a**) Lactic acid bacteria (LAB) that secrete or produce cell-wall-bound microbicidal antibodies against the opportunist pathogen *Candida albicans* have been used to treat experimental infections. (**b**) Single-chain variable fragment (scFv) antibodies that comprise variable regions of the heavy and light chains of an immunoglobulin, which are linked together via a linker peptide, have been expressed on the surface of LAB and used in passive immunotherapy. In the example shown, an scFv that binds to a major adhesion molecule of *Streptococcus mutans* (a causative agent of dental caries) protects against colonization of tooth enamel and dental caries by agglutination and clearance of the bacterium from the mouth. (**c**) ScFv antibodies that are expressed on the surface of LAB could also be considered for immunotherapy or prophylaxis. For example, an scFv might prevent pathogen adherence or neutralize a luminal bacteria toxin, thereby preventing it from reaching the epithelial cells. Reprinted by permission from Macmillan Publishers Ltd: [Nature reviews/Microbiology] (Wells and Mercenier 2008), copyright (2008)

DNA vectors used in DNA vaccines also provide the opportunity to express multiple antigens or epitopes leading to protection against multiple pathogens. Because the expression is carried out by the host cell, this approach also provides the opportunity to carry out essential posttranslational modifications such as glycosilation (required by some antigens). The delivery of DNA vaccines to host cells is facilitated by several means such as viral vectors, addition of chemicals such as PEG and use of liposomes. Although the mechanism has not been clearly stated, LAB have also been bestowed with the ability to transfer DNA into host cells (Guimarães et al. 2005). However, significant immune responses have been detected only from injected *L. lactis* carrying a foot and mouth antigen producing DNA vector (Li et al. 2007). Additionally, mucosal application has also demonstrated the ability to prime a specific immune response albeit with very low delivery, providing enough evidence to warrant investigation into LAB as DNA vaccine delivery systems in the future.

4 Mutant Probiotics, Issues, and Ethics

No matter how brilliant the science adapted in a process, its success and popularity would always depend on safety issues and human perceptions. Hence, the mutant probiotics for food and health applications will be realistic only if the safety is maintained and the issues are properly addressed. The most concerning issues of mutant probiotics are human and animal safety and the adverse effects on the environment.

4.1 Type of Mutation Events in Probiotics

Designing of more effective probiotics can be categorized into three groups. These include the improvement of probiotic tolerance to stresses encountered during food manufacture and storage, improvement of resistance to host-specific stresses for better persistence and colonization, and the improvement of prophylactic and therapeutic effect (Sleator and Hill 2008). Probiotic mutants can be the result of uncontrolled genetic alterations or controlled genetic alteration events. Uncontrolled genetic alterations include spontaneous mutations by natural events such as insertion sequence elements (Visser et al. 2004), radiation, erroneous DNA replication, transcription, and other factors. The frequency of uncontrolled genetics can be further increased by exposing probiotic bacteria to mutagenic conditions such as UV light or chemicals such as ethyl methyl sulphonate, or *N*-methyl-*N*′-nitro-*N*-nitrosoguanidine (NNG) (Sybesma et al. 2006). The controlled genetic alterations involve single base pair substitutions, mutations, addition or deletion of genes/DNA from chromosomes and plasmids.

4.2 Issues of Mutant Probiotics and Possible Solutions

As mentioned earlier the issues of mutant probiotics can be divided into two basic categories: human and animal health risks issues and adverse effects on the environment. The risk associated with mutant probiotics is due to the generation of strains with novel genetic properties and modified cellular behavior. This theoretically could lead to predictable unintended effects as consequences of the novel changes and unexpected effects due to unintended genetic alterations or due to consequences of predictable effects (Sybesma et al. 2006).

4.2.1 Unintended Predictable Effects of Mutants

There are several unintended yet predictable effects that could rise due to directed gene alterations. These include allergenecity and toxicity, chimeric or fusion proteins, and effect on the expression of adjacent genes.

Allergenecity and Toxicity

Out of all the possible unintended yet predictable effects, allergenecity and toxicity are the most possible and the most debated. The production of a novel protein in a host could lead to allergenecity and toxicity. This might be aggravated if there are differences in posttranslational modifications between the original host and the recipient. Production of chimeric proteins could also contribute to this due to changes in their structures. Currently, there is no tool available to predict allergenecity of a novel protein. However, the newly expressed protein could be assessed to find out whether some individuals of the population are already sensitive to it. Although it might be difficult to accurately predict allergic responses or toxicity of such novel products, lessons can be learned by comparing these to already recognized allergens through calculating the percentage identity. In fact Codex Alimentarius Commission, a regulatory system of genetically modified food (GMF), proposes that the IgE cross-reactions between the novel protein and the known allergen should be considered, if there is more than 30% identity in a segment of 80 or more amino acids. Potential toxicity of a substance, which has not been safely consumed before, needs to be assessed on a case-by-case basis of identity, concentration, biological function, and dietary exposure (Sybesma et al. 2006).

Fusion or Chimeric Proteins

The fusion proteins are generated when the foreign DNA inserts are integrated in-frame with a native coding region. In addition to being allergenic or toxic, the fusion proteins can also lead to different types of cellular behavior. Hence, it would be beneficial to clearly confirm that such combinations have not occurred in the product because this additional information may lead to better consumer acceptance.

Effect on Adjacent Gene Expression

When the genetic modification is carried out to a gene located in an operon, the expression of neighboring genes might be affected. The outcome could be a change that may lead to changed metabolism, production of inefficient strain or toxicity.

However, with the advent of novel molecular techniques that concentrate on genomics, transcriptomics, proteomics and metabolomics, it is possible to assess the directed mutations and predict for any alarming changes. The ability to test at various strata such as DNA, mRNA, proteins, and metabolites helps to find out any changes that might have an effect on the final product. High throughput techniques such as genome sequencing and expression analysis (chip arrays) have made it possible to detect such changes across the whole genome, making the screening task less time-consuming.

4.2.2 The Unexpected Effects of Mutants

Although the unexpected effects are more related to uncontrolled genetic alterations, obtained naturally or induced via radiation and chemicals, the directed mutagenesis also could contribute due to unintended consequences of the process and material adopted or the interference of other pathways. However, the unexpected effects of directed gene alteration are more traceable and hence solvable than uncontrolled mutagenesis with low or no information on the mechanism of change. The possible unexpected effects are the undesired metabolic outcomes, horizontal gene transfer, and change of microbial population in vivo and outside environment.

Undesired Metabolic Outcomes

Overexpression of desired metabolites or blocking pathways that produce specific metabolites may lead to undesirable cellular responses. When some metabolites are overexpressed they might cause feedback inhibition of its own pathway or other metabolic pathways. Over expression can also interfere with other metabolite production if both share a common starting or intermediate substrate. For instance, biosynthesis of vitamins riboflavin and folate requires GTP as the starting substrate. So any attempt to increase either of the vitamins may reduce the available GTP for the other.

Horizontal Gene Transfer

Although transfer of genes across species is a natural phenomenon that happens regularly, horizontal gene transfer is the most concerned safety issue with regard to many genetic alterations. The fear is, of course justifiable when it comes to microbes, because there are several methods by which microbes uptake DNA. These DNA transfer methods include transformation, conjugation, transduction, and transposon-mediated transfer. Although conjugation and transduction take place among closely related bacterial species, transformation, which is the uptake of naked DNA, provides the theoretical possibility of transfer of any DNA sequence, be it chromosomal or plasmid to most bacterial species which are naturally competent for DNA uptake. The horizontal gene transfer may facilitate transfer of antibiotic-resistant markers and transfer of traits that lead to improved resilience against harsh conditions.

Transfer of Antibiotic-Resistant Markers

Antibiotic resistance can be an inherent property of the strains used or a consequence of the gene transfer process. Some of the inherent antibiotic-resistant properties of bacteria are not transmissible or have a very low probability of transfer. These include spontaneous point mutations in rRNA, RNA polymerase genes, and some metabolite coding genes. Strains with such antibiotic resistance should not be excluded from GM programs as they do not contribute to transmission of these genes. However, antibiotic-resistant markers which are incorporated into vector plasmids for screening purposes have a risk of transferring into human gut microflora and subsequently to microbes in the environment. Therefore, such mutants should be avoided as a precaution. Alternatively, the marker should either be removed during latter stages of the production or replaced by food grade markers. Some of the food grade markers that could be considered for screening are a marker that forces of D-alanine-dependent growth (Bron et al. 2004), lacG- and lacF-based lactose dependence (Takala et al. 2003).

Transfer of Genes That Offer Resilience to Harsh Conditions

The idea of improving probiotic bacteria in terms of persistence and tolerance of harsh conditions in the GI tract also has a disadvantage. There is a possibility that these mutations could be horizontally transferred to the pathogenic bacteria. Although most such systems already exist in the environment, these are governed by natural selection. Hence, any distantly related gene transfer that may not be possible through nature should be closely scrutinized for its risk. Clearly, such a gene transfer should only be allowed when the character does not give selective advantage to a pathogen.

Change of Microbial Population In Vivo and Outside Environment

This issue is in fact a combinatory outcome of release and persistence of mutant probiotics in the environment and the horizontal gene transfer. In mixed culture environments such as the GI tract and outer environment, there exists a fine balance among bacteria and other organisms. Once a selective advantage is gained through modification or acquisition, it could disrupt this intricate balance and lead to adverse effects. This can be curtailed by containment of such mutants. Mutant probiotics can be made to depend on growth factors, temperature regimes, or pH for their survival, thus limiting their survival in the outside environment. As for the effects in vivo, all mutant probiotics should be properly assessed for their impact on the normal gut flora.

Regulatory Issues

Although regulation of GM probiotics is a must to safeguard the humans, animals and the environment, it is imperative that these should be sensible enough to accommodate differences, rather than rigid categorization. Currently, there are no regulations specific for GM probiotics and these are being governed by the regulations which are applied to GMF and GMO in general. Strangely enough, the awarding of GM status, a major taboo in consumer preference, is very inconsistent. Some of these inconsistencies arise from the fact that the legislation focuses the process rather than the product. Hence, uncontrolled mutants arising from natural events and induced events are accepted without profound safety analysis while directed mutants with deletions are considered as GMO. However, the possibility is that the deleted products are much safer than uncontrolled mutants, due to their predictability. One other example is that self-cloned mutants are not considered as GMOs if they are confined, but the same mutant will be considered a GMO if included in food (Sybesma et al. 2006). On the other hand, it is also important to assess the natural counterpart for its safety rather than taking it for granted that they are safe, especially when there is a possibility of uncontrolled spontaneous mutations.

4.3 Ethics Concerning Mutant Probiotics

The major ethical concern with respect to probiotics will be labeling and the opportunity for consumer preference. Since the probiotics are introduced as food supplements, any food containing a mutant probiotic culture invariably becomes a GMF. Hence, it becomes a requirement to indicate the presence of the mutant probiotic in the label of the food along with possible side effects if any. On the other hand, the use of labeling to promote false claims should not be entertained. However, there should be provisions to allow labeling and promoting for confirmed health benefits without any curtailing. Still for all, these modified products should not be forced upon consumers where there is no option but to consume them. The competitive natural product should be available and easily accessible for the consumer to encourage natural consumer preference. Furthermore, aid programs should not make it an opportunity to popularize the mutant probiotics as such instances make people accept products at all costs. The liability of the manufacturer, in case of proven damage directly related to the mutant product, is also desirable. While it is not reasonable to request for unlimited liability, reasonable level of liability acceptance would probably lead to consumer support for both the product and industry.

5 Conclusion

Currently, probiotics fall to a gray area between food and health industries (Sybesma et al. 2006). However, with increasing awareness of the relationship between probiotic bacteria and the host, the future of probiotics may be dominated by pharmaceutical applications such as vaccines. Such applications may well be oral administration of probiotics rather than food supplements. Hence, the emphasis on the status of probiotics would change from being a GMF to GM vaccines. This would be a welcome change as GM medicine and vaccines are already in use and the consumer is not overtly concerned about therapy when compared to food. So the future of GM probiotics may involve the search for more efficient and cheap oral or mucosal vaccines and modifications of probiotics that allow them to survive harsh conditions of the human and animal GI tract.

References

Altermann E, Buck BL, Cano R, Klaenhammer TR (2004) Identification and phenotypic characterization of the cell division protein CdpA. Gene 342:189–197

Altermann E, Russell WM, Azcarate-Peril MA, Barrangou R, Buck BL, McAuliffe O, Souther N, Dobson A, Duong T, Callanan M, Lick S, Hamrick A, Cano R, Klaenhammer TR (2005) Complete genome sequence of the probiotic lactic acid bacterium *Lactobacillus acidophilus* NCFM. Proc Natl Acad Sci USA 102:3906–3912

Ascarte-Peril MA, Altermann E, Hoover-Fitzula RL, Cano RJ, Klaenhammer TR (2004) Identification and Inactivation of Genetic Loci Involved withLactobacillus acidophilus Acid Tolerance Appl Environ Microbiol 70(9):5315–5322

Barrangou R, Altermann E, Hutkins R, Cano R, Klaenhammer TR (2003) Functional and comparative genomic analyses of an operon involved in fructooligosaccharide utilization by *Lactobacillus acidophilus*. Proc Natl Acad Sci USA 100:8957–8962

Begley M, Hill C, Gahan CGM (2006) Bile salt hydrolase activity in probiotics. Appl Environ Microbiol 72(3):1729–1738

Bermúdez-Humarán LG, Cortes-Perez NG, Le Loir Y, Alcocer-Gonzalez JM, Tamez-Guerra RS, de Oca-Luna RM, Langella P (2004) An inducible surface presentation system improves cellular immunity against human papillomavirus type 16 E7 antigen in mice after nasal administration with recombinant lactococci. J Med Microbiol 53:427–33

Birri DJ, Brede DA, Forberg T, Holo H, Nes IF (2010) Molecular and genetic characterization of a novel Bacteriocin Locusin *Enterococcus avium* isolates from infants. Appl Environ Microbiol 76:483–492

Boekhorst J, Siezen RJ, Zwahlen M-C, Vilanova D, Pridmore RD, Mercenier A, Kleerebezem M, de Vos WM, Brüssow H, Desiere F (2004) The complete genomes of Lactobacillus plantarum and Lactobacillus johnsonii reveal extensive differences in chromosome organization and gene content. Microbiology 150:3601–3611

Boyd MR, Gustafson KR, Mcmahon JB, Shoemaker RH, O'keefe BR, Mori K, Gulakowski RJ, Wu L, Rivera MI, Laurencot CM, Currens MJ, Cardellina JH, Buckheit RW Jr, Nara PL, Panell LK, Sowder RC, Henderson LE (1997) Discovery of cyanovirin-N, a novel human immunodeficiency virus-inactivating protein that binds viral surface envelope glycoprotein gp120: potential applications to microbicide development. Antimicrob Agents Chemother 41:1521–1530

Bron PA, Hoffer SM, Van Swam II, De Vos WM, Kleerebezem M (2004) Selection and characterization of conditionally active promoters in *Lactobacillus plantarum*, using alanine racemase as a promoter probe. Appl Environ Microbiol 70(1):310–317

Bron PA, Molenaar D, Vos WM, Kleerebezem M (2006) DNA micro-array-based identification of bile-responsive genes in *Lactobacillus plantarum*. J Appl Microbiol 100:728–738

Chamberlain LM, Wells JM, Robinson K, Schofield KM, Le Page RWF (1997) Mucosal immunisation with recombinant Lactococcus lactis. In: Pozzi G, Wells JM (eds) Gram-positive bacteria as vaccine vehicles for mucosal immunisation. Biotechnology Intelligence Unit. Landes, Austin, USA, pp 83–106

Cortes-Perez NG, Lefevre F, Corthier G, Adel-Patient K, Langella P, Bermudez-Humaran LG (2007) Influence of the route of immunization and the nature of the bacterial vector on immunogenecity of mucosal vaccines based on lactic acid bacteria. Vaccine 25:6581–6588

Denou E, Pridmore RD, Berger B, Panoff JM, Arigoni F, Brussow H (2008) Identification of genes associated with the long gut persistence phenotype of the probiotic *Lactobacillus johnsonii* strain NCC533 using a combination of genomics and transcriptome analysis. J Bacteriol 190:3161–3168

Drouault S, Juste C, Marteau P, Renault P, Corthier G (2002) Oral treatment with *Lactococcus lactis* expressing *Staphylococcus hyicus* lipase enhances lipid digestion in pigs with induced pancreatic insufficiency. Appl Environ Microbiol 68:3166–3168

Ehrlich K, Plate S, Stroff T, Gretzer B, Respondek M, Peskar BM (1998) Peptidergic and cholinergic neurons and mediators in peptone-induced gastroprotection: role of cyclooxygenase-2. AJP Gastrointest Liver Physiol 274:G955–G964

Fleischmann RD, Adams MD, White O, Clayton RA, Kirkness EF, Kerlavage AR, Bult CJ, Tomb JF, Dougherty BA, Merrick JM (1995) Whole-genome random sequencing and assembly of *Haemophilus influenzae* Rd. Science 269:496–512

Flynn S, van Sindere D, Thornton GM, Holo H, Nes IF, Collins K (2002) Characterization of the genetic locus responsible for the production of ABP-118, a novel bacteriocin produced by the probiotic bacterium *Lactobacillus salivarius* subsp. *salivarius* UCC118. Microbiology 148:973–984

Gahan CG, Hill C (2005) Gastrointestinal phase of *Listeria monocytogenes* infection. J Appl Microbiol 98:1345–1353

Gilbert C, Robinson K, Le Page RW, Wells JM (2000) Heterologous expression of an immunogenic pneumococcal type 3 capsular polysaccharide in *Lactococcus lactis*. Infect Immun 68:3251–3260

Giomarelli B, Provvedi R, Meacci F, Maggi T, Medaglini D, Pozzi G, Mori T, McMahon JB, Gardella R, Boyd MR (2002) The microbicide cyanovirin-N expressed on the surface of commensal bacterium *Streptococcus gordonii* captures HIV-1. Aids 16:1351–1356

Glaasker E, Heuberger EHML, Konings WN, Poolman B (1998) Mechanism of osmotic activation of the quaternary ammonium compound transporter (QacT) of *Lactobacillus plantarum*. J Bacteriol 180:5540–5546

Grangette C, Grangette C, Müller-Alouf H, Hols P, Goudercourt D, Delcour J, Turneer M, Mercenier A (2004) Enhanced mucosal delivery of antigen with cell wall mutants of lactic acid bacteria. Infect Immun 72:2731–2737

Grill JPC, Cayuela J, Antoine M, Schneider F (2000) Isolation and characterization of a *Lactobacillus amylovorus* mutant depleted in conjugated bile salt hydrolase activity: relation between activity and bile salt resistance. J Appl Microbiol 89:553–563

Guimarães VD, Gabriel JE, Lefèvre F, Cabanes D, Gruss A, Cossart P, Azevedo V, Langella P (2005) Internalin-expressing *Lactococcus lactis* is able to invade small intestine of guinea pigs and deliver DNA into mammalian epithelial cells. Microbes Infect 7:836–844

Hanniffy S, Wiedermann U, Repa A, Mercenier A, Daniel C, Fioramonti J, Tlaskolova H, Kozakova H, Israelsen H, Madsen S, Vrang A, Hols P, Delcour J, Bron P, Kleerebezem M, Wells J (2004) Potential and opportunities for use of recombinant lactic acid bacteria in human health. Adv Appl Microbiol 56:1–64

Hanniffy SB, Carter AT, Hitchin E, Wells JM (2007) Mucosal delivery of a pneumococcal vaccine using *Lactococcus lactis* affords protection against respiratory infection. J Infect Dis 195:185–193

Hardy J, Francis KP, De Boer M, Chu P, Gibbs K, Contag CH (2004) Extracellular replication of *Listeria monocytogenes* in the murine gall bladder. Science 303:851–853

Hofmann AF (1999) Bile acids: the good, the bad, and the ugly. News Physiol Sci 14:24–29

Hvalbye BK, Aaberge IS, Lovik M, Haneberg B (1999) Intranasal immunization with heat-inactivated *Streptococcus pneumoniae* protects mice against systemic pneumococcal infection. Infect Immun 67:4320–4325

Jensen PR, Hammer K (1998) The sequence of spacers between the consensus sequences modulates the strength of prokaryotic promoters. Appl Environ Microbiol 63:82–87

Kim GBS, Yi H, Lee BH (2004) Purification and characterization of three different types of bile salt hydrolase from *Bifidobacterium* strains. J Dairy Sci 87:258–266

Kleerebezem M, Boekhorst J, van Kranenburg R, Molenaar D, Kuipers OP, Leer R, Tarchini R, Peters SA, Sandbrink HM, Fiers MW, Stiekema W, Lankhorst RM, Bron PA, Kleerebezem M, Beerthuyzen MM, Vaughan EE, de Vos WM, Kuipers OP (1997) Controlled gene expression systems for lactic acid bacteria: transferable nisin-inducible expression cassettes for Lactococcus, Leuconostoc, and Lactobacillus spp. Appl Environ Microbiol 63:4581–4584

Kleerebezem M, Boekhorst J, van Kranenburg R, Molenaar D, Kuipers OP, Leer R, Tarchini R, Peters SA, Sandbrink HM, Fiers MWEJ, Stiekema W, Lankhorst RMK, Bron PA, Hoffer SM, Groot MNN, Kerkhoven R, de Vries M, Ursing B, de Vos WM, Siezen RJ (2003) Complete genome sequence of *Lactobacillus plantarum*WCFS1. Proc Natl Acad Sci USA 100:1990–1995

Klijn A, Mercenier A, Arigoni F (2005) Lessons from the genomes of bifidobacteria. FEMS Microbiol Rev 29:491–509

Li YG, Tian FL, Gao FS, Tang XS, Xia C (2007) Immune responses generated by *Lactobacillus* as a carrier in DNA immunization against foot and mouth disease virus. Vaccine 25:902–911

Lundeen SG, Savage DC (1992) Multiple forms of bile salt hydrolase from *Lactobacillus* sp. strain 100-100. J Bacteriol 174:7217–7220

Mc Auliffe O, Cano RJ, Klaenhammer TR (2005) Genetic analysis of two bile salt hydrolase activities in *Lactobacillus acidophilus* NCFM. Appl Environ Microbiol 71:4925–4929

Molenaar D, Boingel F, Schuren FH, de Vos WM, Siezen RJ, Kleerebezem M (2005) Exploring *Lactobaccilus plantarum* genomic diversity by using microarrays. J Bacteriol 187:6119–6127

Morita H, Toh H, Fukuda S, Horikawa H, Oshima K, Suzuki T, Murakami M, Hasamatau S, Kato Y, Takizawa T, Fukuoka H, Yashimura T, Itoh K, O'Sullivan DJ, McKay LL, Ohno H, Kibuchi J, Masaoka T, Hattori M (2008) Comparative genome analysis of *Lactobacillus reuteri* and *Lactobacillus fermentum* reveal a genomic island for reuterin and cobalamin production. DNA Res 15:151–161

Moser SA, Savage DC (2001) Bile salt hydrolase activity and resistance to toxicity of conjugated bile salts are unrelated properties in lactobacilli. Appl Environ Microbiol 67:3476–3480

NCBI n.d. http://www.ncbi.nlm.nih.gov, accessed on October 10, 2010

Obis D, Guillot A, Gripon JC, Renault P, Bolotin A, Mistou MY (1999) Genetic and biochemical characterization of a high-affinity betaine uptake system (BusA) in *Lactococcus lactis* reveals a new functional organization within bacterial ABC transporters. J Bacteriol 181:6238–6246

Oliveira ML, Areas AP, Campos IB, Monedero V, Perez-Martinez G, Miyaji EN, Leite LC, Aires KA, Lee Ho P (2006) Induction of systemic and mucosal immune response and decrease in *Streptococcus pneumoniae* colonization by nasal inoculation of mice with recombinant lactic acid bacteria expressing pneumococcal surface antigen A. Microbes Infect 8:1016–1024

Pfeiler EA, Azcarate-Peril MA, Klaenhammer TR (2007) Characterization of a novel bile-inducible operon encoding a two component regulatory system in *Lactobacillus acidophilus*. J bacteriol 189:4624–4634

Pieterse B, Leer RJ, Schuren FH, van der Werf MJ (2005) Unravelling the multiple effects of lactic acid stress on *Lactobacillus plantarum* by transcription profiling. Microbiology 151: 3881–3894

Polonelli L, Morace G (1987) Production and characterization of yeast killer toxin monoclonal antibodies. J Clin Microbiol 25(2):460–462

Pouwels PH, Leer RJ, Shaw M, den Bak-Glashouwer MJH, Tielen SFDE, Martinez B, Jore J, Conway PL (1998) Lactic acid bacteria as antigen delivery vehicles for oral immunization purposes. Int J Food Microbiol 41:155–167

Pridmore RD, Berger B, Desiere F, Vilanova D, Barretto C, Pittet AC, Zwahlen MC, Rouvet M, Altermann E, Barrangou R, Mollet B, Mercenier A, Klaenhammer T, Arigoni F, Schell MA (2004) The genome sequence of the probiotic intestinal bacterium *Lactobacillus johnsonii* NCC 533. Proc Natl Acad Sci USA 101:2512–2517

Ramasamy R, Yasawardena S, Zomer A, Venema G, Kok J, Leenhouts K (2006) Immunogenicity of a malaria parasite antigen displayed by *Lactococcus lactis* in oral immunisations. Vaccine 24:3900–3908

Robinson K, Chamberlain LM, Schofield KM, Wells JM, Le Page RWF (1997) Oral vaccination of mice against tetanus with recombinant Lactococcus lactis. Nat Biotechnol 15:653–657

Sánchez B, Champomier-Vergès MC, Anglade P, Baraige F, de los Reyes-Gavilán CG, Margolles A, Zagores M (2005) Proteomic analysis of global changes in protein expression during bile salt exposure of Bifidobacterium longum NCIMB 8809. J Bacteriol 187:5799–5808

Schell MA, Karmirantzou M, Snel B, Vilanova D, Berger B, Pessi G, Zwahlen MC, Desiere F, Bork P, Delley M, Pridmore RD, Arigoni F (2002) The genome sequence of *Bifidobacterium longum* reflects its adaptation to the human gastrointestinal tract. Proc Natl Acad Sci USA 99:14422–14427

Schotte L, Steidler L, Vandekerckhove J, Remaut E (2000) Secretion of biologically active murine interleukin-10 by *Lactococcus lactis*. Enzyme Microb Technol 27(10):761–765

Sheehan VM, Sleater RD, Fitzgerald GF, Hill C (2006) Heterologous expression of BetL, a betaine uptake system, enhances the stress tolerance of *Lactobacillus salivarius* UCC118. Appl Environ Microbiol 72(3):2170–2177

Sleator RD, Hill C (2001) Bacterial osmoadaptation: the role of osmolytes in bacterial stress and virulence. FEMS Microbiol Rev 26:49–71

Sleator RD, Hill C (2005) A novel role for the LisRK two component regulatory system in listerial osmotolerance. Clin Microbiol Infect 11:599–601

Sleator RD, Hill C (2006) Patho-biotechnology; using bad bugs to do good things. Curr Opin Biotechnol 17:211–226

Sleator RD, Hill C (2008) New frontiers in probiotic research. Lett Appl Microbiol 46:143–147

Sleator RD, Gahan CGM, O'Driscoll B, Hill C (2000) Analysis of the role of *betL* in contributing to the growth and survival of *Listeria monocytogenes* LO28. Int J Food Microbiol 60:261–268

Sleator RD, Gahan CGM, Hill C (2003a) A postgenomic appraisal of osmotolerance in *Listeria monocytogenes*. Appl Environ Microbiol 69:1–9

Sleator RD, Wood JM, Hill C (2003b) Transcriptional regulation and posttranslational activity of the betaine transporter BetL in *Listeria monocytogenes* are controlled by environmental salinity. J Bacteriol 185:7140–7144

Sleator RD, Wemekamp-Kamphuis HH, Gahan CGM, Hill C, Abee T (2005) A PrfA-regulated bile exclusion system is a novel virulence factor in *Listeria monocytogenes*. Mol Microbiol 55:1183–1195

Steidler L, Robinson K, Chamberlain L, Schofield KM, Remaut E, Le Page RW, Wells JM (1998) Mucosal delivery of murine interleukin-2 (IL-2) and IL-6 by recombinant strains of *Lactococcus lactis* coexpressing antigen and cytokine. Infect Immun 66(7):3183–3189

Steidler L, Hans W, Schotte L, Neirynck S, Obermeier F, Falk W, Fiers W, Remaut E (2000) Treatment of murine colitis by Lactococcus lactis secreting interleukin-10. Science 289 (5483):1352–1355

Sybesma W, Hugenholtz J, de Vos WM, Smid EJ (2006) Safe use of genetically modified lactic acid bacteria in food. Bridging the gap between consumers, green groups, and industry. Electronic J Biotechnol 9:4. doi:10.2225/vol9-issue4-fulltext-12

Tacket CO, Levine MM (2007) CVD 908, CVD 908- htrA, and CVD 909 live oral typhoid vaccines: a logical progression. Clin Infect Dis 45:S20–S23

Takala T, Saris P, Tynkkynen S (2003) Foodgrade host/vector expression system for *Lactobacillus casei* based on complementation of plasmid-associated phosphobeta- galactosidase gene *lacG*. Appl Microbiol Biotech 60(5):564–570

Taranto MP, Fernandez Murga ML, Lorca G, Font de Valdez G (2003) Bile salts and cholesterol induce changes in the lipid cell membrane of *Lactobacillus reuteri*. J Appl Microbiol 95:86–91

Taupin D, Podolsky DK (2003) Trefoil factors: initiators of mucosal healing. Nat Rev Mol Cell Biol 4:721–732

Thim L, Mortz E (2000) Isolation and characterization of putative trefoil peptide receptors. Regul Pept 90:61–68

Thompson C, Powrie F (2004) Regulatory T cells. Curr Opin Pharmacol 4:408–414

Tilg H, Ulmer H, Kaser A, Weis G (2002) Role of IL 10 for induction of anemia during inflammation. J Immunol 169:2204–2209

Toba T, Virkola R, Westerlund B, Björkman Y, SillanpääJ VT, Kalkkinen N, Korhonen TK (1995) A collagen binding S-layer protein in Lactobacillus crispatus. Appl Environ Microbiol 61:2467–2471

Vandenbroucke K, Hans W, Van Huysse J, Neirynck S, Demetter P, Remaut E, Rottiers P, Steidler L (2004) Active delivery of trefoil factors by genetically modified *Lactococcus lactis* prevents and heals acute colitis in mice. Gastroenterology 127:502 513

Ventura M, O'Flaherty S, Claesson MJ, Turroni F, Klaenhammer TR, Sinderen DV, O'Toole PW (2009) Genome-scale analyses of health-promoting bacteria: probiogenomics. Nat Rev Microbiol 7:61–71

Visser JA, Akkermans GM, Hoekstra RF, De Vos WM (2004) Insertionsequence- mediated mutations isolated during adaptation to growth and starvation in *Lactococcus lactis*. Genetics 168:1145–1157

Wall T, Bath M, Britton RA, Jonsson H, Versalovic J, Roos S (2007) The early response to acid shock in *Lactobacillus reuteri* involves the ClpL chaperone and a putative cell wall-altering esterase. Appl Environ Microbiol 73:3924–3935

Watsen D, Sleater RD, Hill C, Gahan CG (2008) Enhancing bile tolerance and improved survival and persistence of Bifidobacterium and lactococcus in the murine gastrointestinal tract. BMC Microbiol. doi:10.1186/1471-2180-8-176

Wells JM, Mercenier A (2008) Mucosal delivery of therapeutic and prophylactic molecules using lactic acid bacteria. Nat Rev Microbiol 6:349–362

Whitehead K, Versalovic J, Roos S, Britton RA (2008) Genomic and genetic characterization of the bile stress response of probiotic *Lactobacillus reuteri* ATCC 55730. Appl Environ Microbiol 74:1812–1819

Zhang JG, Matthews JM, Ward LD, Simpson RJ (1997) Disruption of the disulfide bonds of recombinant murine interleukin-6 induces formation of a partially unfolded state. Biochemistry 36(9):2380–2389

Roles of Probiotic on Gut Health

Wai-Yee Fung, Huey-Shi Lye, Ting-Jin Lim, Chiu-Yin Kuan, and Min-Tze Liong

Contents

1 Introduction 140
2 Alleviation of Diarrhea 140
2.1 Acute Diarrhea 141
2.2 Antibiotic-Associated Diarrhea 141
2.3 Radiation-Induced Diarrhea 144
3 Lactose Intolerance 146
4 Irritable Bowel Syndrome 147
5 Inflammatory Bowel Disease 148
6 Antibacterial Activities 150
7 Anticolorectal Cancer 154
8 Postoperative Gut Maintenance 157
8.1 Bariatric Surgery 157
8.2 Short Bowel Syndrome 158
8.3 Other Postoperative Complications 159
9 Conclusions 160
References 160

Abstract There is an increasing interest and demand for probiotics today, after a long history of safe use in fermented dairy products, due to greater recognition of its beneficial effects to the human gut health. The maintenance and promotion of gut health is critical, since 70% of the human immune system is located in the gut, thus directly implying the role of probiotics in human health. This chapter documents the strong in vivo and in vitro evidences of probiotics in reestablishing the intestinal ecosystem balance, and alleviating gut and malabsorption disorders such as diarrhea,

W.-Y. Fung (✉) • C.-Y. Kuan
School of Pharmaceutical Sciences, Universiti Sains Malaysia, 11800, USM Penang, Malaysia
e-mail: waiyee_fung@yahoo.com

H.-S. Lye • T.-J. Lim • M.-T. Liong
School of Industrial Technology, Universiti Sains Malaysia, 11800, USM Penang, Malaysia

M.-T. Liong (ed.), *Probiotics*, Microbiology Monographs 21,
DOI 10.1007/978-3-642-20838-6_6,

lactose intolerance, and irritable bowel syndrome. Probiotics are also therapeutic against postoperative complications and inflammatory bowel diseases, in addition to exerting antibacterial and anticancer properties in the gut, due to its ability to attenuate the immune system. Studies suggest mechanisms include competitive exclusion of pathogenic bacteria for nutrients and adhesion sites, production of antimicrobial bacteriocins and metabolites, and gut immunomodulation.

1 Introduction

Probiotics are defined as "live microorganisms that beneficially affect the host once consumed in adequate amounts" (Lye et al. 2009). Microorganisms identified as probiotics include *Streptococcus thermophilus*, *Enterococcus*, *Saccharomyces*, *Lactobacillus*, and *Bifidobacterium* species. Large populations of lactobacilli and bifidobacteria in the intestinal tract are generally an indication of a healthy gut. Thus, *Lactobacillus* and *Bifidobacterium* strains have been the focus of studies investigating the role of probiotics in maintenance of gut health. Probiotics are also increasingly used as functional adjuncts in dairy and new products such as soymilk, fruit juices, meat derivatives, and cereal-based foods, with documented physiological benefits. Probiotics have been reported to suppress diarrhea, alleviate lactose intolerance and postoperative complications, reduce irritable bowel symptoms, prevent inflammatory bowel disease (IBD), exhibit antimicrobial and anticolorectal cancer activities, via various proposed mechanisms. However, in order to exert their beneficial effects, probiotics must be able to survive within the gastrointestinal tract and be sustained at sufficiently high levels in the intestine. The generally observed dose of probiotics via oral administration is in excess of 10^9 CFU/day. Most administered probiotic cells exert health benefits upon successful adherence to intestinal cells. However, it has been recommended that oral administration of probiotics is continued on a daily basis, as cells are constantly eliminated from the digestive track into feces upon physiological changes of the host and/or administration of antimicrobial agents such as antibiotics. *Sacchoromyces boulardii* has been shown able to adhere to colonic cells but were eliminated from stools after 2–5 days of discontinued consumption or upon administration of antifungal agents such as nystatin (Czerucka and Rampal 2002).

2 Alleviation of Diarrhea

Diarrhea, defined as the increase of liquidity or decreased consistency of stools usually associated with increased frequency of stools and fecal weight (de Vrese and Marteau 2007), is a major cause of morbidity among young children in developing countries. Microorganisms responsible for the majority of infections are transmitted mostly by fecal–oral route through water and food (Farthing and Kelly 2007).

In recent years, there has been a growing interest in the use of probiotics for the prevention and treatment of diarrhea such as acute diarrhea, antibiotic-associated diarrhea (AAD), and radiation-induced diarrhea (RID). Probiotics confer health benefits primarily by promoting the proliferation of beneficial gastrointestinal indigenous microflora (Lye et al. 2009).

2.1 Acute Diarrhea

Acute diarrhea is one of the most common diseases in children. In the United States, 16.5 million children develop acute diarrhea annually, while children in developing nations commonly experience several episodes per year (Huang et al. 2002). Gastroenteritis, commonly due to several viral and bacterial pathogens, is the main cause of acute diarrhea, but rotaviruses infection is the major cause of acute diarrhea in infants and children (Marteau et al. 2001).

The alleviation of symptoms and shortening of acute infections are the most prominent probiotic effects against diarrhea. Clinical studies have demonstrated these beneficial effects using different interventions such as fermented milk product, oral rehydration-probiotic solution, probiotic capsules, and nonfat dry milk powder (Table 1).

The mechanisms and the efficacy of a probiotic effect often depend on the stimulation of specific and nonspecific immunity system. The reduction in duration of acute diarrhea by *Lactobacillus GG* has been attributed to the elicitation of local immune response. *Lactobacillus GG* potentiated gut immune response to rotavirus, shortening the duration of acute diarrhea (Kaila et al. 1992). *Lactobacillus GG* also promotes recovery from acute rotavirus diarrhea via augmentation of the local immune defense mechanisms. *Lactobacillus GG* enhanced the immunoglobulin (IgA) response to rotavirus and influenced the permeability of antigens through Peyer's patches, thus stimulating numerous IgA-committed B-cell populations.

Competitive inhibition for bacterial adhesion sites on intestinal epithelial surfaces is another mechanism of probiotic action. Simakachorn et al. (2000) showed that *L. acidophilus* LB possessed specific properties, including the reinforcement of intestinal microflora that reduces shedding of pathogens. Probiotics adhere to cultured human intestinal absorptive and secretory cells to form a protective biofilm, thus inhibiting invasion of enterocytes by diarrhea-causing microorganisms. Probiotics also produce a variety of substances that are inhibitory to pathogens. Shornikova et al. (1997) suggested that the production of antimicrobial reuterin by *L. reuteri* suppressed the overgrowth of urease-producing bacteria, thus inhibiting and alleviating rotavirus infection.

2.2 Antibiotic-Associated Diarrhea

Antibiotic-associated diarrhea (AAD) is a common complication raised from the consumption of antibiotic, especially for broad-spectrum antibiotics such as

Table 1 Effect of probiotics on acute diarrhea

Probiotic strain	Dose (CFU)	Intervention	Subjects	Therapeutic effect[a]	References
L. casei strain *GG*	3×10^9	200 mL milk; twice daily; 5 days	100 children; 3–36 months with acute diarrhea	Diarrhea duration reduced by 50% ($p < 0.01$)	Guarino et al. (1997)
Lactobacillus GG	$>10^{10}$	250 mL oral rehydration solution; once daily; until diarrhea stopped	287 children; 1 month to 3 years with acute diarrhea	58.3 ± 27.6 h vs. 71.9 ± 35.8 h*	Guandalini et al. (2000)
L. acidophilus; *B. infantis*	10^9	Capsule (Infloran Berna); once daily; 4 days	100 children; 6–60 months	3.1 ± 0.7 days vs. 3.6 ± 0.8 days*	Lee et al. (2001)
L. bulgaricus and *Strep. thermophilus*	10^8	180 mL yoghurt; once daily; 5 days	112 children; 3–24 months; with acute watery diarrhea	44.1 ± 34.6 h vs. 61.7 ± 32.5 h*	Boudraa et al. (2001)
L. acidophilus LB	2×10^{10}	Lyophilized heat-killed probiotic; 5 doses; 48 h	73 children; 3–24 months with acute watery diarrhea	43.4 ± 25.9 h vs. 57.0 ± 36.3 h*	Simakachorn et al. (2000)
L. rhamnosus 19070-2; *L. reuteri* DSM 12246	1.7×10^{10}; 0.5×10^{10}	2.5–5.0 mL mixture; twice daily; 5 days	69 children 6–36 months with acute diarrhea	81.5 ± 37.3 h vs. 101.1 ± 47.6 h	Rosenfeldt et al. (2002)
L. reuteri SD 2112	10^{10}–10^{11}	Nonfat dry milk powder; once daily; 5 days	40 children 6–36 months with acute diarrhea	1.7 ± 1.6 h vs. 2.9 ± 2.3 h	Shornikova et al. (1997)

[a]Duration of acute diarrhea in the treatment group versus (vs.) placebo group*$P < 0.05$ compared to controls

ampicillin, amoxicillin, cephalosporins, and clindamycin (McFarland 2009). AAD varies in incidence but can occur in up to 39% of hospitalized patients receiving antibiotics. The consequences of AAD include extended hospital stays, increased medical costs, increased incidence of other nosocomial infections, and mortality (Surawicz 2003). Probiotics evaluated for the prevention of AAD include various strains of *Lactobacillus* spp., *Bifidobacterium* spp., *Streptococcus* spp., *Enterococcus faecium*, and yeasts (*Saccharomyces boulardii*) (Table 2).

Clostridium difficile colitis is a distinct complication of antibiotic administration, often resulting in AAD. The administration of antibiotics allows the proliferation of *C. difficile*, leading to colonization of intestinal pathogens and excretion of exotoxins. *C. difficile* produces toxins A and B that cause mucosal damage and inflammation of the colon, thus accelerates the pathogenesis of colonic lesions (Marteau et al. 2001). Probiotics produce proteases that directly degrade *C. difficile* toxins and increase the immune response to the toxins (McFarland 2009). Castagliuolo et al. (1996, 1999) showed that *S. boulardii* reduced toxin-induced intestinal secretion and permeability, produced toxin-degrading proteases and inhibited binding of toxins to the brush border membrane. Growth supernatant of *S. boulardii* was filtered and chromatographically fractionated, yielding an active 54-kDa serine protease fraction that decreased toxin A-induced rat ileal secretion by 46% and intestinal permeability by 74%, prevented toxin A-mediated inflammation and villus damage (Fig. 1), and also exerted a proteolytic activity against

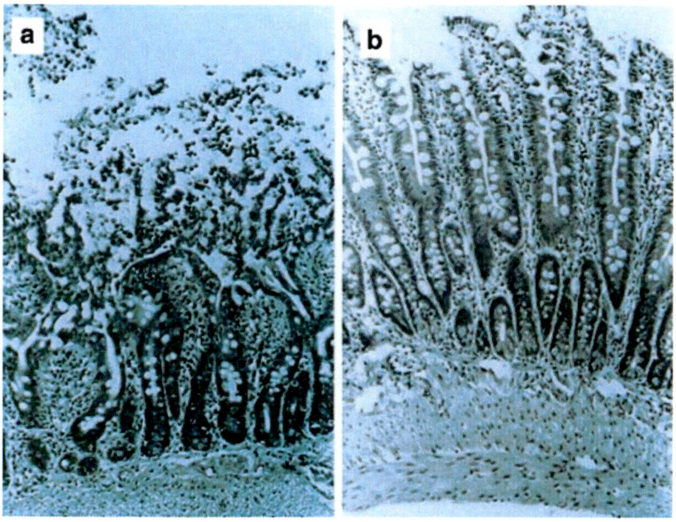

Fig. 1 Protective effect of *S. boulardii* protease against toxin A-induced enteritis in rats. (**a**) Villus necrosis and infiltration of the lamina propria by neutrophils in rat ileal loops injected with toxin A and (**b**) ileal loop exposed to toxin A after preincubation with *S. boulardii* protease, showing prevention of toxin A-induced changes (**b**); Castagliuolo et al. (1996). Reprinted from Czerucka and Rampal (2002), with permission from Elsevier (License number: 2611150041785)

the toxin B molecule. Interestingly, when anti-*S. boulardii* protease antibody and toxin A were administered, the inhibitory effect of *S. boulardii*-conditioned medium on fluid secretion and mucosal permeability in rat ileum was reversed, strengthening the claims that the *S. boulardii* serine protease mediated inhibitory effects against *C. diffcile*. Additionally, *S. boulardii* has also been found to exhibit steric hindrance that inhibited the adhesion of *C. difficile* (Surawicz 2003), and thus often used as a preventive and therapeutic agent for AAD.

The therapeutic role of probiotics in AAD has been critically assessed in meta-analyses of clinical trials. Meta-analyses are used to obtain a pooled estimate of efficacy for prevention of AAD, using studies of different probiotic strains (McFarland 2009). The most comprehensive meta-analysis was conducted by McFarland (2006) which included 25 randomized controlled trials of probiotics for the prevention of AAD and *C. difficile* infections (CDI) in 2,810 subjects. Results revealed that 67% of trials, using a probiotic dose more than 10^{10} CFU/day, were significantly protected, compared to only 17% if the dose was less than 10^{10} CFU/day (McFarland 2009). However, the action of probiotics is found to be strain- and disease-specific, thus, extrapolation of selected results would not give an accurate representation of predicted behavior for all probiotics. Nevertheless, the positive and consistent evidence of probiotics' therapeutic role in alleviating AAD and gut imbalance warrants further research to determine the specific actions of potential strains and establish the suitable levels of administration for adult, geriatric, and pediatric use.

2.3 Radiation-Induced Diarrhea

Chemotherapy and radiotherapy frequently cause severe disturbances of the immune system and the indigenous intestinal microflora. RID is common in radio-oncology patients (de Vrese and Marteau 2007) and several attempts to treat this complication with probiotics have been examined (Table 2). Studies have demonstrated that bacteriotherapy with the use of probiotics could protect patients against the risk of RID. Patients supplemented with probiotics may be allowed to receive a greater dose of radiation to enhance the efficacy of radiotherapy.

Clinical trials have provided insights into the mechanisms of probiotics to protect patients against RID. Probiotics have been reported to lower the production of pro-inflammatory cytokines and effectors of inflammation (nitric oxide and metallo-proteinases), interfere with the pro-inflammatory signal transduced by toll-like receptors, and thus exert a significant protection upon the integrity of the intestinal epithelial barrier. Additionally, probiotics has the ability to down-modulate the process of apoptosis, the main factor responsible for the radiation-induced injury of the intestinal epithelium (Delia et al. 2007).

Table 2 Effect of probiotics on antibiotic-associated diarrhea (AAD) and radiation-induced diarrhea (RID)

Probiotic	Dose (CFU)	Intervention	Subjects	Therapeutic effect[a]	References
Antibiotic-associated diarrhea (AAD):					
Lactobacillus GG	1×10^{10}–2×10^{10}	Capsule, once daily 10 days	188 children	7.5% vs. 26.3%[a,*]	Vanderhoof et al. (1999)
L. casei DN-114001; *Strep. thermophilus*; *L. bulgaricus*	10^8; 10^8; 10^7	Yoghurt, twice daily 14 days	135 patients	12.3% vs. 33.9%[a,*]	Hickson et al. (2007)
S. boulardii	10^{10}	Twice daily, 7–9 days	269 children	3.4% vs. 17.3%[a,*]	Kotowska et al. (2005)
Radiation-induced diarrhea (RID):					
Antibiophilus (*L. rhamnosus*)	1.5×10^9	3 sachets once daily 1 week	206 cancer patients, radiotherapy in the abdominal region	138 ± 5 h vs. 125 ± 5 h[b,*]	Urbancsek et al. (2001)
VSL#3[c]	4.5×10^{11}	1 sachet three times daily until end of radiation therapy	190 cancer patients, radiotherapy of adjuvant postoperative	118 ± 6 h vs. 97 ± 4 h[b,*]	Delia et al. (2002)
VSL#3[c]	4.5×10^{11}	1 sachet three times daily until end of radiation therapy	490 cancer patients, radiotherapy of adjuvant postoperative	122 ± 8 h vs. 86 ± 6 h[b,*]	Delia et al. (2007)

[a]Percentage of subjects with antibiotic-associated diarrhea (AAD) in the treatment group versus (vs.) placebo group. [b]Time of relief medication administration in subjects with RID in the treatment group versus (vs.) placebo group. [c]*L. casei, L. plantarum, L. acidophilus, L. delbruekii bulgaricus, B. longum, B. breve, B. infantis* and *Strep. salivarius thermophilus*. $^*P < 0.05$ compared to controls

3 Lactose Intolerance

Lactase insufficiency refers to low concentration of lactase in the brush border membrane of the mucosa of the small intestine. Hypolactasia leads to incomplete digestion of lactose, termed as lactose maldigestion. Lactose maldigestion is defined as an increase in concentration of blood glucose of less than 1.12 mmol/L or in breath hydrogen of more than 20 ppm after intake of 1 g/kg body weight (de Vrese et al. 2001). The ingestion of 50 g of lactose by lactose maldigesters could cause symptoms such as abdominal pain, diarrhea, bloating and flatulence, and up to 50% of them suffer symptoms after intake of 200–250 mL of milk (Vesa et al. 2000).

Lactose must be hydrolyzed by β-galactosidase or lactase into its monomers, glucose and galactose, before it can be absorbed by intestinal enterocytes into the bloodstream, where glucose is used as a source of energy and galactose becomes a component of glycolipids and glycoproteins (Lomer et al. 2008). The enzyme, normally found in the jejunum or the beginning of the small intestine (Vesa et al. 2000), has optimal activity at pH 6–8 in the small intestines. When the pH drops to as low as 4 in the colon, bacterial lactase activity declines, leading to lactose fermentation and maldigestion (Lomer et al. 2008). Undigested lactose is used by the gut bacteria to produce short-chain fatty acids (SCFAs) and gases (hydrogen, carbon dioxide, and methane) which cause discomfort (Honda et al. 2007) and contribute to the development of diarrhea.

Unabsorbed lactose in the ileum and colon cause acidification of colonic contents and increased osmotic load (Lomer et al. 2008). The osmotic load triggers secretion of fluid and electrolytes until osmotic equilibrium is reached, leading to dilatation of the intestines which induces rapid small intestinal transit time and increases the degree of maldigestion. These conditions lead to excretion of loose stools (Vesa et al. 2000). The gases produced will be removed either through consumption by intestinal bacteria or absorption into the bloodstream. However, the excessive rectal gas or abdominal distension may occur when the removal mechanisms are overwhelmed.

Some lactic acid bacteria (LAB) with the ability to utilize lactose such as *Lactobacillus*, *Bifidobacterium*, and *Streptococcus* can alleviate symptoms of lactose intolerance, and are generally added into dairy products to increase the digestibility of lactose present (Rolfe 2000). Lactobacilli such as *L. casei*, *L. acidophilus*, and *L. lactis* utilize two lactose transportation mechanisms:

(a) Lactose-permease transportation and hydrolysis by β-galactosidase (β-gal, EC 3.2.1.23);
(b) Lactose-specific phosphoenolpyruvate-dependent phosphotransferase system (lac-PTS) and hydrolysis by phosphor-β-galactosidase (P-β-gal, LacG, EC 3.2.1.85)

(Honda et al. 2007)

Lactase activities of 65 LAB isolated from human feces were examined. The authors found that *Lactobacillus mucosae* OLL 2848 was able to exhibit β-galactosidase activity, while *L. gasseri* OLL 2863 and OLL 2948 were able to exhibit phospho-β-galactosidase and phospho-β-glucosidase activities. The expression of these enzymes was due to the presence of lactose, suggesting the importance of these enzymes in utilization of lactose in human intestine (Honda et al. 2007). In a study evaluating the ability of *L. acidophilus* LA-1 (1.5×10^{10} cells) to ferment lactose in vitro, authors observed a significant decrease in concentration of lactose, while production of SCFAs (acetate and propionate) increased with increasing β-galactosidase activity compared to the control (Jiang and Savaiano 1997). Pelletier et al. (2001) conducted a double-blind, randomized, and crossover trial with 24 male lactose malabsorbers to evaluate the effect of probiotic-fermented yoghurt on hydrogen production and symptoms of lactose intolerance. The excretion of hydrogen by bacteria reflects lactose maldigestion. The authors found that ingestion of yoghurt with *Strep. thermophilus* and *L. bulgaricus* (10^8 CFU/ml) led to reduced excretion of hydrogen and decreased symptoms of lactose intolerance in lactose malabsorbers. These studies are among many that have corroborated the role of probiotics in alleviation of lactose intolerance, and justified its continuous use as a treatment for lactose intolerance.

4 Irritable Bowel Syndrome

Irritable bowel syndrome (IBS) is a chronic functional gastrointestinal disorder where the common diagnosis is mainly based on self-reported symptoms according to Rome Consensus Diagnosis Criteria for IBS (Longstreth et al. 2006). IBS normally involves the middle and lower gastrointestinal tract, in the absence of structural and metabolic disorders. Common symptoms include abdominal discomfort and abnormal bowel habits such as irregular stool form and stool passage frequency. The current treatment for IBS is symptom-oriented therapy with prescriptions based on self-described symptoms, an approach with limited efficacy and controversial side effects. The common pathophysiology of IBS include small intestine bacterial overgrowth (SIBO), altered gut motility, visceral hypersensitivity and overproduction of hydrogen due to food malfermentation, all of which can be modulated by probiotics to alleviate IBS.

Probiotics alleviate SIBO in IBS by maintaining a healthy colonic microflora population via production of bacteriocins and SCFAs, which inhibit pathogenic bacteria, and improving intestinal barrier functions by competitively inhibiting the attachment of pathogenic bacteria. SCFAs also serve as an important energy source for the maintenance of normal colonic cell functions. Gut inflammation is known to increase visceral hypersensitivity which causes abdominal discomfort in IBS. *L. paracasei* (NCC2461) therapy reportedly inhibited inflammation-associated visceral hypersensitivity and sensory neurotransmitter expression in murine model, suggesting a possible probiotic mechanism in inhibiting visceral hypersensitivity (Verdu et al. 2006).

Overproduction of fermentation gases, particularly hydrogen, from malfermentation of food in the small bowel results in abdominal bloating and flatulence, as observed in IBS patients. Probiotics has been reported to modulate the fermentation process, subsequently affecting gas production, colonic transit, and fluid fluxes. Most probiotics do not produce gas from carbohydrate fermentation, while some pathogenic bacteria such as *Clostridia* spp. are gas-producing. A double-blind, placebo-controlled, crossover study, reported significant decrease in breath hydrogen levels in *L. plantarum 299v* supplemented group at 120 min after ingestion of lactulose, suggesting the potential beneficial effect of probiotics in the reduction of flatulence in IBS (Sen et al. 2002).

Excess intracolonic bile acids cause diarrhea in IBS by promoting secretion of colonic fluid, stimulating gut and fluid motility. Propulsive colonic contraction responsible for defecation has been induced by bile acids infusion (Bampton et al. 2002). Probiotics possess the ability to absorb bile acids (Lye et al. 2009), thus alleviating abnormal gut and fluid motility, subsequently bile acid-induced diarrhea. Verdu et al. (2004) reported a significant decrease in intestinal muscle hypercontractility after *L. paracasei* treatment in a murine model with postinfective IBS.

Reduction of pathogenic bacteria by probiotics enhances intestinal immune response, thus reducing inflammation in IBS. Immunomodulatory actions of probiotics include stimulation of anti-inflammatory regulatory T cells, reduction of pro-inflammatory cytokines, and increase of anti-inflammatory cytokines composition (Sartor 2004). The administration of *B. infantis* 35624 normalized the ratio of anti-inflammatory cytokine to pro-inflammatory cytokine (IL-10/IL-12), resulting in a significant improvement in IBS symptoms in a randomized, placebo-controlled study (O'Mahony et al. 2005).

A meta-analysis that systematically reviewed 20 randomized, controlled, blinded trials involving 1,404 subjects in 23 probiotic treatments over periods ranging from 2 to 24 weeks (McFarland and Dublin 2008) concluded a significant reduction of common IBS symptoms in probiotics-treated group as compared to the placebo group at the end of the studies. Moayyedi et al. (2010) also reported beneficial role of probiotics in the treatment of IBS from systematic review of 19 randomized, controlled, blinded trials that involved 1,650 IBS patients. Probiotic therapy resulted in significant alleviation of IBS symptoms in all trials, while combinations of different types of probiotics resulted in significant improvements in pain and flatulence scores as compared to the placebo group.

5 Inflammatory Bowel Disease

Inflammatory bowel diseases (Crohn's disease, ulcerative colitis, and pouchitis) are immune-mediated diseases with chronic recurrent inflammatory conditions in gut of unknown multifactorial etiology (Marteau 2000). Crohn's disease is an acute inflammatory disease involving mainly the ileum and other parts along gastrointestinal tract with symptoms like diarrhea, abdominal pain, and weight loss.

Inflammation of ulcerative colitis is confined to the colon and causes mucosal inflammation, erosion, and ulceration. Pouchitis manifests as nonspecific inflammation of the ileal pouch following ileal pouch-anal anastomosis surgery in patients with refractory ulcerative colitis. Symptoms include abdominal pain, diarrhea, and rectal bleeding with possibly systemic features such as fever.

Probiotics reportedly alleviate IBD via inhibition of pathogenic bacterial growth via secretion of bactericidal compounds, decreasing luminal pH by formation of SCFAs and colonization resistance. Probiotics also influence gut immuno-regulatory activities via modulation of immunocytes apoptosis, reduction of phagocytic activity, down-regulation of T-cell responsiveness, induction of antimicrobial substances secretion from specialized intestinal cells such as immunoglobulin A, and alteration of cytokines profile by up-regulation of anti-inflammatory cytokines such as IL-10 and down-regulation of pro-inflammatory cytokines. Probiotics protect and improve intestinal epithelial barrier function, via production of SCFAs and other metabolites, maintenance of tight junction integrity, enhancement of mucus production, enhancement of transepithelial resistance affecting intestinal permeability, enhancement of epithelial cell glycosylation, and stimulation of epithelial self repair.

Feeding of *B. subtilis* to chickens improved histological alterations of intestinal villi such as villus height, cell area and cell mitosis, attributed to better maintenance of normal intestinal microflora, increased digestive enzyme activity, and decreased ammonia production (Fig. 2) (Samanya and Yamauchi 2002). In vitro experiment using Caco-2/TC-7 cells revealed that spent culture supernatants of *Lactobacillus*

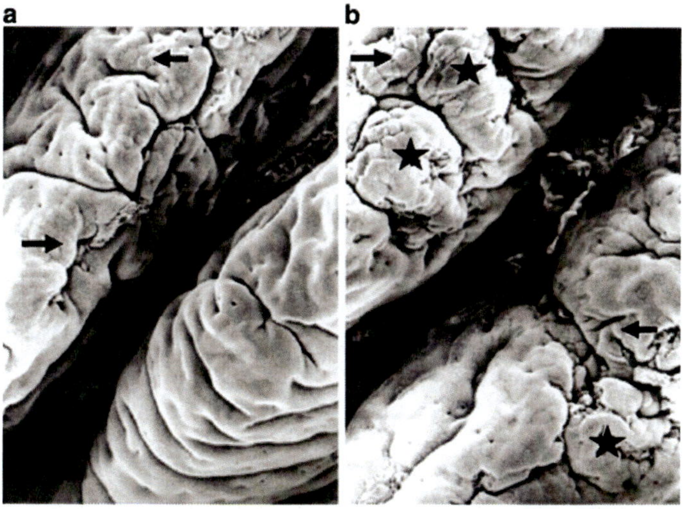

Fig. 2 Duodenal villus tip surface of control chickens (**a**) and chickens fed dried *B. subtilis* (0.2%; **b**). Control chickens developed flat cell outline on the duodenal villus surface (*arrows* in **a**), while chickens fed *B. subtilis* developed large, protruded cell clusters (*stars* in **b**) and cell protuberances (*arrows* in **b**), indicating activation of intestinal functions. Reprinted from Samanya and Yamauchi (2002), with permission from Elsevier (License number: 2546350699365)

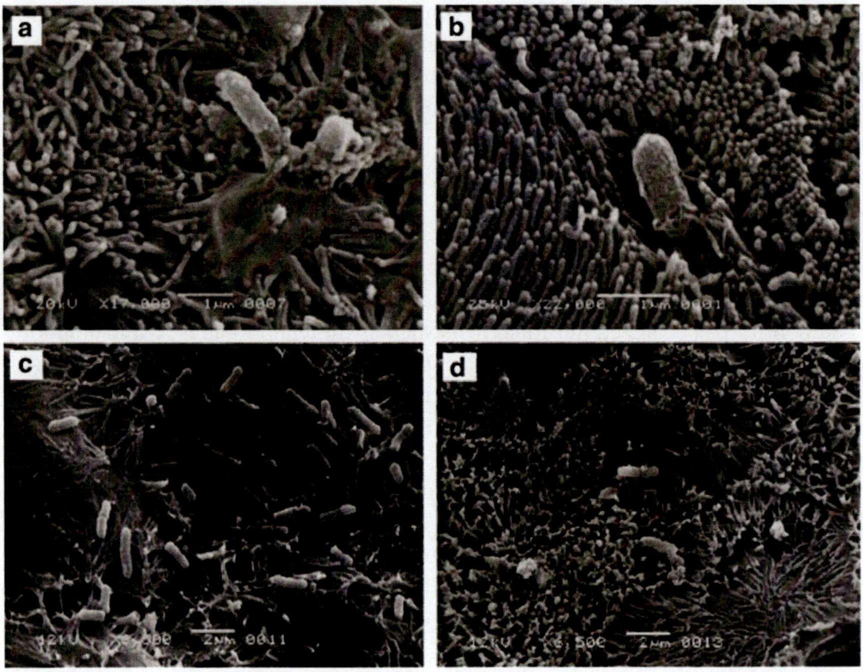

Fig. 3 Caco-2/TC7cells infected with *Salmonella enteritidis* after (**a**) 40 min and (**c**) 120 min, infected with *S. enteritidis* pretreated with spent culture supernatants of *L. kefir* CIDCA 8321 after 40 min (**b**) and 120 min (**d**). Reprinted from Golowczyc et al. (2007), with permission from Elsevier (License number: 2546370389722)

kefir strains could protect epithelial cells against damage induced by *Salmonella enteritidis* (Golowczyc et al. 2007). *Salmonella* induced microvilli disorganization in Caco-2/TC-7 cells, such as shortened and coalesced microvilli (Figs. 3a and c). But, Caco-2/TC-7 cells infected with *Salmonella* that were pretreated with spent culture supernatants of *L. kefir* 8321 showed a confluent monolayer of normal microvilli (Figs. 3b and d).

The therapeutic efficacy of probiotics in preventing the development, induction, and maintenance of the remission of IBD in human models has been vigorously evaluated, with most clinical trials showing promising effects in ulcerative colitis and pouchitis patients (Table 3).

6 Antibacterial Activities

Probiotics exhibit antibacterial activities via several mechanisms namely the production of inhibitory substances, inhibition of pathogen adhesion to surfaces, and production of iron-siderophore.

Table 3 Clinical trials on the effects of probiotics in treating IBD

Study	Design	Treatment	No. of subjects (n)	Duration (t)	Results in probiotics-treated group
Crohn's disease:					
Plein and Hotz (1993)	Randomized, double-blind, placebo-controlled, single center	*S. boulardii* vs placebo	20	9 weeks	Significant reduction in relapse (CDAI) and frequency of diarrhea
Guslandi et al. (2000)	Randomized, single center	*S. boulardii* + mesalamine vs mesalamine	32	6 months	Significant efficiency in maintenance of remission (CDAI)
Ulcerative colitis:					
Kato et al. (2004)	Randomized, placebo controlled, single center	Bifidobacteria-fermented milk vs placebo	20	12 weeks	Significantly improvement (CAI, endoscopic activity index and histological score)
Kruis et al. (2004)	Randomized, double-blind, double dummy, multicenter	*E. coli Nissle 1917* vs mesalazine	327	12 months	As efficient as mesalazine in maintenance of remission.
Sood et al. (2009)	Randomized, double-blind, placebo-controlled, multicenter	VSL#3 vs placebo	144	12 weeks	Significantly higher induction of remission: 42.9% vs 15.7%. Significant greater decreases in UCDAI scores and individual symptoms
Tursi et al. (2010)	Randomized, double-blind, placebo-controlled, multicenter	VSL#3 supplementation vs placebo with concomitant standard treatment	131	8 weeks	Significantly improved disease activity (UCDAI)
Pouchitis:					
Gionchetti et al.(2000)	Randomized, double-blind, placebo-controlled, single center	VSL#3 vs placebo	40	9 months	Significant higher maintenance of remission (PDAI): 85% vs 0%. Significant increase fecal concentration of lactobacilli, bifidobacteria, and *S. thermophilus*
Mimura et al. (2004)	Randomized, double-blind, placebo-controlled, two centers	VSL#3 vs placebo	36	12 months	Significantly higher maintenance of remission (PDAI): 85 % vs. 6 %
Pronio et al. (2008)	Randomized, controlled, single center	VSL#3 vs no probiotic supplementation	31	12 months	Significant reduction in PDAI

CDAI Crohn's disease activity index, *CAI* Clinical activity index, *UCDAI* Ulcerative colitis disease activity index, *PDAI* Pouchitis disease activity index, VSL#3: *L. casei*, *L. plantarum*, *L. acidophilus*, *L. delbrueckii* subsp. *Bulgaricus*, *B. longum*, *B. breve*, *B. infantis*, *Strep. salivarius* subsp. *thermophilus* (300 billion viable, lyophilized bacteria per gram)

Inhibitory substances such as hydrogen peroxide, bacteriocins, and SCFAs (lactic and acetic acids) produced by *Lactobacillus* spp., *Bifidobacterium* spp., *Enterococcus* spp., *E. coli*, *Leuconostoc* spp., *Pediococcus* spp., *S. cerevisiae*, and *Streptococcus* spp. (Voravuthikunchai et al. 2006) are effective in inhibiting the growth of food pathogens, thus spurring increased use of bacteriocins as food preservatives, such as enterocin AS-48 in vegetable sauces, sakacin in meat sausages, and nisin in cheese and liquid whole egg (Galvez et al. 2007).

Bacteriocins produced by LAB are generally small, heterogeneous, and cationic proteins consisting of 30–60 amino acid residues (Lima et al. 2007). They are divided into three classes: class I (bacteriocins or lantibiotics), class II (nonlantibiotics), and class III (large protein bacteriocins). Most bacteriocins act by creating pores in the membrane of their target cells that cause the dissipation of proton motive force, ATP depletion, and leakage of nutrients that subsequently lead to cell damage or cell death (Muthaiyan et al. 2010). Bacteriocins exhibit a broad-spectrum of action against numerous genera of pathogenic and nonpathogenic bacteria. The efficacy of probiotic against pathogens depends on the combined action of bacteriocins and antimicrobial substances such as hydrogen peroxide, organic acids, and bacteriophages (Lima et al. 2007).

Hutt et al. (2006) evaluated the antimicrobial activity of five lactobacilli (*L. rhamnosus* GG, *L. fermentum* ME-3, *L. acidophilus* La5, *L. plantarum* 299v, and *L. paracasei* 8700:2) and two bifidobacteria (*B. lactis* Bb12 and *B. longum* 46) against six target pathogens using different assays. *L. rhamnosus* GG and both bifidobacteria suppressed the growth of *E. coli*. *Lactobacillus* strains 8700:2, 299v and ME-3 reduced the growth of *Salmonella enterica* ssp. *enterica* in microaerobic assays. *Lactobacillus* strain ME-3 and both bifidobacteria showed high antagonistic effect against *Shigella sonnei* in anaerobic condition, while *L. paracasei*, *L. rhamnosus*, and *L. plantarum* showed antimicrobial activity against *Helicobacter pylori* under microaerobic conditions on solid media. Lima et al. (2007) observed a similar trend where *Lactobacillus* strains (*L. reuteri*, *L. salivarius*, and *Lactobacillus* spp.) isolated from chicken inhibited the growth of *Enterococcus faecalis*, *E. faecium*, *L. monocytogens*, and *Salmonella* spp.

Karska-Wysocki et al. (2010) investigated the antimicrobial effects of *L. acidophilus* CL 1285 s and *L. casei* LBC80R extracted from commercial food products on methicillin-resistant *Staphylococcus aureus* (MRSA). Direct interaction of lactobacilli strains and MRSA in solid or liquid medium caused a 99% reduction of MRSA cells after incubation at 37°C for 24 h. The reduction in MRSA cells was attributed to the production of bacteriocins, hydrogen peroxide, and organic acids (lactic and acetic acids). There is a synergistic action between lactic acid and secreted nonlactic acid molecules, where lactic acid acts as a permeabilizer of the outer membrane of pathogens in favor of penetration of nonlactic acid molecules (Zhang et al. 2010a). The inhibitive effect of lactic acid was also demonstrated in a recent study by Bernardeau et al. (2009), which showed that cell-free *Lactobacillus* supernatant containing mainly lactic acid exerted growth inhibiting activity against pig pathogenic intestinal spirochaetes, *Brachyspira hyodysenteriae* and *Brachyspira pilosicoli*. The exposure of spirochaetal cells

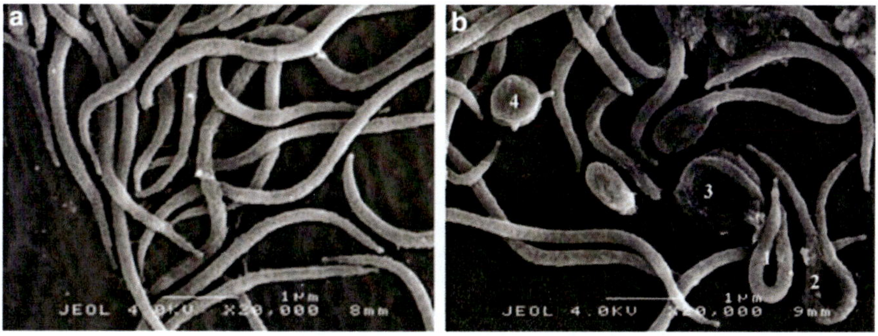

Fig. 4 Scanning electron micrographs of *B. hyodysenteriae* 9828978 incubated 1 h with phosphate-buffer saline (control) (**a**) and with *Lactobacillus* culture supernatant (**b**). Various stages of stress injury (1) cell retracts, (2) cell folds in two, probably from synthesis of exopolymers between the two parallel strands of the cell, (3) cell curls up, (4) cell form spherical body. Reprinted from Bernardeau et al. (2009), with permission from Elsevier (License number: 2546301150301)

to the lactic acid supernatant for 1 h caused cellular morphological variations, attributed to a progressive degree of injury ranging from stress to lethality (Fig. 4). Morphological injuries include retraction of cells, folding of cells (probably due to stress-induced synthesis of exopolymers), curling of cells, and formation of spherical body.

The ability to adhere to mucosal surface is an important factor in the competitive exclusion of pathogenic bacteria (Balcazar et al. 2008), as the colonization of the gastrointestinal tract is based on their ability to adhere to the intestinal epithelium (Del Re et al. 2000). However, not all bacteria have the ability to adhere to intestinal cells. Probiotics such as lactobacilli and bifidobacteria exhibit adhesive properties which allow them to interact with the brush border of human intestinal epithelium cell lines in culture (Bernet et al. 1993). The adhesion to the intestinal mucosa also stimulates gut immune system (Balcazar et al. 2008), as probiotics competitively inhibit colonization by pathogenic bacteria and subsequently prevent their infection (Voravuthikunchai et al. 2006).

Bifidobacteria isolated from human (*B. breve* and *B. infantis*) were able to adhere to Caco-2 cells by possessing a proteinaceous adhesion-promoting factor that was present in bacterial whole cells and extracellularly in the spent supernatant. The adherence of bifidobacteria inhibited cell association and affected the adherence of pathogenic *E. coli* and *Salmonella typhimurium* to enterocystic Caco-2 cells (Bernet et al. 1993). The competition for specific sites on the enterocystic cell surface and formation of bacterial biofilm prevented the binding of enterovirulent organisms. Balcazar et al. (2008) observed that *Lactococcus lactis* CLEP 101 reduced the adhesion activities of fish pathogens (*Aeromonas hydrophila*, *Aeromonas salmonicida*, *Yersinia ruckeri*, and *Vibrio anguillarum*), while *L. fermentum* CLEP 242 and probiotics mixture (*Lactococcus lactis* CLFP 101, *L. plantarum* CLFP 238, and *L. fermentum* CLFP 242) reduced the adhesion activities of *Aeromonas* strains and *Yersinia ruckeri*. The adherence ability of LAB has been attributed to the

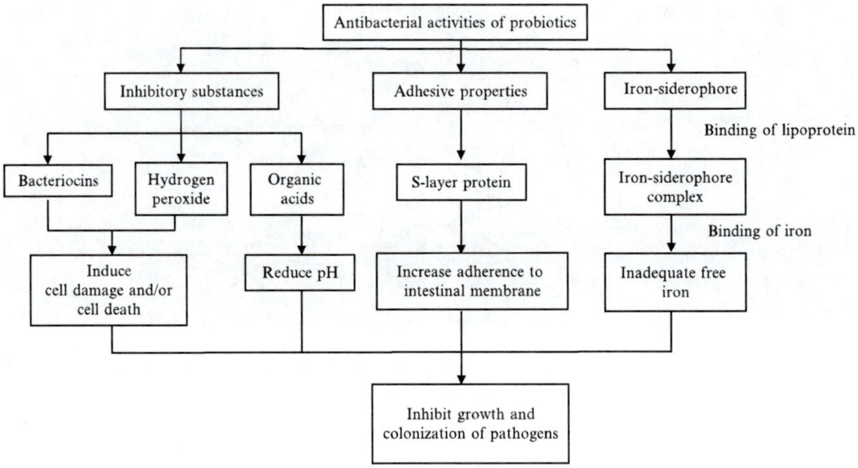

Fig. 5 Summary of antibacterial activities of probiotics

carbohydrate-specific molecules of bacterial cell surface components, which allow specific interaction between the cell surface and intestinal epithelial cells. In addition, S-layer proteins in the cell surface of lactobacilli strains has been associated with adhesion to host epithelial cells, as decreased adhesion ability was observed when the proteins were removed or disrupted (Zhang et al. 2010b).

Antimicrobial ability of probiotics has also been attributed to the production of iron-siderophore, used for the uptake of transferrin or lactoferrin (iron-binding glycoproteins). Generally, all microbes require iron for growth as it plays a vital role in metabolism and acts as a global regulator of gene expression in microbes (O'Sullivan 2000). Thus, the iron-acquiring system of pathogenic bacteria is imperative in bacterial virulence in human intestines (iron-limited environments). Iron-siderophore binding lipoproteins are crucial in gram-positive bacteria, as iron-siderophore complexes (ferrisiderophores) are first identified and bound by lipoproteins (Moon 2008). Moon (2008) found that *Bifidobacterium pseudocatanulatum* BP1 was able to secrete putative iron-siderophore binding lipoprotein which enabled competitive acquisition of iron leaving inadequate sources for pathogenic bacteria in the human gut. All the postulated mechanisms for antimicrobial activity of probiotics are as summarized in Fig. 5.

7 Anticolorectal Cancer

About 80% of the body's immune system is localized in the gastrointestinal tract (Kim et al. 2006), thus the general health of human is strongly influenced by gut health, which can be attenuated by probiotics. Potential protective effects of probiotics against cancer has been established, with most evidences in colon cancer.

Colorectal cancer is the third leading cause of cancer mortality in the United States (American Cancer Society 2008). Some strains of bacteria (*Strep. bovis*, *Bacteroides*, and clostridia) have been implicated in the pathogenesis of colon cancer, while others (*L. acidophilus* and *B. longum*) demonstrated protective effects (Ewaschuk et al. 2006). Thus, dietary interventions which can enhance the balance of colonic microflora are being studied to reduce risks of colon cancer, as a mean of prevention rather than cure (Liong 2008).

Probiotics suppress putrefactive intestinal bacteria with deleterious enzymatic activity which produce carcinogenic substances from dietary components and convert procarcinogens into carcinogens (Rafter et al. 2007). Supplementation with synbiotic consisting of prebiotic oligofructose-enriched inulin and probiotics (*L. rhamnosus* GG and *B. lactis* Bb12) increased fecal counts of *Lactobacillus* and *Bifidobacterium*, reduced counts of *Clostridium perfringens*, and reduced necrosis and DNA damage in the colon cells, in polypectomized and colon cancer patients. Synbiotic intervention produced significantly favorable alterations of colorectal cancer biomarkers, including improved composition of colonic bacterial ecosystem, decreased exposure of the epithelium to cytotoxins and genotoxins, and improved mucosa structure (Rafter et al. 2007).

Anticolorectal cancer property of probiotics can manifest via alteration of intestinal microflora metabolic activities. Increased LAB population decreased bacterial enzymes implicated in the synthesis and activation of carcinogens, tumor promoters, and genotoxins (Burns and Rowland 2000). Xenobiotic-metabolizing enzymes include NADPH dehydrogenase (azoreductase), nitroreductase, and β-glucuronidase which are purportedly produced by *Bacteroides*, *Clostridium*, and *Enterobacteriaceae* (Wollowski et al. 2001). *Enterobacteriaceae* such as enteropathogenic *E. coli* (EPEC) has also been associated with the increased risks of colorectal cancer. EPEC has been found able to attach to the gastrointestinal epithelium and forming lesions, causing worn microvilli and adversely altered the cellular cytoskeleton (Frankel et al. 1998). Upon adherence and colonization, EPEC also caused changes in epithelial cell function and changes in paracellular permeability. Philpott et al. (1996) previously reported that the human colonic epithelial cell line T84 form tight junctions and desmosomes. However, once infected with EPEC, these T84 cell lines showed a decrease in transepithelial resistance, increased flux of inert molecules, and modification of the distribution of the tight-junction-associated protein ZO-1. The probiotic yeast, *Saccharomyces boulardii* was found able to preserve the tight-junction structure in infected cells (Fig. 6), and delayed EPEC-induced cell apoptosis (Czerucka et al. 2000).

L. acidophilus-fermented milk product (3×10^{11} lactobacilli per day) administered for 6 weeks in 14 colon cancer patients, resulted in a 14% decrease in β-glucuronidase, increased counts of lactobacilli, and decreased counts of *E. coli* (Lidbeck et al. 1991). Alteration in enzyme activity and metabolite concentration decreased preneoplastic lesions such as aberrant crypt foci (ACF) and tumors in carcinogen-treated rats (Burns and Rowland 2000). Administration of *B. longum* (4×10^8 viable cells/g diet) decreased small ACF induced by colon carcinogen (azoxymethane) in rats by 26% (Rowland et al. 1998). Probiotics also

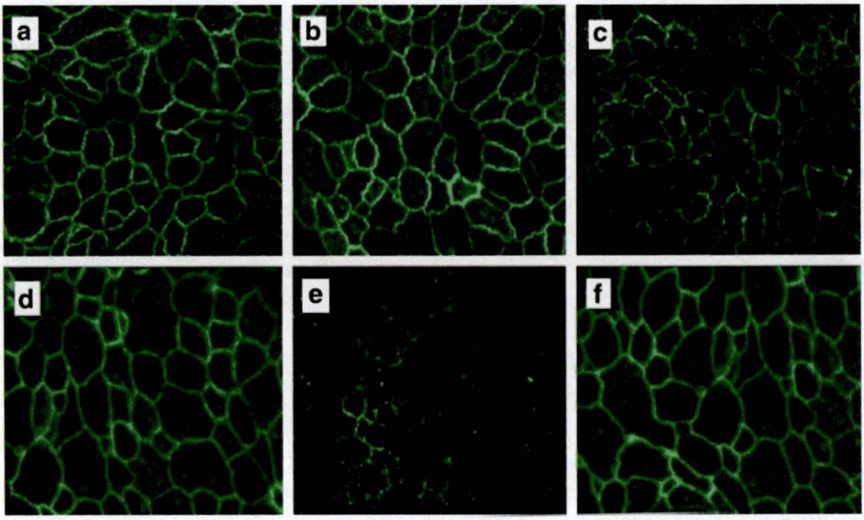

Fig. 6 Effect of *S. boulardii* on T84 cells infected with EPEC. (**a**) and (**b**) are control cells; (**a**) uninfected cells and (**b**) uninfected cells that were exposed to *S. boulardii*. (**c**) ZO-1 disappeared in cells infected for 6 h or (**e**) 9 h with EPEC. ZO-1 distribution was not modified when *S. boulardii* was present during EPEC infection after (**d**) 6 h and (**f**) 9 h (Castagliuolo et al. 1996). Reprinted from Czerucka and Rampal (2002), with permission from Elsevier (License number: 2611150041785)

stimulate protective enzyme – glutathione transferase (GSH) – which inactivates food-borne carcinogen such as heterocyclic amines and polycyclic aromatic hydrocarbons. The activity of GSH in colonic mucosa was inversely related to the ACF numbers in carcinogen-treated rats fed with *B. longum* and lactulose (Challa et al. 1997).

Anticolorectal cancer effect is also achieved by enhancing host's immune response, manifesting in increased phagocytic activity of monocytes and granulocytes, and increased levels of antibody-secreting cells (Burns and Rowland 2000). Yoghurt administration inhibited the development of 1,2-dimethylhydrazine (DMH)-induced colorectal carcinoma in mice (Perdigón et al. 2002), with increased inflammatory cytokines such as tumor necrosis factor (TNF)-α, interferon-γ (IFN-gamma), interleukin-10 (IL-10), capable of inducing apoptosis. Apoptosis provides an innate cellular defense against oncogenesis by eliminating mutated cells that might otherwise progress to malignancy (Le Leu et al. 2010). Antitumor activity in the colon is exerted by apoptosis induction and decreased inflammatory immune response mediated by immunoglobulin IgA(+).

Antitumorigenic effect is achieved via suppression of cell proliferation and angiogenesis in the colon. The extracellular extract of a commercial probiotic, *Bacillus polyfermenticus* inhibited growth of human colon cancer cells including HT-29, DLD-1 and Caco-2 cells, reduced carcinogen-induced colony formation of normal colonocytes, and reduced tumor size in mouse xenograft model of human colon cancer cells (Ma et al. 2010). The inhibition of tumor growth was partially attributed to reduced expression levels of ErbB2 and ErbB3 protein and mRNA, and subsequently their downstream signaling molecules E2F-1 and cyclin D1, which play critical roles in tumor proliferation (Ma et al. 2010).

Food-borne mutagenic compounds such as fungal toxin aflatoxin B1 and heterocyclic amines can be bound by intestinal LAB, thus reducing mutagenicity (Wollowski et al. 2001). Mutagen-binding is attributed to the cation-exchange mechanism, which involves cell wall peptidoglycans and polysaccharides (Hirayama and Rafter 2000). Probiotics have shown species- and dose-dependent protective effects against DNA damage induced by colon carcinogens such as *N*-methyl-*N*-nitro-*N*-nitrosoguanidine (MNNG) and DMH (Wollowski et al. 2001). *L. acidophilus*, *L. gasseri*, *L. confusus*, *B. breve* and *B. longum* strongly inhibited colonic genotoxicity in MNNG- and DMH-treated rats (Pool-Zobel et al. 1996).

A study suggests a new anticolorectal cancer mechanism via the production of conjugated linoleic acid (CLA). Selected probiotics such as VSL#3 could convert linoleic acid to CLA, which exerted anticarcinogenic effects in animal models. Conditioned medium containing probiotic-produced CLA reduced viability and induced apoptosis of HT-29 and Caco-2 colon cancer cells (Ewaschuk et al. 2006). Although there is limited information on this mechanism, the positive results warrant further research to provide more evidence on the significance and specificity of different probiotics strains, especially in human models.

8 Postoperative Gut Maintenance

Various postoperative complications such as bacteremia, infections, and recurrence of diseases are attributed to bacteria overgrowth and translocation, loss of intestinal epithelial integrity, and compromised immunological defense in patients after gut surgeries. Probiotics have been reported as potential candidates in combating these postoperative complications via inhibition of pathogen growth and modulation of the immune system.

8.1 Bariatric Surgery

Bariatric surgery refers to gastrointestinal bypass surgery for effective treatment of morbid obesity in patients with body mass index (BMI) exceeding 40. Bariatric surgery, which includes gastric transection, Roux-en-Y gastric bypass (RNYGB), vertical gastric division and vertical-banded gastroplasty (Fig. 7), uses restriction in stomach size and intestinal malabsorption to achieve substantial weight loss (Gratz et al. 2010). However, gastroenterological surgery and subsequent antibiotic therapy can significantly influence gut microflora balance, causing infectious complications, bacterial overgrowth, and dysmotility (Folwarski and Dobosz 2010). Postoperative effects of bariatric surgery such as creation of a blind pouch,

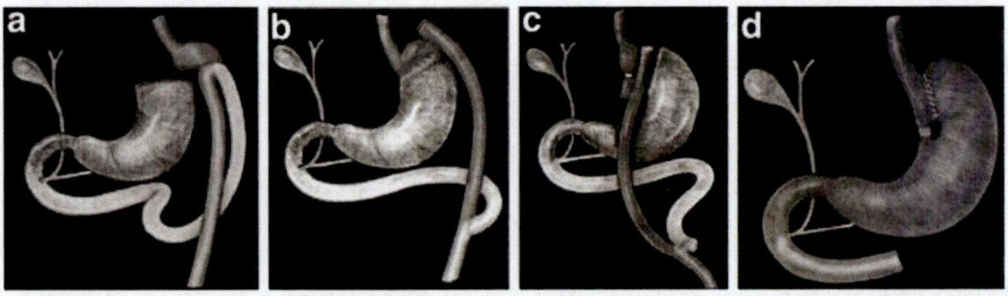

Fig. 7 Types of bariatric surgery: (**a**) Gastric transection with loop gastrojejunostomy; (**b**) horizontal gastric stapling with Roux-en-Y gastrojejunostomy; (**c**) vertical gastric division with interposed Roux gastrojejunostomy and proximal silastic ring; (**d**) vertical-banded gastroplasty. Reprinted from Buchwald and Buchwald (2002) with permission from Springer (License number: 2556360622753)

decreased gastric acid, declined intestinal motility, and Vitamin B12 malabsorption can contribute to bacterial overgrowth as the stasis of digestive material and unbound vitamin is catabolized by intestinal bacteria (Woodard et al. 2009). Bariatric surgery could also unmask previously unidentified lactose intolerance, which can be alleviated with probiotic administration.

Complications of bariatric surgery include abdominal pain, diarrhea, gastrointestinal bleeding, or wound infections (Lee et al. 2007), which are alleviated by restoring the natural bacterial milieu toward the normal state (ASBS Public/Professional Education Committee 2008). Woodard et al. (2009) conducted a double-blinded, randomized, controlled study involving 44 patients undergoing RNYGB. Daily probiotic supplementation (2.4 billion live cells of *Lactobacillus* species per pill) for 6 months postoperatively resulted in significantly lower bacterial overgrowth, greater percent excess weight loss, and significantly higher postoperative vitamin B12 levels.

8.2 *Short Bowel Syndrome*

Short bowel syndrome (SBS) is a malabsorption disorder, presented inborn or after surgical resection of a significant part of the intestine (Uchida et al. 2004). Bacterial translocation is often associated with SBS due to intestinal dysmotility, bacterial overgrowth, and immunocompetent intestinal tissue loss (Eizaguirre et al. 2002).

Probiotics could reduce complications of SBS such as bacterial translocation and D-lactic acidosis via production of antimicrobial substances and metabolites, competitive inhibition of bacteria and immunostimulation. A randomized and placebo-controlled trial evaluating the effect of probiotics after 80% gut resection in 128 Wistrar rats showed that administration of *Bifidobacterium lactis* (7.8×10^8 CFU/day) reduced the risk of bacterial translocation by 43% compared to control (Eizaguirre et al. 2002). The administration of *Lactobacillus GG* (5×10^8 CFU/day) decreased bacterial translocation by 20% in the liver and peripheral blood compared to the control, in 30 male Sprague–Dawley rats after

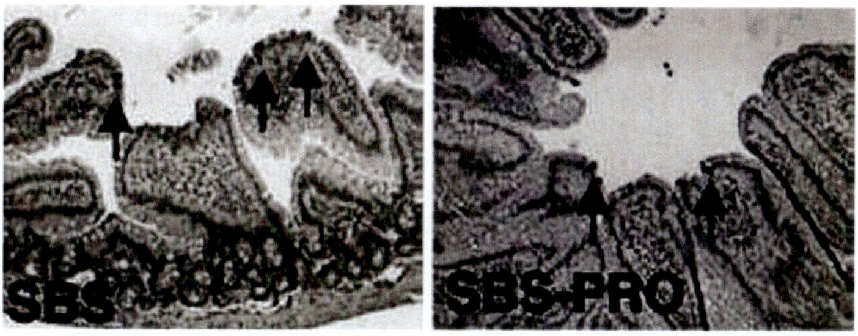

Fig. 8 Effect of bowel resection and treatment with *Lactobacillus GG* on enterocyte apoptosis in ileum. Immunohistochemistry slides show enterocyte apoptosis (*arrows*). SBS: short bowel syndrome male Sprague–Dawley rats; SBS-PRO: short bowel syndrome male Sprague–Dawley rats treated with *Lactobacillus GG* (5×10^8 CFU/day) after bowel resection. Reprinted from Mogilner et al. (2007) with permission from Elsevier (License number: 2556350743796)

bowel resection (Mogilner et al. 2007). Probiotics reportedly decreased apoptosis of enterocytes in the ileum (Fig. 8), and contributed a decrease in the apoptotic index of jejunum, all indicating a stimulating effect on gut regrowth (Mogilner et al. 2007).

8.3 Other Postoperative Complications

Patients undergoing other gut surgeries such as biliary cancer surgery and pancreaticoduodenectomy frequently suffer from postoperative infections (Pitsouni et al. 2009). Viable probiotics and galacto-oligosaccharides can achieve a significant reduction in infectious complications after high-risk hepatectomy (Kanazawa et al. 2005) by improving intestinal microbial imbalance induced by surgical stress, thus lowering postoperative infection rate. Pancreatic resection has relatively high postoperative infection rates caused by gut bacteria especially *Enterococci* and *Escherichia coli*. Rayes et al. (2007) showed that thc intervention of probiotics with fibers significantly decreased the incidence of postoperative bacterial nosocomial infections among patients undergoing pancreas resection.

Curative intestinal surgeries such as ileostomy and ileoanal anastomosis in the treatment of chronic Crohn's disease and ulcerative colitis have high postoperative recurrence rates of ileal Crohn's disease (40–90%) and the development of pouchitis, respectively (Gionchetti et al. 2003a). Probiotics have been suggested as a prophylactic agent in view of the role of endogenous intestinal microflora in the pathogenesis of these diseases and the efficacy of probiotics in restoration of healthy gut microbial population.

Combination of probiotic cocktail (VSL#3) and antibiotics was effective in lowering the incidence of endoscopic recurrence of Crohn's disease after surgical resection (Campieri et al. 2000). A randomized, double-blind, placebo-controlled trial illustrated the efficiency of VSL#3 in the prevention of postoperative occurrence of pouchitis (10% of probiotic-supplemented group versus 40% of placebo group, $p < 0.01$) in 40 ulcerative colitis patients after ileo-pouch anal anastomosis (Gionchetti et al. 2003b). Daily administration of *Lactobacillus GG* decreased postoperative occurrence of pouchitis in an open-label, controlled study involving 117 patients after ileal pouch-anal anastomosis (7% in probiotics group versus 29% in historical controls, $p = 0.01$) (Gosselink et al. 2004).

9 Conclusions

To this end, past experimental data and clinical trials have provided strong evidences to support the positive influences of probiotics on gut health. Probiotics have been shown to alleviate diarrhea, lactose intolerance, and reduce postoperative complications. Clinical trials have also shown that probiotics improved symptoms of IBS and IBD. Antibacterial activities of probiotics effectively reduce putative pathogens and infectious disease in the gastrointestinal tract. In vitro and in vivo research have also revealed many possible mechanisms for probiotics' anti-colorectal cancer properties and postoperative gut health improvement. These strengthened the suggestions that probiotics could possibly serve as an alternative therapy to the current available treatment regimes in enhancing gastrointestinal health. However, as demonstrated in various clinical trials, the efficacy of probiotics varies with different probiotics strains, formulation, dosage, and subjects. Thus, further controlled studies and clinical trials are required to elucidate the exact underlying mechanisms and establish the specific effective dosage.

Acknowledgement The authors would like to acknowledge the USM RU grants (1001/PTEKIND/815056, 1001/PTEKIND/811141, 1001/PTEKIND/811089, 1001/PFARM/843004, 1001/PFARMASI/843071) and the USM fellowship for supporting this work.

References

American Cancer Society (2008) Colorectal cancer facts & figures 2008-2010. American Cancer Society, Atlanta

American Society for Bariatric Surgery (ASBS) Public/Professional Education Committee (2008) Bariatric surgery: postoperative concerns. http://www.asbs.org/html/pdf/asbs_bspc.pdf. Cited 19 Nov 2010

Balcazar JL, Vendrell D, de Blas I, Ruiz-Zarzuela I, Muzquiz JL, Girones O (2008) Characterization of probiotic properties of lactic acid bacteria isolated from intestinal microbiota of fish. Aquaculture 278:188–191

Bampton PA, Dinning PG, Kennedy ML, Lubowski DZ, Cook IJ (2002) The proximal colonic motor response to rectal mechanical and chemical stimulation. Am J Physiol Gastrointest Liver Physiol 282:G443–G449

Bernardeau M, Gueguen M, Smith DGE, Corona-Barrera E, Vernoux JP (2009) In-vitro antagonistic activities of *Lactobacillus* spp. against *Brachyspira hyodysenteriae* and *Brachyspira pilosicoli*. Vet Microbiol 138:184–190

Bernet M-F, Brassart D, Neeser J-R, Servin AL (1993) Adhesion of human bifidobacterial strains to cultured human intestinal epithelial cells and inhibition of enteropathogen-cell interactions. Appl Environ Microbiol 59:4121–4128
Boudraa G, Benbouabdellah M, Hachelaf W, Boisset M, Desjeux JF, Touhami M (2001) Effect of feeding yogurt versus milk in children with acute diarrhea and carbohydrate malabsorption. J Pediatr Gastroenterol Nutr 33:307–313
Buchwald H, Buchwald JN (2002) Evolution of operative procedures for the management of morbid obesity 1950–2000. Obesity Surgery 12:705–717
Burns AJ, Rowland IR (2000) Anti-carcinogenicity of probiotics and prebiotics. Curr Issues Intest Microbiol 1(1):13–24
Campieri M, Rizzello F, Venturi A, Poggioli G, Ugolini F, Helwig U, Amadini C, Romboli E, Gionchetti P (2000) Combination of antibiotic and probiotic treatment is efficacious in prophylaxis of post-operative recurrence of Crohn's disease: a randomized controlled study vs mesalazine. Gastroenterol A118:A781
Castagliuolo I, Lamont JT, Nikulasson ST, Pothoulakis C (1996) *Saccharomyces boulardii* protease inhibits *Clostridium diffcile* toxin A effects in the rat ileum. Infect Immun 64:5225–5232
Castagliuolo I, Riegle MF, Valenick L, LaMont JT, Pothoulakis C (1999) *Saccharomyces boulardii* protease inhibits the effects of *Clostridium difficile* toxins A and B in human colonic mucosa. Infect Immun 67:302–307
Challa A, Rao DR, Chawan CB, Shackelford L (1997) *Bifidobacterium longum* and lactulose suppress azoxymethane-induced colonic aberrant crypt foci in rats. Carcinogenesis 207:517–521
Czerucka D, Rampal P (2002) Experimental effects of *Saccharomyces boulardii* on diarrheal pathogens. Microb Infect 4:733–739
Czerucka D, Dahan S, Mograbi B, Rossi B, Rampal P (2000) *Saccharomyces boulardii* preserves the barrier function and modulates the transduction pathway induced in Enteropathogenic *Escherichia coli*-infected T84 cells. Infect Immun 68:5998–6004
de Vrese M, Marteau PR (2007) Probiotics and prebiotics: effects on diarrhea. J Nutr 137:803S–811S
de Vrese M, Stegelmann A, Richter B, Fenselau S, Laue C, Schrezenmeir J (2001) Probiotics-compensation for lactase insufficiency. Am J Clin Nutr 73:421S–429S
Del Re B, Sgorbati B, Miglioli M, Palenzona D (2000) Adhesion, autoaggregation and hydrophobicity of 13 strains of *Bifidobacterium longum*. Lett Appl Microbiol 31:438–442
Delia P, Sansotta G, Donato V, Messina G, Frosina P, Pergolizzi S, de Renzis C (2002) Prevention of radiation-induced diarrhea with the use of VSL#3, a new high-potency probiotic preparation. Am J Gastroenterol 97(8):2150–2152
Delia P, Sansotta G, Donato V, Frosina P, Messina G, De Renzis C, Famularo G (2007) Use of probiotics for prevention of radiation-induced diarrhea. World J Gastroenterol 13(6):912–915
Eizaguirre I, Urkia GN, Asensio AB, Zubillaga I, Zubillaga P, Vidales C, Garcia-Arenzana JM, Aldazabal P (2002) Probiotic supplementation reduces the risk of bacterial translocation in experimental short bowel syndrome. J Pediatr Surgery 37:699–702
Ewaschuk JB, Walker JW, Diaz H, Madsen KL (2006) Nutrient physiology, metabolism, and nutrient-nutrient interactions bioproduction of conjugated linoleic acid by probiotic bacteria occurs in-vitro and in-vivo in mice. J Nutr 136:1483–1487
Farthing MJ, Kelly P (2007) Infectious diarrhea. Med 35(5):251–256
Folwarski M, Dobosz M (2010) Probiotics in surgery. Polski Przegląd Chirurgiczny 82 (8):489–495
Frankel G, Phillips AD, Rosenshine I, Dougan G, Kaper JB, Knutton S (1998) Enteropathogenic and enterohaemorrhagic *Escherichia coli*: more subsertive elements. Mol Microbiol 30: 911–921
Galvez A, Abriouel H, Lopez RL, Omar NB (2007) Bacteriocin-based strategies for food biopreservation. Int J Food Microbiol 120:51–70

Gionchetti P, Rizzello F, Venturi A, Brigidi P, Matteuzzi D, Bazzocchi G, Poggioli G, Miglioli M, Campieri M (2000) Oral bacteriotherapy as maintenance treatment in patients with chronic pouchitis: a double-blind, placebo-controlled trial. Gastroenterol 119(2):305–309

Gionchetti P, Amadini C, Rizzello F, Venturi A, Poggioli G, Campieri M (2003a) Probiotics for the treatment of postoperative complications following intestinal surgery. Best Pract Res Clin Gastroenterol 17(5):821–831

Gionchetti P, Rizzello F, Helwig U, Venturi A, Lammers KM, Brigidi P, Vitali B, Poggioli G, Miglioli M, Campieri M (2003b) Prophylaxis of pouchitis onset with probiotic therapy: a double-blind, placebo-controlled trial. Gastroenterol 124(5):1202–1209

Golowczyc MA, Mobili P, Garrote GL, Abraham AG, De Antoni GL (2007) Protective action of *Lactobacillus* kefir carrying S-layer protein against *Salmonella enterica* serovar enteritidis. Int J Food Microbiol 118:264–273

Gosselink MP, Schouten W, van Lieshout LM, Hop WC, Laman JD, Ruseler-van Embden JG (2004) Delay of the first onset of pouchitis by oral intake of the probiotic strain *Lactobacillus rhamnosus GG*. Dis Colon Rectum 47(6):876–884

Gratz SW, Mykkanen H, El-Nezami HS (2010) Probiotics and gut health: a special focus on liver diseases. World J Gastroenterol 16(4):403–410

Guandalini S, Pensabene L, Zikri MA, Dias JA, Casali LG, Hoekstra H, Kolacek S, Massar K, Micetic-Turk D, Papadopoulou A, de Sousa JS, Sandhu B, Szajewska H, Weizman Z (2000) *Lactobacillus GG* administered in oral rehydration solution to children with acute diarrhea: a multicenter European trial. J Pediatr Gastroenterol Nutr 30:54–60

Guarino A, Canani RB, Spagnuolo MI, Albano F, Di Benedetto L (1997) Oral bacterial therapy reduces the duration of symptoms and of viral excretion in children with mild diarrhea. J Pediatr Gastroenterol Nutr 25(5):516–519

Guslandi M, Mezzi G, Sorghi M, Testoni PA (2000) *Saccharomyces boulardii* in maintenance treatment of Crohn's disease. Dig Dis Sci 45(7):1462–1464

Hickson M, D'Souza AL, Muthu N, Rogers TR, Want S, Rajkumar C, Bulpitt C (2007) Use of probiotic *Lactobacillus* preparation to prevent diarrhea associated with antibiotics: randomized double blind placebo controlled trial. Br Med J 335(7610):80–83

Hirayama K, Rafter J (2000) The role of probiotic bacteria in cancer prevention. Microb Infection 2(6):681–686

Honda H, Kataoka F, Nagaoka S, Kawai Y, Kitazawa H, Itoh H, Kimura K, Taketomo N, Yamazaki Y, Tateno Y, Saito T (2007) β-galactosidase, phospho- β-galactosidase and phosphor- β-glucosidase activities in lactobacilli strains isolated from human faeces. Lett Appl Microbiol 45:461–466

Huang JS, Bousvaros A, Lee JW, Diaz A, Davidson EJ (2002) Efficacy of probiotic use in acute diarrhea in children: a meta-analysis. Digest Dis Sci 47(11):2625–2634

Hutt P, Shchepetova J, Loivukene K, Kullisaar T, Mikelsaar M (2006) Antagonistic activity of probiotic lactobacilli and bifidobacteria against entero- and uropathogens. J Appl Microbiol 100:1324–1332

Jiang T, Savaiano DA (1997) In-vitro lactose fermentation by human colonic bacteria is modified by *Lactobacillus acidophilus* supplementation. J Nutr 127:1489–1495

Kaila M, Isolauri E, Soppi E, Virtanen E, Laine S, Arvilommi H (1992) Enhancement of the circulating antibody secreting cell response in human diarrhea by a human *Lactobacillus* strain. Pediatr Res 32(2):141–144

Kanazawa H, Nagino M, Kamiya S, Komatsu S, Mayumi T, Takagi K, Asahara T, Nomoto K, Tanaka R, Nimura Y (2005) Synbiotic reduce postoperative infectious complications: a randomized controlled trial in biliary cancer patients undergoing hepatectomy. Langenbecks Arch Surg 390:104–113

Karska-Wysocki B, Bazo M, Smoragiewicz W (2010) Antibacterial activity of *Lactobacillus acidophilus* and *Lactobacillus casei* against methicillin-resistant *Staphylococcus aureus* (MRSA). Microbiol Res 165(8):674–686

Kato K, Mizuno S, Umesaki Y, Ishii Y, Sugitani M, Imaoka A, Otsuka M, Hasunuma O, Kurihara R, Iwasaki A, Arakawa Y (2004) Randomized placebo-controlled trial assessing the effect of bifidobacteria-fermented milk on active ulcerative colitis. Aliment Pharmacol Ther 20(10):1133–1141

Kim H-S, Park H, Cho I-Y, Paik H-D, Park E (2006) Dietary supplementation of probiotic *Bacillus polyfermenticus*, Bispan strain, modulates natural killer cell and T cell subset populations and Immunoglobulin G levels in human subjects. J Med Food 9(3):321–327

Kotowska M, Albrecht P, Szajewska H (2005) *Saccharomyces boulardii* in the prevention of antibiotic-associated diarrhea in children: a randomized double-blind placebo-controlled trial. Aliment Pharmacol Ther 21:583–590

Kruis W, Fric P, Pokrotnieks J, Lukas M, Fixa B, Kascak M, Kamm MA, Weismueller J, Beglinger C, Stolte M, Wolff C, Schulze J (2004) Maintaining remission of ulcerative colitis with the probiotic *Escherichia coli* Nissle 1917 is as effective as with standard mesalazine. Gut 53 (11):1617–1623

Le Leu RK, Hu Y, Brown IL, Woodman RJ, Young GP (2010) Synbiotic intervention of *Bifidobacterium lactis* and resistant starch protects against colorectal cancer development in rats. Carcinogenesis 31(2):246–251

Lee MC, Lin LH, Hung KL, Wu HY (2001) Oral bacterial therapy promotes recovery from acute diarrhea in children. Acta Paediatr Taiwan 42(5):301–305

Lee CW, Kelly JJ, Wassef WY (2007) Complications of Bariatric surgery. Curr Opin Gastroenterol 23(6):636–643

Lidbeck A, Geltner-Allinger U, Orrhage KM, Ottava L, Brismar B, Gustafsson JA, Rafter JJ, Nord CE (1991) Impact of *Lactobacillus acidophilus* supplements on the faecal microflora and soluble faecal bile acids in colon cancer patients. Microbial Ecol Health Dis 38:81–88

Lima ET, Andreatti Filho RL, Okamoto AS, Noujaim JC, Barros MR, Crocci AJ (2007) Evaluation in vitro of the antagonistic substances produced by *Lactobacillus* spp. isolated from chickens. Can J Vet Res 71:103–107

Liong MT (2008) Roles of probiotics and prebiotics in colon cancer prevention: postulated mechanisms and in-vivo evidence. Int J Mol Sci 9:854–863

Lomer MCE, Parkers GC, Sanderson JD (2008) Review article: lactose intolerance in clinical practice-myths and realities. Aliment Pharmacol Ther 27:93–103

Longstreth GF, Thompson WG, Chey WD, Houghton LA, Mearin F, Spiller RC (2006) Functional bowel disorders. Gastroenterology 130:1480–1491

Lye HS, Kuan CY, Ewe JA, Fung WY, Liong MT (2009) The improvement of hypertension by probiotics: effects on cholesterol, diabetes, renin and phytoestrogens. Int J Mol Sci 10:3755–3775

Ma EL, Choi YJ, Choi J, Pothoulakis C, Rhee SH, Im E (2010) The anticancer effect of probiotic *Bacillus polyfermenticus* on human colon cancer cells is mediated through ErbB2 and ErbB3 inhibition. Int J Cancer 127:780–790

Marteau P (2000) Inflammatory bowel disease. Endoscopy 32:131–137

Marteau PR, de Vrese M, Cellier CJ, Schrezenmeir J (2001) Protection from gastrointestinal diseases with the use of probiotics. Am J Clin Nutr 73:430S–436S

McFarland LV (2006) Meta-analysis of probiotics for prevention of antibiotic associated diarrhea and treatment of *Clostridium difficile* disease. Am J Gastroenterol 101:812–822

McFarland LV (2009) Evidence-based review of probiotics for antibiotic-associated diarrhea and *Clostridium difficile* infections. Anaerobe 15:274–280

McFarland LV, Dublin S (2008) Meta-analysis of probiotics for the treatment of irritable bowel syndrome. World J Gastroenterol 14(17):2650–2661

Mimura T, Rizzello F, Helwig U, Poggioli G, Schreiber S, Talbot IC, Nicholls RJ, Gionchetti P, Campieri M, Kamm MA (2004) Once daily high dose probiotic therapy (VSL#3) for maintaining remission in recurrent or refractory pouchitis. Gut 53(1):108–114

Moayyedi P, Ford AC, Talley NJ, Cremonini F, Foxx-Orenstein AE, Brandt LJ, Quigley EMM (2010) The efficacy of probiotics in the treatment of irritable bowel syndrome: a systematic review. Gut 59:325–332

Mogilner JG, Srugo I, Lurie M, Shaoul R, Coran AG, Shiloni E, Sukhotnik I (2007) Effect of probiotics on intestinal regrowth and bacterial translocation after massive small bowel resection in a rat. J Pediatr Surgery 42:1365–1371

Moon G-S (2008) A preliminary analysis of secreted proteins from *Bifidobacterium pseudocatanulatum* BP1 by two-dimensional gel electrophoresis. J Food Sci Nutr 13:366–369

Muthaiyan A, Limayem A, Ricke SC (2010) Antimicrobial strategies for limiting bacterial contaminants in fuel bioethanol fermentations. Progr Energ Combust Sci doi:10.1016/j.pecs.2010.06.005

O'Mahony L, McCarthy J, Kelly P, Hurley G, Luo F, Chen K, O'Sullivan GC, Kiely B, Collins JK, Shanahan F, Quigley EM (2005) *Lactobacillus* and *bifidobacterium* in irritable bowel syndrome: symptom responses and relationship to cytokine profiles. Gastroenterol 128 (3):541–551

O'Sullivan DJ (2000) Bifidobacteria and siderophores produced thereby and methods of use. US 2004/0191233 A1 U.S. Provisional Application Serial No. 60/212,273, filed 19 June, 2000

Pelletier X, Laure-Boussuge S, Donazzolo Y (2001) Hydrogen excretion upon ingestion of dairy products in lactose intolerant male subjects: importance of the live flora. Eur J Clin Nutr 55:509–512

Perdigón G, de Moreno de LeBlanc A, Valdez J, Rachid M (2002) Role of yoghurt in the prevention of colon cancer. Eur J Clin Nutr 56(Suppl 3):S65–S68

Philpott DJ, McKay DM, Sherman PM, Perdue MH (1996) Infection of T84 cells enteropathogenic *Escherichia coli* alters barrier and transport functions. Am J Physiol 270:G634–G645

Pitsouni E, Alexiou V, Saridakis V, Peppas G, Falagas ME (2009) Does the use of probiotics/sunbiotics prevent postoperative infections in patients undergoing abdominal surgery? A meta analysis of randomized controlled trials. Eur J Clin Pharmacol 65:561–570

Plein K, Hotz J (1993) Therapeutic effects of *Saccharomyces boulardii* on mild residual symptoms in a stable phase of Crohn's disease with special respect to chronic diarrhea – a pilot study. Z Gastroenterol 31:129–134

Pool-Zobel BL, Neudecker C, Domizlaff I, Ji S, Schillinger U, Rumney C, Moretti M, Vilarini I, Scassellati-Sforzolini R, Rowland I (1996) *Lactobacillus*- and *bifidobacterium*-mediated antigenotoxicity in the colon of rats. Nutr Cancer 26(3):365–380

Pronio A, Montesani C, Butteroni C, Vecchione S, Mumolo G, Vestri A, Vitolo D, Boirivant M (2008) Probiotic administration in patients with ileal pouch-anal anastomosis for ulcerative colitis is associated with expansion of mucosal regulatory cells. Inflamm Bowel Dis 14(5):662–668

Rafter J, Bennett M, Caderni G, Clune Y, Hughes R, Karlsson PC, Klinder A, O'Riordan M, O'Sullivan GC, Pool-Zobel B, Rechkemmer G, Roller M, Rowland I, Salvadori M, Thijs H, Loo JV, Watzl B, Collins JK (2007) Dietary synbiotics reduce cancer risk factors in polypectomized and colon cancer patients. Am J Clin Nutr 85(2):488–496

Rayes N, Seehofer D, Theruvath T, Mogl M, Langrehr JM, Nüssler NC, Bengmark S, Neuhaus P (2007) Effect of enteral nutrition and synbiotics on bacterial infection rates after pylorus-preserving pancreatoduodenectomy. Ann Surg 246(1):36–41

Rolfe RD (2000) The role of probiotic cultures in the control of gastrointestinal health. J Nutr 130:396S–402S

Rosenfeldt V, Michaelsen KF, Jakobsen M, Larsen CN, Møller PL, Pedersen P, Tvede M, Weyrehter H, Valerius NH, Paerregaard A (2002) Effect of probiotic *Lactobacillus* strains in young children hospitalized with acute diarrhea. Pediatr Infect Dis J 21:411–416

Rowland IR, Rumney CJ, Coutts JT, Lievense LC (1998) Effect of *Bifidobacterium longum* and inulin on gut bacterial metabolism and carcinogen-induced aberrant crypt foci in rats. Carcinogenesis 19:281–285

Samanya M, Yamauchi KE (2002) Histological alterations of intestinal villi in chickens fed dried *Bacillus subtilis* var. natto. Comp Biochem Physiol A Mol Integr Physiol 133:95–104
Sartor RB (2004) Therapeutic manipulation of the enteric microflora in inflammatory bowel diseases: antibiotics, probiotics, and prebiotics. Gastroenterol 126:1620–1633
Sen S, Mullan MM, Parker TJ, Woolner JT, Tarry SA, Hunter JO (2002) Effect of *Lactobacillus plantarum* 299v on colonic fermentation and symptoms of irritable bowel syndrome. Dig Dis Sci 47(11):2615–2620
Shornikova AV, Casas IA, Isolauri E, Mykkanen H, Vesikari T (1997) *Lactobacillus reuteri* as a therapeutic agent in acute diarrhea in young children. J Pediatr Gastroenterol Nutr 24(4):399–404
Simakachorn N, Pichaipat V, Rithipornpaisarn P, Kongkaew C, Tongpradit P, Varavithya W (2000) Clinical evaluation of the addition of lyophilized, heat-killed *Lactobacillus acidophilus* LB to oral rehydration therapy in the treatment of acute diarrhea in children. J Pediatr Gastroenterol Nutr 30(1):68–72
Sood A, Midha V, Makharia GK, Ahuja V, Singal D, Goswami P, Tandon RK (2009) The probiotic preparation, VSL#3 induces remission in patients with mild-to-moderately active ulcerative colitis. Clin Gastroenterol Hepatol 7(11):1202–1209
Surawicz CM (2003) Probiotics, antibiotic-associated diarrhea and *Clostridium difficile* diarrhea in humans. Best Pract Res Clin Gastroenterol 17(5):775–783
Tursi A, Brandimarte G, Papa A, Giglio A, Elisei W, Giorgetti GM, Forti G, Morini S, Hassan C, Pistoia MA, Modeo ME, Rodino' S, D'Amico T, Sebkova L, Sacca' N, Di Giulio E, Luzza F, Imeneo M, Larussa T, Di Rosa S, Annese V, Danese S, Gasbarrini A (2010) Treatment of relapsing mild-to-moderate ulcerative colitis with the probiotic VSL#3 as adjunctive to a standard pharmaceutical treatment: a double-blind, randomized, placebo-controlled study. Am J Gastroenterol 105(10):2218–2227
Uchida H, Yamamoto H, Kisaki Y, Fujino J, Ishimaru Y, Ikeda H (2004) D-lactic acidosis in short bowel syndrome managed with antibiotics and probiotics. J Pediatr Surgery 39:634–636
Urbancsek H, Kazar T, Mezes I, Neumann K (2001) Results of a double-blind, randomized study to evaluate the efficacy and safety of Antibiophilus® in patients with radiation-induced diarrhea. Eur J Gastroenterol Hepatol 13:391–396
Vanderhoof JA, Whitney DB, Antonson DL, Hanner TL, Lupo JV, Young RJ (1999) *Lactobacillus GG* in the prevention of antibiotic-associated diarrhea in children. J Pediatr 135(5):564–568
Verdu EFBP, Bergonzelli GE, Huang XX, Blennerhasset P, Rochat F, Fiaux M, Mansourian R, Corthésy-Theulaz I, Collins SM (2004) *Lactobacillus paracasei* normalizes muscle hypercontractility in a murine model of postinfective gut dysfunction. Gastroenterol 127(3):826–837
Verdu EFBP, Verma-Gandhu M, Huang XX, Blennerhassett P, Jackson W, Mao Y, Wang L, Rochat F, Collins SM (2006) Specific probiotic therapy attenuates antibiotic induced visceral hypersensitivity in mice. Gut 55(2):182–190
Vesa TH, Marteau P, Korpela R (2000) Lactose intolerance. J Am Coll Nutr 19:165S–175S
Voravuthikunchai SP, Bilasoi S, Supamala O (2006) Antagonistic activity against pathogenic bacteria by human vaginal lactobacilli. Anaerobe 12:221–226
Wollowski I, Rechkemmer G, Pool-Zobel BL (2001) Protective role of probiotics and prebiotics in colon cancer. Am J Clin Nutr 73(2):451S–455S
Woodard GA, Encarnacion B, Downey JR, Peraza J, Chong K, Hernandez-Boussard T, Morton JM (2009) Probiotics improve outcomes after Roux-En-Y Gastric Bypass surgery: a prospective randomized trial. J Gastrointest Surg 13(7):1198–1204
Zhang Y, Liu Y, Bao Y, Zhang H-P (2010a) Influence of pH, heat and enzymatic treatments on the activity of antibacterial substance in MRS and milk media produced by *Lactobacillus fermentum* F6. Agric Sci China (ASC) 9:911–920
Zhang Y-C, Zhang L-W, Tuo Y-F, Guo C-F, Yi H-X (2010b) Inhibition of *Shigella sonnei* adherence to HT-29 cells by lactobacilli from Chinese fermented food and preliminary characterization of S-layer protein involvement. Res Microbiol 161(8):667–672

New Health Potentials of Orally Consumed Probiotic Microorganisms

Vivekananda Mandal and Narayan C. Mandal

Contents

1 Introduction 168
2 Immunomodulatory Effects of Probiotic 169
2.1 Mechanism of Immunostimulatory Activity of Probiotics 170
3 Hypocholesteromic Effects of Probiotics 172
3.1 Some Proof of Concepts on Hypocholesteromic Probiotic Study 173
3.2 Mechanisms of Action 173
4 Antihypertensive Properties of Probiotics 177
4.1 Antihypertensive Properties of Probiotic Organisms: Some Proof of Concepts ... 178
5 Alleviation of Postmenopausal Symptoms 178
5.1 Treatment of Menopause: Alternative Medicines and Prebiotics 179
5.2 Probiotics for Treatment and Prophylaxis to Postmenopausal Women 179
5.3 Mechanisms of Actions of Probiotics 180
6 Antiallergic Effects of Probiotics 181
6.1 Probiotics Prophylaxis and Management to Allergic Diseases 181
6.2 Mechanisms of Antiallergic Probiotic Action 182
7 Protection Against Lung Emphysema 184
7.1 Probiotics Prophylaxis for Lung Emphysema and Other Lung Complications 184
8 Conclusions 185
References 185

V. Mandal
Plant Physiology and Biochemistry Laboratory, Post Graduate Department of Botany, Darjeeling Government College, Darjeeling 734101, West Bengal, India
e-mail: mandalvivek@gmail.com

N.C. Mandal (✉)
Mycology and Plant Pathology Laboratory, Microbiology Division, Department of Botany, Siksha Bhavan, Visva-Bharati University, Santiniketan 731235, West Bengal, India
e-mail: mandalnc@rediffmail.com

M.-T. Liong (ed.), *Probiotics*, Microbiology Monographs 21,
DOI 10.1007/978-3-642-20838-6_7,

Abstract The microorganism and their live formulations upon consumption contribute to intestinal microbial balance are known as probiotics. They generally live in our gut mucosal layer and also in some other body parts like mouth, vagina and offer a number of health benefits to the host. They fight with the invading pathogens by producing a number of active principles like bacteriocins or by other mechanisms and kill or outnumber them. These probiotic organisms not only fight with the pathogens of the host but also contribute a lot in improving the immunological and physiological state of the host by interfering their metabolic processes. As they colonize in vital parts of human intestine, they intricately involve with different systems of human body and alleviate the problem associated with them. The major probiotic organisms used today are the species of *Lactobacillus*, *Bifidobacterium*, *Enterococcus*, *Pediococcus*, etc. In this chapter how different strains or species of such orally consumed probiotic organisms could offer different kinds of health benefits particularly in relation to modulation of immunological parameters, allergy, and lung emphysema are discussed with very recent references.

1 Introduction

Human body is a good ecological niche for different types of microbes, many of which are either beneficial to the host or commensal taking the body system as shelter for survival. The interaction between these two diverse living organisms is a great ecological scenario. Diverse group of bacteria are the part of this ecological system, each group of microbe has a specific role and site of colonization building a faithful chemistry in-between the microbes and the harboring body systems. For example, *Escherichia coli* is a predominating colonizer in the intestine. Likewise, lactic acid bacteria (LAB) is also a well-known colonizer in different parts of our body system such as mouth cavity, intestine, secretory organs like breast, urogenital tracts, etc. Mucous membranes are the unique environment where different bacterial species are able to survive. About 10^{14} bacteria of 200 species belonging to 40–50 genera of the families *Bacteroides*, *Fusobacterium*, *Butyrivibrio*, *Clostridium*, *Bifidobacterium*, *Eubacterium*, and *Lactobacillus* live on these surfaces. Mucous epithelial surfaces of the gastrointestinal tract and respiratory tract are the areas where the host is confronted with a range of different microorganisms from the outside environment. There they are confronted with both beneficial and harmful microorganisms. Therefore, the host must have some defense mechanisms to combat the harmful effects of the incoming microbes. The integrity of the epithelial layer, the production of mucus, glycolipids, cytoprotective peptides, and antibiotic-like substances are some protection systems of the host. The coexistence of microbes within the body is essential as they play a crucial role in the anatomical, physiological, and immunological development of the host. Therefore, a selection between beneficial and harmful microbes is needed and this vital role is solely controlled by the well-organized immune system of the host.

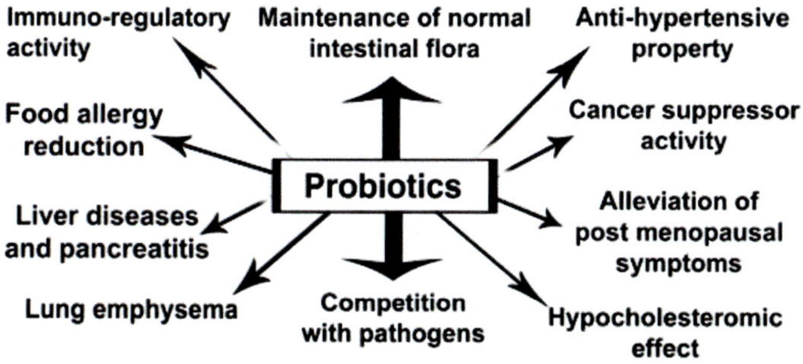

Fig. 1 Schematic presentation of proposed benefits of probiotics uses

To maintain and improve the balance between beneficial and harmful microbes in our body system, a new idea in microbial application has developed in the last few years that is the probiotic therapy, i.e., therapeutic application of potentially beneficial microorganisms. A probiotic has been defined as a live microbial feed supplement that beneficially affects the host by improving its intestinal microbial balance (Fuller 1991). Among the several parameters for probiotic selection, the adherence of probiotic microbes to host epithelial tissue, at least transiently, is one of the prime parameter in selection of probiotic microorganisms because until and unless they survive and adhere to the colonization site it is impossible to reflect the host beneficial effects for long-time basis. Among the probiotic microbes reported to have health beneficial effect, LAB constitute a major fraction though nonlactic bacteria such as the species of *Bacillus* (*Bacillus cereus*, *B. clausii*, *B. pumilus*, and *B. subtilis*) and also yeasts like *Saccharomyces boulardii* and *S. cerevisiae* have also been proved to have probiotic effects on the host (Kimmey et al. 1990; Agarwal et al. 2000; Pinchuk et al. 2001; Le Duc et al. 2004). The great advantage of LAB as probiotic organism is the "GRAS status" (Generally Regarded as Safe) attached with this group of bacteria.

Several scientists have reported the multifaceted beneficial aspects of orally consumed probiotics but the detailed mechanisms are still unclear to us. In this presentation, we have tried to explain the immunomodulatory effects, hypocholesterolemic effects, antihypertensive properties, alleviation of postmenopausal symptoms, antiallergic effects, and protection against lung emphysema activities of the orally consumed probiotics. The multidimensional uses of probiotic microbes is presented schematically in Fig. 1.

2 Immunomodulatory Effects of Probiotic

Probiotic bacteria are claimed to have immunomodulatory activity in the host body. They do this effect either by the stimulation or by the alteration of the normal immune response against the antigens. In general, probiotic bacteria exert their immunomodulatory activity in two ways: nonspecific immunomodulation and specific immunomodulation. Nonspecific immune response constitutes the first line of defense for the host. The cell line of nonspecific immunity is composed of mononuclear phagocytic cells (monocytes, macrophages), polymorphonuclear lymphocytes (PML) (mainly neutrophils), and natural killer (NK) cells. Oral introduction of *Lactobacillus casei* and *Lactobacillus bulgaricus* was found to enhance nonspecific host resistance to microbial pathogens and thereby facilitate the exclusion of pathogens in the gut. Several strains of live LAB have been shown to induce in vitro release of the proinflammatory cytokines, tumor necrosis factor α, and interleukin 6, and also activates the production of macrophages and phagocytosis in mice reflecting the stimulation of nonspecific immunity (Perdigon et al. 1986, 1998). Enhanced phagocytosis was also reported in humans by *Lactobacillus acidophilus* Lal (Schiffrin et al. 1994). Phagocytosis is responsible for early activation of the inflammatory response before antibody production. Thus, probiotic therapy may help stabilize the gut microbial environment and thereby prevent the generation of inflammatory mediators. Oral bacteriotherapy with *Lactobacillus rhamnosus* GG (ATCC 53103) was shown to reduce elevated fecal concentrations of tumor necrosis factor α in patients with atopic dermatitis and cow milk allergy (Majarmaa and Isolauri 1997). It is remarkable that fermented milk products expressed better stimulation of the nonspecific immune system than nonfermented one probably due to immune-active peptides formed during fermentation from milk proteins (Fiat et al. 1993). In a study with rats feeding yoghurt showed a greater in vitro proliferative response, greater interferon-γ production in the lymphoid compartments in response to yoghurt bacteria (Aattouri et al. 2002). Probiotic *L. rhamnosus* GG prevents cytokine-induced apoptosis in two different intestinal epithelial cell models (Yan and Polk 2002).

On the other hand, specific immunomodulatory activity by probiotic bacteria is achieved by modulation of the host's immune responses toward harmful antigens. Oral introduction of *Bifidobacterium bifidum* was shown to enhance antibody response to ovalbumin (Moreau et al. 1990) and *Bifidobacterium breve* was shown to stimulate IgA response to cholera toxin in mice (Yasui et al. 1992). An increase in rotavirus-specific antibody-secreting cells in the IgA class was detected in children with acute rotavirus diarrhea who received *L. rhamnosus* GG during the acute phase of diarrhea (Kaila et al. 1992). Dietary supply of probiotic bacteria stimulates IgA immune response and the transport of target antigens through Peyer's patches (Isolauri et al. 1993). Pessi et al. (1998) showed that the probiotics influence mucous permeability by rebuilding damaged macromolecule transportation and they are able to positively influence the lesions caused by an inflammatory response on the membrane.

2.1 Mechanism of Immunostimulatory Activity of Probiotics

The detailed mechanisms by which LAB affect the immune system and produce immunostimulative effects are still unknown. Probably, LAB alone or their products are absorbed by M cells and transported to deeper lying lymphatic follicles where they are checked by immunocompetent cells. LAB enter the body possibly through unspecific pathways mediated by receptor mechanisms. Interactions of LAB and their products with immunocompetent cells such as macrophages and T cells can lead to the production of cytokines which have a manifold effect on immune and nonimmune cells. Epithelial cells can differentiate between pathogenic and nonpathogenic bacteria, probably by means of the recognition of conserved structure components in the bacteria. Nonpathogenic probiotic bacteria present contact with the epithelial cells and with the immune cells in Peyer's patches of the intestine, the associated immune cells such as monocytes/macrophages and dendritic cells and then initiate the innate immune response against these antigenic stimulations. Galdeano and Perdigon (2006) reported the changes of pattern recognition receptors like cluster of differentiation (CD-206) and toll-like receptor (TLR-2) in the positive cells in the small intestine of mice that fed with or without probiotic *L. casei* treatment. They found that for the animals with *L. casei* administration for 2, 5, or 7 consecutive days, the number of CD-206 cells in the small intestine increased significantly in comparison with the control group. The numbers of TLR-2 cells in the lamina propria of mice were also increased significantly. Macrophages and dendritic cells obtained from Peyer's patches after *L. casei* administration showed an important increase in the number of CD-206 positive cells. This increase in the number of TLR-2 positive cells could indicate an activation of the myeloid dendritic cells (mDC) population in Peyer's patches. T cells in the presence of mDC secrete a large amount of IFN-γ with little IL-4, IL-5, and IL-10 and that, in contrast, when T cells are cultured with plasma cytoid DC, they secrete a large amount of IL-4, IL-5, and IL-10.

The indigenous microflora of gut-associated lymphoid tissue (GALT) helps in immune exclusion and protects the host from the adhesion of pathogens through competition for substrates and places of adhesion. These bacteria produce antibacterial substances and they stimulate the production of specific antibodies. Therefore, disturbances in the ecological balance in the gut lead to the growth of harmful bacteria and their possible translocation to internal organs ultimately inducing diseases. The application of some probiotic strains particularly LAB strains were found to produce enhanced activity of peritoneal and pulmonary macrophages and blood leukocytes; and increased secretion of lysozomal enzymes, increased production of reactive oxygen, nitrogen radicals, and monokines of phagocytic cells (Kaur et al. 2002). Therefore, ingestion of probiotic bacteria may stabilize the immunologic barrier of the gut mucosa by reducing the generation of local proinflammatory tumor necrosis factor α and by reinforcing the systemic production of interferon γ with physiologic protective effects in the gut.

The immunomodulating effects of probiotic LAB depend on some factors like (1) the ability of LAB to influence the immune system differs largely from strain to strain; (2) depends on the dose; (3) depends on the medium of probiotic intake, and also (4) it depends on the state of probiotic organism. At last it could be concluded that the probiotic bacteria would be a good choice to improve the mucosal immune system if the mechanisms through which they work are known.

3 Hypocholesteromic Effects of Probiotics

Cholesterol, the most abundant sterol in animal tissues, occurs in the plasma membranes and in the lipoproteins of blood plasma, is often deposited on the inner wall of blood vessels together with other lipids leading to occlusion of blood vessels, a condition known as atherosclerosis, in the heart and the brain resulting in heart attack and strokes, respectively. Cholesterol in association with bile can also crystallize to form gall stones that may block the bile ducts. Though its negative impact is today's catchy topic, but it is essential to life. The human body contains about 100 g of cholesterol. To a great number of biomolecules cholesterol is the starting ingredient for synthesis of many vital compounds such as progesterone, estrogens, androgens, mineralocorticoids, and is also the precursor of vitamin D. Total cholesterol content in human blood is the sum of high-density lipoproteins (HDLs) cholesterol, low-density lipoproteins (LDLs) cholesterol, and 20% of the triglyceride value. Of these fractions, the major culprit seems to be levels of LDLs that are in excess of the body's needs and fastens arteriosclerosis by accumulating cholesterol in the blood vessel. Most of the rest is carried by so-called HDLs. High HDL values are good for health because HDL cholesterol transports cholesterol from the tissues back to the liver where it is secreted in the bile and thus known to prevent arteriosclerosis by removing cholesterol from blood stream. Because of their relationship to cardiovascular disease, the regular analysis of serum lipids has become an important health measure.

Coronary heart disease (CHD) is one of the major causes of death and disability in industrialized countries. The development of atherosclerosis is strongly correlated with the level of plasma cholesterol. Most of the cholesterol in the tissues of higher organisms is esterified at the 3-hydroxyl group with long-chain fatty acids by the enzyme, cholesterol acyltransferase in the liver or phosphatidyl-cholesterol acyltransferase in the blood plasma. The amount of cholesterol and its esters content of human blood plasma ranges between 150 and 250 mg/100ml. The plasma cholesterol concentration can be regulated by various ways such as by the inhibition of cholesterol biosynthesis from saturated fat, removal of cholesterol from the circulation, absorption of dietary cholesterol, and excretion of cholesterol via bile and feces. Therefore, cellular cholesterol homeostasis is very important for the prevention of cardiovascular disease. Some cholesterol-lowering drugs generally act by blocking its biosynthesis through inhibition of enzymes 3-hydroxy-3-methylglutaryl coenzyme A (HMG-CoA) reductase viz., statins, by binding bile acids with insoluble powders of colestipol-like compounds followed by their elimination through feces, nicotinic acid, and fibric acids. The main drawback to these drugs is that all these have some side effects if they are taken on long-term basis. The current dietary strategies for prevention of CHD implicate adherence to a low-fat/low saturated fat diet. Although such diets may present an effective therapy they are difficult to maintain on a long-term basis. Therefore, new approaches toward the identification of other dietary means of reducing blood cholesterol levels have been evaluated. These include the use of soluble fibers, soy protein, plant sterols, probiotic bacteria, and prebiotic compounds.

For probiotic bacteria to exert hypocholesteromic effect, the strain must have to colonize in the intestine. This survivability is seriously challenged by several parameter of the gastrointestinal tract such as variations in pH, low oxygen levels, nutrient limitation and elevated osmolarity and bile salts. Liver secretes as much as a liter of bile into the intestinal tract each day. It has some antimicrobial activity due to the ability to affect the phospholipids and proteins of cell membranes and disrupt cellular homeostasis. Therefore, the ability of pathogens and commensals to tolerate bile is likely to be important for their survival and subsequent colonization in the gastrointestinal tract. A prebiotic substance is a nondigestible food ingredient that beneficially affects the host by selectively stimulating the growth and/or activity of one or a limited number of bacteria in the colon (Gibson and Roberfroid 1995). To be effective, prebiotics should escape digestion in the upper gut by pancreatic and brush-border enzymes, reach the large bowel, and be utilized selectively by a restricted group of microorganisms that have clearly identified as having health-promoting properties, i.e., probiotic bacteria (Cummings et al. 2001; Roberfroid 2001). In practice a range of dietary nonstarch polysaccharides, resistant starches, undigested sugars, oligosaccharides and proteins are used as prebiotics, and of these it is the nondigested dietary carbohydrates that provide the principal substrates for colonic bacterial growth. Fructooligosaccharide (FOS) such as inulin and oligofructose are naturally occurring indigestible carbohydrates found in many plant foods like chicory, onions and asparagus, and found to have good prebiotics activity (Roberfroid 1996).

3.1 Some Proof of Concepts on Hypocholesteromic Probiotic Study

Shaper et al. (1963) and later on Mann and Spoerry (1974) were the first who observed that men from the tribes of Samburu and Maasai warriors in Africa showed a reduced serum cholesterol after consumption of large amounts of milk fermented with a wild *Lactobacillus* strain. Since then a large number of microbial strains have been evaluated for their hypocholesteromic effects and are now proved to have hypocholesteromic effect both in vitro and in vivo as shown in Table 1. Since every 1% reduction in serum cholesterol concentration is associated with an estimated 2–3% reduction in risk for CHD, therefore, reduction of plasma cholesterol through the biological organisms or products could serve as a viable alternative of reducing risk for CHDs.

3.2 Mechanisms of Action

Based on the results from different probiotic studies four hypothesis have been explained to prove the hypocholesteromic effect of probiotic microorganisms these

Table 1 Hypocholesteromic effects of probiotics and synbiotics both in vitro and in vivo conditions

In vitro studies			
Authors	Probiotic strains	Supplemented with/in the form of	Observations
De Rodas et al. (1996)	*Lactobacillus acidophilus*	–	Hypocholesterolemic effect.
Anderson and Gilliland (1999)	*L. acidophilus* L1	Fermented milk products	2.4% reduction of serum cholesterol concentration.
Kimoto et al. (2002)	*L. lactis* subsp. *lactis* biovar *diacetylactis* N7	–	The cholesterol removal quantity was more by the live cells than that of heat-killed cells.
Pereira and Gibson (2003)	*Lactobacillus fermentum*	In presence of galacto-oligosaccharides	Acetate production decreased by 9–27%, while the propionate and butyrate concentrations increased by 50–90% and 52–157%, respectively.
Psomas et al. (2003)	Eight yeast strains	Growth medium supplemented with 0.3% oxgall	*Saccharomyces cerevisiae* KK1 *S. cerevisiae* 832, and *S. boulardii* and *Isaatchenkia orientalis* KK5.Y.1 removed cholesterol from the growth medium.
Liong and Shah (2005a)	11 strains of lactobacilli	–	Altered fatty acid profiles, especially of hexadecanoic, octadecanoic, total saturated and unsaturated acids.
Liong and Shah (2005b)	*Lactobacillus casei* ASCC 292	Six prebiotics	The synbiotic treatment had removed the highest level of cholesterol.
Mandal et al. (2009)	*Pediococcus acidilactici* LAB 5	*P. acidilactici* LAB 5 with sorbitol	Removed more cholesterol in synbiotics conditions than the control groups.

In vivo studies			
Authors	Animal model	Treatment conditions	Observations
De Rodas et al. (1996)	Placebo-controlled trial involving 33 hypercholesterolemia induced pigs (Yorkshire barrows)	*L. acidophilus* ATCC 43121 (2.5×10^{11} cells per feeding) for 15 days	Pigs showed reduced total blood cholesterol by 11.8% compared to the control.
Taranto et al. (1998)	Swiss Albino mice		

		Fat enriched diet for 7 days with *Lactobacillus reuteri* CRL 1098 (10^4 cells/d)	38% decrease in total cholesterol, triglycerides by 40%, and increased the ratio of HDL to LDL by 20% without bacterial translocation of the native microflora into the spleen and liver.
Kieling et al. (2002)	Randomized, crossover, and placebo-controlled trial involving 29 women	Yoghurt (300 g/day) supplemented with *L. acidophilus* 145 and *B. longum* 913 for 21 weeks'	HDL cholesterol was increased significantly by 0.3 mmol/l and the ratio of LDL/HDL cholesterol decreased from 3.24 to 2.48.
Sindhu and Khetarpaul (2003)	20 young Swiss mice	Probiotic fermented food for 42 days	*L. casei* NCDC-19 (10^9 CFU) and *Saccharomyces boulardii* (10^9 CFU) caused 19% reduction in the total serum cholesterol and 37% decrease in LDL cholesterol levels.
Park et al. (2007)	36 male Sprague–Dawley hypercholesterolemic rats	Supplementation of *L. acidophilus* ATCC 43121 (2×10^6 CFU/day) for 21 days	Reduced total serum cholesterol by 25%, reduced very low and intermediate density lipoprotein and LDL cholesterol, compared to the control.
Mandal et al. (2009)	Swiss albino mice	Synbiotic treatment (*P. acidilactici* LAB 5 with sorbitol) for 1 month	16% reduction of plasma cholesterol level compared to untreated one or only probiotic treatment (7%).

are, first, assimilation or binding of cholesterol to the probiotic strains; second, greater deconjugation of bile salts; third, coprecipitation of cholesterol with free bile salts, and fourth, binding of cholesterol with prebiotics and subsequent release from the host body. First hypothesis is the assimilation of cholesterol to the probiotic strains. Existing evidences from human and animal studies suggest a moderate cholesterol-lowering action of some probiotic strains was due to the ability of some strains to assimilate cholesterol. This hypothesis is supported by the evidence of altered lipid profiles of the probiotic bacteria grown in presence of cholesterol (Kimoto et al. 2002, Liong and Shah 2005a). They also observed that cells grown in the presence of cholesterol micelles and bile salts were more resistant to lysis by sonication than were those grown in their absence. It was found that most of the cholesterol from the medium was incorporated into the cytoplasmic membrane, however the outer membranes of the intact cells are more easily accessed by cholesterol. There was twice the amount of cholesterol in the protoplast membranes than that of intact cells. This indicated cholesterol could preferentially bind to the cytoplasmic membranes (Walker and Gilliland 1993; Pereira and Gibson 2002). Some probiotics could also produce exopolysaccharides (EPS) which adhered to the cell surface and could absorb cholesterol. Kimoto-Nira et al. (2007) suggested cholesterol was bound to bacterial cells due to the chemical and structural properties of their cell wall peptidoglycans, which contain various amino acid compositions that facilitate the attachment of cholesterol to cellular surfaces. Therefore, the assumed hypocholesteromic action of probiotic strains may be due to assimilation or binding of cholesterol on the cells.

Second hypocholesterolemic mechanism was postulated that involves the ability of certain probiotics to enzymatically deconjugate bile acids. Cholesterol is the precursor molecule for bile salts synthesis in vivo and is synthesized in pericentral hepatocytes of the liver. Bile salts have some antimicrobial toxicity. To detoxify such antimicrobial compounds, these microbes have developed detoxification mechanism by producing bile salt hydrolases (BSH; cholyglycine hydrolase; EC 3.5.1.24). Microbial BSH catalyzes the hydrolysis (deconjugation) of glycine- and/or taurine-conjugated bile salts into amino acid residues and free bile acids (De Smet et al. 1995; Ridlon et al. 2006). Free bile acids are less soluble and there are no transporter for free bile salts and thus less efficiently reabsorbed into the intestines compared to conjugated bile salts, and thus is more prone to be excreted with the feces. This will increase the need for the synthesis of new bile acids to replace the lost ones. To compensate this bile salt level and also to meet other needs the liver synthesizes approximately 1,500–2,000 mg of new cholesterol each day. Therefore, GRAS bacteria which have high bile salt tolerance by deconjugating conjugated bile salts help in utilizing more cholesterol for bile salt synthesis in the colonized intestine and help in minimizing the cholesterol-associated problem. Therefore, the ingestion of probiotic bacteria containing active BSH might serve as a "biological" alternative to common medical or surgical interventions to treat hypocholesteromia.

Thirdly, coprecipitation of cholesterol with free bile salts. Klaver and van der Meer (1993), Tahri et al. (1996) and Zhang et al. (2008) proposed that the apparent loss of cholesterol by *L. acidophilus* and *B. bifidum* from the culture medium in presence of conjugated bile salts was not due to bacterial uptake of cholesterol rather due to coprecipitation with deconjugated bile salts in an acidic environment. Klaver and van der Meer (1993) showed that in the absence of bacterial cells, cholesterol was partially removed from the medium at pH values lower than 5.5 when deconjugated bile salts were added but no effect on pH controlled experiments (pH >6.0). This transient phenomenon was due to that the precipitated cholesterol was redissolved when solutions were adjusted to pH 7. Though coprecipitation is considered as one of the hypothesis for hypocholesteromic effect this is not applicable for in vivo effect because the intestinal pH is higher than 6.0.

Fourthly, to get the best effect of added probiotic supplement one important mechanism is by stimulating the growth of that organism by selective nutrients which are known as "prebiotics." Collins and Gibson (1999) and Tuohy et al. (2003) emphasized on using probiotics and prebiotics combined (i.e., as synbiotics) to improve gut health and also for modulating the microbial ecology of the gut. This has attracted considerable attention, mainly because unlike probiotics, prebiotic substances are not subject to viability problems and also have greater possibilities for incorporation into a wide range of common foodstuffs. One hypothesis for a lipid-lowering mechanism is by increasing the synthesis of fermentation byproducts (e.g., propionate) which potentially modulate the hepatic cholesterol synthesis (Levrat et al. 1994). However, the exact mechanism of action needs to be clarified. It is also proved that blood cholesterol synthesis is decreased due to inhibition of HMG-CoA reductase that converts HMG-CoA to mevalonate by organic acids in the fermented food products.

In conclusion, probiotic strains have some hypocholesteromic activity in vivo due to either of the mechanisms. The degree of hypocholesteromic effect is strain-specific. Probiotic supplementation through fermented food products (more particularly dairy products) adds another benefit in both ways as easy intake of live probiotic organisms and the fermented byproducts have direct hypocholesteromic effect by inhibiting cholesterol synthesizing enzymes and lowering circulating cholesterol concentrations, thus diminishing the risk of CHD.

4 Antihypertensive Properties of Probiotics

Hypertension or high blood pressure is a chronic medical condition in which the systemic arterial blood pressure is elevated and is the most important cause of death and disability in the world. It is the most prevalent preventable disease affecting 20–50% of the adult population in developed countries and it is expected that by 2025, almost three-quarters of the worldwide population in developing countries will be with hypertension. Randomized controlled trials have shown that treatment of hypertension reduces the risk of stroke, CHD, congestive heart failure, and mortality.

4.1 Antihypertensive Properties of Probiotic Organisms: Some Proof of Concepts

Studies on spontaneously hypertensive rats (Nakamura et al. 1996) and one human clinical study (Hata et al. 1996) provide the evidence that the bioactive peptides viz. casokinins and lactokinins and two tripeptides, valine-proline-proline and isoleucine-proline-proline resulting from the proteolytic action of probiotic bacteria on casein (αs2-casein, κ-casein and β-casein) and whey during fermentation by *Saccharomyces cerevisiae* and *Lactobacillus helveticus* may suppress the blood pressure of hypertensive individuals (Takano 1998). These tripeptides function as angiotensin-I converting enzyme (ACE) inhibitors and reduce blood pressure. Donkor et al. (2007) reported that soy yogurt from fermenting soy milk containing starter culture of *Lactobacillus delbrueckii* ssp. *bulgaricus* Lb1466, *Streptococcus thermophilus* St1342, and probiotic organisms (*Lactobacillus acidophilus* LAFTI® L10, *L. paracasei* LAFTI® L26, and *Bifidobacterium lactis* LAFTI® B94) showed a substantial increase in in vitro ACE inhibitory activity in comparison to the control produced by yogurt culture only. This improvement of ACE inhibition in soy yogurt is partly due to higher proteolytic activity of probiotics (Woo et al. 2009). Yeo and Liong (2010) reported that probiotic-fermented soymilk containing *Lactobacillus* sp. FTDC 2113, *L. acidophilus* FTDC 8033, *L. acidophilus* ATCC 4356, *L. casei* ATCC 393, *Bifidobacterium* FTDC 8943, and *B. longum* FTDC 8643 supplemented with FOS, inulin, mannitol, maltodextrin and pectin had increased proteolytic activity and in the presence of FOS the ACE-I inhibitory activity had increased. The β-glucosidase activity was also enhanced in the presence of pectin resulting in higher bioconversion of glucosides to aglycones by probiotics, especially genistin and malonyl genistin to genistein. They found that the supplementation of prebiotics enhanced the in vitro antihypertensive effect and production of bioactive aglycones in probiotic-fermented soymilk that could potentially be used as a dietary therapy to reduce the risks of hypertension.

The renin–angiotensin system plays a major role in the pathophysiology of hypertension and heart failure and ACE inhibitors play a pivotal role in the management of hypertension and heart failure. Data from the research on probiotic therapy for hypertension proved that probiotic products have the same efficacy as of synthetic ACE inhibitors. Thus probiotics and their fermented food products will provide the only biological method for future treatment of hypertension.

5 Alleviation of Postmenopausal Symptoms

Menopause is the permanent cessation of ovarian function resulting cessation of ova release, uterine lining creation, and subsequent uterine lining shedding (i.e., menses). Menopause typically occurs in women in midlife, during their late 40s or early 50s, and signals the end of the fertile phase of a woman's life.

The transition from reproductive to nonreproductive is not sudden rather tends to occur over a period of years which is designated as "perimenopause." During perimenopause, the ovarian production of the estrogens and progesterone becomes more irregular. At this stage many women undergo noticeable and clinically observable physical changes; the most well-known of these is the "hot flush," a sudden temporary increase in body temperature. The term postmenopause is applied to women who have not experienced a menstrual bleed for a minimum of 12 months, assuming that they do still have a uterus, and are not pregnant or lactating. Postmenopause affect the women both physically, psychologically, and also in sexual life. The physical symptoms of postmenopause may include hot or cold flashes, and rapid heartbeat, watery discharge, urinary frequency, and increased susceptibility to inflammation and infection to vaginal candidiasis, and urinary tract infections (UTI), decreased breast tenderness and elasticity of the skin. The psychological symptoms include depression, anxiety, fatigue, memory loss, mood and sleep disturbances, and insomnia. The sexual symptoms include dyspareunia or painful intercourse, decreased libido, vaginal dryness and vaginal atrophy.

5.1 Treatment of Menopause: Alternative Medicines and Prebiotics

Apart from conventional synthetic medicines, some popular alternative medications are based on foods or food supplements enriched with phytoestrogens (a group of plant-derived nonsteroidal estrogens that have estrogenic and/or antiestrogenic activity) (Humfrey Charles 1998). They are present in many human foodstuffs including beans, sprouts, cabbage, spinach, soybean, grains, and hops. The main classes of phytoestrogens are the isoflavones, coumestans, and lignans. Hooper et al. (2009) reported that isoflavone-rich soy products decrease follicle-stimulating hormone (FSH) and leutenizing hormone (LH) in premenopausal women and may increase estradiol in postmenopausal women. Fermentation products (lactic acid and acetic acid) and probiotics (*B. breve*) can influence the absorption or metabolism of isoflavones through endogenous estrogen metabolism in postmenopausal women (Xu et al. 2000).

5.2 Probiotics for Treatment and Prophylaxis to Postmenopausal Women

The use of probiotics in obstetrics and gynecology is in early stages of experimentation and study. The recurrent nature of urogenital infection, emergence of multidrug-resistant bacteria, and patient dissatisfaction with side effects of drugs needs better ways to diagnose, treat, and prevent infection. Alternative strategies like probiotics would be a beneficial treatment option and are found effective in the treatment of vaginosis, candidiasis, and recurrent UTI. The yeast *Candida albicans*, a commensal inhabitant of the mouth, genital and intestinal tracts, under certain conditions such as low pH, impaired immune responses, suppression of the normal microflora by antibiotics or due to hormonal changes associated with the menstrual cycle, oral contraceptives and pregnancy, undergoes dramatic genetic and physiological changes and causes opportunistic candidacies infecting the vagina (vaginal thrush), the mouth (oral thrush) or the intestinal tract. Therefore, women usually develop yeast-related disorders much more frequently than men and premenopausal women are often the most susceptible ones. There is evidence that probiotic bacteria taken prophylactically or postinfection can help reduce candidiasis and treat recurrent thrush. A probiotic bacteria *Pediococcus acidilactici* LAB 5 could effectively destroy pathogenic *C. albicans* in vitro (Mandal, Ph.D. Thesis 2010). Locally applied *L. acidophilus* lyophilizate or *L. acidophilus*-yogurt given orally also proved effective in bacterial vaginosis (BV) or candida vaginosis (CV). Lactobacilli are the dominant flora of premenopausal women and of hormonally treated postmenopausal women. Studies have shown that absence or depletion of lactobacillus in vagina is one of the main reasons for BV infections and these results in significantly increased risk of HIV as well as gonorrhea, chlamydia and Herpes simplex viral-like infections. In a controlled crossover study, Hallen et al. (1992) examined that daily ingestion (twice a day) of yogurt-containing hydrogen peroxide producing *L. acidophilus* for 6 days prevented vulvovaginal candida infection and led to 43% improvement compared to none in the placebo group. Neri et al. (1993) reported that intravaginal applications of yogurt to treat woman with BV in the first trimester of pregnancy reduced the disease severity. Randomized, placebo-controlled trial in 64 healthy women for 60 days showed that oral use of *L. rhamnosus* GR-1 and *L. fermentum* RC-14 significantly altered (cure rate of 87.6%) vaginal flora in antibiotic untreated or treated conditions (Reid et al. 2003a, b). Thus lactobacillus as an adjunct to antibiotics may be helpful in increasing cure rates and preventing recurrence. The combination of probiotic *L. rhamnosus* GR-1 and *L. fermentum* RC-14 is not only safe for daily use in healthy women, but it can also reduce colonization of the potential pathogenic bacteria and yeast in the vagina.

5.3 Mechanisms of Actions of Probiotics

Probiotics are believed to protect the host against infections by means of several mechanisms, such as (1) stabilization of low pH and the production of antimicrobial substances like acids, hydrogen peroxide and bacteriocins, (2) tight adhesion to the epithelial surface of the urinary tract, (3) the degradation of polyamines, and (4) the production of surfactants, which have anti-adhesive properties against the adhesion of pathogens (Reid 2001). Probiotic lactobacilli given orally can also act via immunostimulation or inhibition of bacterial translocation. They also modulate the host immunity by decreasing IL-1 and IL-8 levels which are elevated during bacterial vaginosis (BV). Probiotics also lead to production of cell signaling molecules which downregulate pathogen molecule that signals for mucous production which acts as a barrier to anti-inflammatory cytokine production. Lactobacilli have also been shown to produce biosurfactants and collagen binding proteins that inhibit pathogen adhesionness and displace the pathogens.

Pre- or postmenopausal symptoms are inevitable to every woman's life. Successful management to minimize or alleviate the agony of pre- or postmenopausal symptoms in women's health to extend the life expectancy needs judicial application of alternative strategies. Evidences accumulating on the use of potent probiotic strains and manipulation of the host's own intestinal and vaginal or urethral microbiota will provide valuable options to help restore and maintain urogenital health at premenopausal or postmenopausal stages. Thus daily intake of scientifically selected probiotics would provide natural, safe, and effective means of regulating the fluctuating vaginal flora and thereby lower the risk of infection in healthy and sick women and thus alleviating such problems.

6 Antiallergic Effects of Probiotics

Allergy is a hypersensitive disorder of the immune system and is one of the four forms of hypersensitivity and is grouped in type I (or immediate) hypersensitivity. Allergic reactions occur in response to normally harmless environmental substances known as allergens and these reactions are acquired, predictable, and rapid. It is characterized by excessive activation of certain white blood cells called mast cells and basophiles by a type of antibody known as immunoglobulin E (IgE), resulting in an extreme inflammatory response. Some common allergic reactions include asthma, eczema, hay fever, hives, food allergies, allergic conjunctivitis, itchiness, and runny nose and reactions to the venom of stinging insects. In the context of probiotic associations in alleviating the allergic problems, major emphasis is given on food-associated allergenic problem in this discussion.

6.1 Probiotics Prophylaxis and Management to Allergic Diseases

The prevalence of allergic diseases in western societies is increasing at an alarming rate over the last 35–40 years, whereas the frequency of allergic diseases is much lower in developing countries. The breast-fed infants show lower incidence of allergies compared with formula-fed infants. Studies show that the balance between bifidobacteria and clostridia appears to be the key factor in determining predisposition to allergies. Several epidemiological and experimental studies indicated that stimulation of the immune system by certain microbes or microbial products might be effective in the prevention and management of allergic diseases (Furrie 2005). Several well-designed studies have provided evidence that supplementation with specific strains of probiotics could be effective in the management of atopic disorders. Perinatal administration of *L. rhamnosus* GG decreased subsequent occurrence of eczema in at-risk infants by one-half (Isolauri et al. 2000). The effectiveness of probiotic therapy in the prevention of allergic diseases has been demonstrated in several randomized controlled trials as summarized in Table 2. From these studies, it can be assumed that specific probiotic strains might be effective in the management of allergies.

6.2 Mechanisms of Antiallergic Probiotic Action

The exact mechanisms behind the favorable effects of probiotics on allergy are not entirely known, but it is expected that they may exert a beneficial effect by improving mucosal barrier function and microbial stimulation of immune system (MacFarlane and Cummings 2002). In addition to modulation of the intestinal microbiota, probiotics have been observed to improve the barrier function of the intestinal mucosa and thus reducing leakage of antigens through the mucosa and thereby exposure to them (Qin et al. 2005). There are two pathways viz. transcellular movement and paracellular route through which antigenic proteins traverse the epithelial barrier of the gastrointestinal tract. Direct modulation of the immune system by probiotics is mediated through the induction of anti-inflammatory cytokines or through increased production of secretory IgA. Secretory IgA contributes to exclusion of antigens from the intestinal mucosa (Kalliomaki and Isolauri 2004). Further, probiotics help in enzymatic degradation of dietary antigens and thus reduce the load and exposure to antigens. Probiotics may also be helpful in alleviating some of the symptoms of food allergies such as those associated with milk protein possibly by degrading these proteins to smaller peptides and amino acids (Majarmaa and Isolauri 1997). Probiotics have also been found to up regulate anti-inflammatory cytokines, such as interleukin-10, in atopic children (Pessi et al. 2000).

Food allergy affects ~2% of the general US population, and its prevalence seems to be increasing with the increasing prevalence of the extrinsic asthma and environmental allergies. However, unlike other allergic diseases, food allergy is the only disorder for which there is no specific therapy. Dietary avoidance is the current standard of care, but accidental ingestions of food allergens are common because of the ubiquitous presence of certain foods, such as peanut, soy, milk and egg, and cross-contamination during processing. Furthermore, in cases of multiple food allergies, restricted diets may result in unbalanced nutrition and pose a considerable hardship to the family of a food-allergic child. Several probiotic organisms have been proved to have antiallergic effects in ameliorating the food allergen-associated disorders. The efficacy of antiallergic effects of probiotics is strain-specific. Therefore, successful prophylaxis of allergy disorders can be achieved with a good probiotic bacterium and thus, probiotic therapy appears to be a promising approach in the treatment and prevention of allergy.

Table 2 Some evidences of prophylactic antiallergic probiotic efficacies in randomized, double-blind, and placebo-controlled study

Authors	Probiotics strains	Target group	Treatment conditions	Observations
Isolauri et al. (2000)	*Lactobacillus* GG or *Bifidobacterium lactis* BB-12	Breast fed infants with atopic eczema	Whey formula containing with or without probiotics for 2 months	Probiotics-treated infants showed significant improvements in the extent and severity of atopic eczema and reduction in inflammatory responses compared with the placebo groups.
Gill et al. (2001)	*L. rhamnosus* GG and *B. lactis* BB-12	–	–	Proved useful in infants allergic to cow's milk.
Kalliomäki et al. (2001)	*Lactobacillus* GG	High risk children at early atopic disease stage	Pre or postnatal supplementation of *Lactobacillus* GG to mothers	Significantly reduced incidence of atopic eczema compared with placebo group given formula without probiotics.
Rautava et al. (2002)	*L. rhamnosus* GG	Pregnant women and newborn children at high risk of allergy	For 4 weeks prior to delivery and then 6 months to newborn children	A significant reduction in early atopic disease
Lodinova-Zadnikova et al. (2003)	*E. coli* (nonenteropathogenic)	The preventive efficacy of the strain at birth colonization	Incidence of allergies was assessed both after 20 years and 10 years after colonization	Subjects colonized with *E. coli* showed significantly lower incidence of allergies compared to control subjects both after 10 and 20 years

7 Protection Against Lung Emphysema

Emphysema is a long-term progressive disease of the lung that primarily causes shortness of breath. In people with emphysema, the tissues necessary to support the physical shape and function of the lungs are destroyed. It is included in a group of diseases called chronic obstructive pulmonary disease (COPD). Emphysema is an irreversible degenerative condition and is the fourth leading cause of death in the United States. In addition to the normal aging process, certain habits like cigarette smoking fasten the process at early age (Pelkonen 2008). Emphysema may also develop from genetic disorderness. About 2% of all emphysema seen in young people is due to a mutation in the alpha-1 antitrypsin gene. Alpha-1-antitrypsin (also known as alpha-1-antiprotease) is a substance that fights a destructive enzyme in the lungs called trypsin (or protease). Trypsin, a digestive enzyme, is also released by immune cells in their attempt to destroy bacteria and other material. People with alpha-1-antitrypsin deficiency (A1AD) cannot fight the destructive effects of trypsin once it is released in the lung and thus the imbalance develops between trypsin and antitrypsin. Nowadays, gene therapy for A1AD has developed by Dr. Darrell Kotton through lung's alveolar macrophages (Wilson et al. 2008) but this is yet to be applied in human beings and also it matters of cost for such treatment. Therefore, more cost-effective treatment strategies are needed for large fraction of world population and here probiotics might be an option.

7.1 *Probiotics Prophylaxis for Lung Emphysema and Other Lung Complications*

Probiotics prophylaxis with *L. acidophilus* NCFM and *Bifidobacterium animalis* subsp. *lactis* Bi-07 twice daily against cold and influenza-like symptoms in children (3–5 years) for 6 months resulted in lower incidence of fever, coughing, and rhinorrhea compared with the placebo group (de Vrese et al. 2005, 2006; Leyer et al. 2009). Weizman et al. (2005) found that infant formula preparation with two probiotic organism have positive influencing results on infections in child care centers. In another study, daily use of probiotics reduced ventilator-associated pneumonia (VAP) in critically ill patients by almost half (Morrow et al. 2010). Patients who received *L. rhamnosus* (2×10^9 colony-forming units) twice daily-half the dose as slurry to the oropharynx and others through nasogastric tube recovered quickly from the critical stage. The researchers also found that daily use of probiotics not only decreased VAP infections by about 50% compared to placebo, but also reduced the amount of antibiotics needed in comparison to placebo-treated patients (Morrow et al. 2010). These data suggest that probiotics may represent a novel, inexpensive, and nonantibiotic approach to prevention of nosocomial infections in properly selected ICU patients. Larger, multicenter clinical trials with more liberal inclusion criteria are needed to establish efficacy of probiotics and also for extrapolation to a larger risk population. Recently, BBC News stated that probiotics containing "benign bacteria" could be used to protect critically ill patients from developing pneumonia. It claimed the probiotic bacterium *Lactobacillus plantarum* 299, did as well as the currently used antiseptic, chlorhexidine, in "keeping pneumonia-causing bacteria at bay." Thus this probiotic is a viable alternative to chlorhexidine in intensive care.

8 Conclusions

Probiotic microorganisms are the new age tool in medical interventions to solve or alleviate many health problems encountered by human beings. Several challenge experiments have proved the utility and efficacy of several species or strains of probiotic organisms. These beneficial attributes either directly involve the organism itself or their contribution in modifying from harsh to friendly or converting into less biologically active to highly potent bioactive compounds. The resultant efficacy is longstanding. Therefore, in near future, probiotics organisms will be a biological alternative to synthetic drugs that will subsiding their harmful sides of drug reaction and achieving an era of prophylactic probiotics for all round diseases. Hope our present-day scientists will explore the new and new areas of probiotic therapeutic uses to expand the dimension of biotherapeutics.

References

Aattouri N, Bouras M, Tome D, Marcos A, Lemonnier D (2002) Oral ingestion of lactic-acid bacteria by rats increases lymphocyte proliferation and interferon-g production. Br J Nutr 87:367–373

Agarwal N, Kamra DN, Chaudhary LC, Sahoo A, Pathak NN (2000) Selection of *Saccharomyces cerevisiae* strains for use as a microbial feed additive. Lett Appl Microbiol 31:270–273

Anderson JW, Gilliland SE (1999) Effect of fermented milk (yogurt) containing *Lactobacillus acidophilus* L1 on serum cholesterol in hypercholesterolemic humans. J Am Coll Nutr 18(1):43–50

Collins MD, Gibson GR (1999) Probiotics, prebiotics, and synbiotics: approaches for modulating the microbial ecology of the gut. Am J Clin Nutr 69(suppl):1052S–1057S

Cummings JH, Macfarlane GT, Englyst HN (2001) Prebiotic digestion and fermentation. Am J Clin Nutr 73:415S–420S

De Rodas BZ, Gilliland SE, Maxweell CV (1996) Hypocholesterolemic action of *Lactobacillus acidophilus* ATCC 43121 and calcium in swine with hypocholesterolemia induced by diet. J Dairy Sci 79:2121–2128

De Smet I, van Hoorde L, vande Woestyne M, Christiaens H, Verstraete W (1995) Significance of bile-salt hydrolytic activities of lactobacilli. J Appl Bacteriol 79:292–301

de Vrese M, Winkler P, Rautenberg P, Harder T, Noah C, Laue C (2006) Probiotic bacteria reduced duration and severity but not the incidence of common cold episodes in a double blind, randomized, controlled trial. Vaccine 10:6670–6674 de Vrese M, Winkler P, Rautenberg P, Harder T, Noah C, Laue C, Ott S, Hampe J, Schreiber S, Heller K, Schrezenmeir J (2005) Effect of *Lactobacillus gasseri* PA 16/8, *Bifidobacterium longum* SP 07/3, *B bifidum* MF 20/5 on common cold episodes: a double blind, randomized, controlled trial. Clin Nutr 24(4):481–491

Donkor ON, Henriksson A, Singh TK, Vasiljevic T, Shah NP (2007) ACE-inhibitory activity of probiotic yoghurt. Int Dairy J 17(11):1321–1331

Fiat AM, Migliore-Samour D, Jolles P, Drouet L, Bal Dit Sollier C, Caen J (1993) Biologically active peptides from milk proteins with emphasis on two examples concerning antithrombotic and immunomodulating activities. J Dairy Sci 76:301–310

Fuller R (1991) Probiotics in human medicine. Gut 32:439–442

Furrie E (2005) Probiotics and allergy. Proc Nutr Soc 64(4):465–469

Galdeano CM, Perdigon G (2006) The probiotic bacterium *Lactobacillus casei* induces activation of the gut mucosal immune system through innate immunity. Clin Vaccine Immun 13:219–226

Gibson GR, Roberfroid MB (1995) Dietary modulation of the human colonic microbiota: introducing the concept of prebiotics. J Nutr 125:1401–1412

Gill HS, Rutherfurd KJ, Cross ML (2001) Dietary probiotic supplementation enhances natural killer cell activity in the elderly: an investigation of age related immunological changes. J Clin Immunol 21:264–271

Hallen A, Jarstrand C, Pahlsson C (1992) Treatment of bacterial vaginosis with lactobacilli. Sex Transm Dis 19:146–148

Hata Y, Yamamoto M, Ohni M, Nakajima K, Nakamura Y, Takano T (1996) A placebo-controlled study of the effect of sour milk on blood pressure in hypertensive subjects. Am J Clin Nutr 64:767–771

Hooper L, Ryder JJ, Kurzer MS, Lampe JW, Messina MJ, Phipps WR, Cassidy A (2009) Effects of soy protein and isoflavones on circulating hormone concentrations in pre- and post-menopausal women: a systematic review and meta-analysis. Hum Reprod Update 15:423–440

Humfrey Charles DN (1998) Phytoestrogens and human health effects: weighing up the current evidence. Nat Toxins 6:51–59

Isolauri E, Arvola T, Sutas Y, Moilanen E, Salminen S (2000) Probiotics in the management of atopic eczema. Clin Exp Allergy 30:1604–1610

Isolauri E, Majamea H, Arvola T, Rantela I, Arvilommi H (1993) *Lactobacillus* strain GG reverses increases intestinal permeability induced by cow milk in suckling rats. Gastroenterology 105:1643–1650

Kaila M, Isolauri E, Soppi E, Virtanen E, Laine S, Arvilommi H (1992) Enhancement of the circulating antibody secreting cell response in human diarrhea by a human lactobacillus strain. Pediatr Res 32:141–144

Kalliomaki MA, Isolauri E (2004) Probiotics and down regulation of the allergic response. Immunol Allergy Clin North Am 24:739–752

Kalliomäki M, Salminen S, Arvilommi H, Kero P, Koskinen P, Isolauri E (2001) Probiotics in primary prevention of atopic disease: a randomized placebo-controlled trial. Lancet 357:1076–1079

Kaur IP, Chopra K, Saini A (2002) Probiotics: potential pharmaceutical applications. Eur J Pharm Sci 15:1–9

Kieling G, Schneider J, Jahreis G (2002) Long-term consumption of fermented dairy products over 6 months increases HDL cholesterol. Eur J Clin Nutr 56:843–849

Kimmey MB, Elmer GW, Surawicz CM, Mc Farland LV (1990) Prevention of further recurrences of *Clostridium difficile* colitis with *Saccharomyces boulardii*. Dig Dis Sci 35:897–901

Kimoto H, Ohmomo S, Okamoto T (2002) Cholesterol removal from media by Lactococci. J Dairy Sci 85:3182–3188

Kimoto-Nira H, Mizumachi K, Nomura M, Kobayashi M, Fujita Y, Okamoto T, Suzuki I, Tsuji NM, Kurisaki JI, Ohmomo S (2007) *Lactococcus* sp. as potential probiotic lactic acid bacteria. Jpn Agr Res Q 41:181–189

Klaver FAM, Van Der Meer R (1993) The assumed assimilation of cholesterol by Lactobacilli and *Bifidobacterium bifidum* is due to their bile salt-deconjugating activity. Appl Environ Microbiol 59:1120–1124

Le Duc H, Hong HA, Barbosa TM, Henriques AO, Cutting SM (2004) Characterization of *Bacillus* probiotics available for human use. Appl Environ Microbiol 70(4):2161–2171

Levrat MA, Favier ML, Moundras C, Rémésy C, Demigné C, Morand C (1994) Role of dietary propionic acid and bile acid excretion in the hypocholesterolemic effects of oligosaccharides in rats. J Nutr 124:531–538

Leyer GJ, Li S, Md Mubasher E, Reifer C, Ouwehand AC (2009) Probiotic effects on cold and influenza-like symptom incidence and duration in children. Pediatrics 124:e172–e179

Liong MT, Shah NP (2005a) Acid and bile tolerance and cholesterol removal ability of *Lactobacillus* strains. J Dairy Sci 88:55–66

Liong MT, Shah NP (2005b) Roles of probiotics and prebiotics on cholesterol: the hypothesized mechanisms. Nutrafoods 4:45–57

Lodinova-Zadnikova R, Cukrowska B, Tlaskalova-Hagenova H (2003) Oral adminstration of probiotic *Escherichia coli* after birth reduces frequency of allergies and repeated infections later in life (after 10 and 20 years). Int Arch Allergy Immunol 131:209–211

Macfarlane GT, Cummings JH (2002) Probiotics, infection and immunity. Curr Opin Infect Dis 15:501–506

Majarmaa H, Isolauri E (1997) Probiotics: a novel approach in the management of food allergy. J Allergy Clin Immunol 99:179–186

Mandal V (2010) Production of bacteriocin by *Pediococcus acidilactici* for application as topical therapeutics and probiotics. (Unpublished doctoral dissertation) Visva-Bharati University, Santiniketan, India

Mandal V, Sen SK, Mandal NC (2009) Effect of prebiotics on bacteriocin production and cholesterol lowering activity of *Pediococcus acidilactici* LAB 5. World J Microbiol Biotech 25:1837–1847

Mann GV, Spoerry A (1974) Studies of a surfactant and cholesteremia in the Maasai. Am J Clin Nutr 27:464–469

Moreau MC, Hudault S, Bridonneau C (1990) Systemic antibody response to ovalbumin in gnotobiotic C3H/HeJ mice with *Bifidobacterium bifidum* or *Escherichia coli*. Microecol Ther 20:309–312

Morrow LE, Kollef MH, Casale TB (2010) Probiotic prophylaxis of ventilator-associated pneumonia: a blinded, randomized, controlled trial. Am J Respir Crit Care Med 182:1058–1064

Nakamura Y, Masuda O, Takano T (1996) Decrease of tissue angiotensin- I-converting enzyme activity upon feeding sour milk in spontaneously hypertensive rats. Biosci Biotechnol Biochem 60:488–489

Neri A, Sabha G, Samra Z (1993) Bacterial vaginosis in pregnancy treated with yogurt. Acta Obstet Gynaecol Scand 72:17–19

Park YH, Kim JG, Shin YW, Kim SH, Whang KY (2007) Effect of dietary inclusion of *Lactobacillus acidophilus* ATCC 43121 on cholesterol metabolism in rats. J Microbiol Biotechnol 17:655–662

Pelkonen M (2008) Smoking: relationship to chronic bronchitis, chronic obstructive pulmonary disease and mortality. Curr Opin Pulm Med 14(2):105–109

Perdigón G, de Macías ME, Alvarez S, Oliver G, de Ruiz Holgado AP (1998) Systemic augmentation of the immune response in mice by feeding fermented milks with *Lactobacillus casei* and *Lactobacillus acidophilus*. Immuno 63:17–23

Perdigon G, Nader De Macias ME, Alvarez S, Oliver G, Pesce De Ruiz Holgado AA (1986) Effect of perorally administered lactobacilli on macrophage activation in Mice. Infect Immun 53:404–410

Pereira DI, Gibson GR (2002) Cholesterol assimilation by lactic acid bacteria and Bifidobacteria isolated from the human gut. Appl Environ Microbiol 68:4689–4693

Pereira DI, Gibson GR (2003) Effects of consumption of probiotics and prebiotics on serum lipid levels in humans. Crit Rev Biochem Mol Biol 37:259–281

Pessi T, Sutas Y, Hurme M, Isolauri E (2000) Interleukin-10 generation in atopic children following oral *Lactobacillus rhamnosus* GG. Clin Exp Allergy 30:1804–1808

Pessi T, Sutas Y, Marttine NA, Isolauri E (1998) Probiotics reinforce mucosal degradation of antigens in rats: Implications for therapeutic use of probiotics. Am Soc Nutr Sci 210:2313–2318

Pinchuk IV, Bressollier P, Verneuil B, Fenet B, Sorokulova IB, MéGraud F, Urdaci MC (2001) In vitro anti-*Helicobacter pylori* activity of the probiotic strain *Bacillus subtilis* 3 is due to secretion of antibiotics. Antimicrob Agents Chemother 45(11):3156–3161

Psomas EI, Fletouris DJ, Litopoulou-Tzanetaki E, Tzanetakis N (2003) Assimilation of cholesterol by yeast strains isolated from infant feces and feta cheese. J Dairy Sci 86:3416–3422

Qin HL, Shen TY, Gao ZG, Fan XB, Hang XM, Jiang YQ, Zhang HZ (2005) Effect of lactobacillus on the gut microflora and barrier function of the rats with abdominal infection. World J Gastroenterol 11:2591–2596

Rautava S, Kalliomaki M, Isolauri E (2002) Probiotics during pregnancy and breast-feeding might confer immunomodulatory protection against atopic disease in the infant. J Allergy Clin Immunol 109:119–121

Reid G (2001) Probiotic agents to protect the urogenital tract against infection. Am J Clin Nutr 73(suppl):437S–443S

Reid G, Charbonneau D, Erb J, Kochanowski B, Beuerman D, Poehner R, Bruce AW (2003a) Oral use of *Lactobacillus rhamnosus* GR-1 and *L. fermentum* RC-14 significantly alters vaginal fora: randomized, placebo-controlled trial in 64 healthy women. FEMS Immunol Med Microbiol 35:131–134

Reid G, Hammond JA, Bruce AW (2003b) Effect of lactobacilli oral supplement on the vaginal microflora of antibiotic treated patients: randomized placebo controlled study. Nutraceut Food 8:145–148

Ridlon JM, Kang DJ, Hylemon PB (2006) Bile salt biotransformations by human intestinal bacteria. J Lipid Res 47:241–259

Roberfroid MB (1996) Functional effects of food components and the gastrointestinal system: chicory fructooligosaccharides. Nutr Rev 54:S38–S42

Roberfroid MB (2001) Prebiotics: preferential substrates for specific germs. Am J Clin Nutr 73:406S–409S

Schiffrin EJ, Rochat F, Link-Amster H, Aeschlimann JM, Donnet-Hughes A (1994) Immunomodulation of human blood cells following the ingestion of lactic acid bacteria. J Dairy Sci 78:491–497

Shaper AG, Jones KW, Jones M, Kyobe J (1963) Serum lipids in three nomadic tribes of northern Kenya. Am J Clin Nutr 13:135–146

Sindhu SC, Khetarpaul N (2003) Effect of feeding probiotic fermented indigenous food mixture on serum cholesterol levels in mice. Nutr Res 23:1071–1080

Tahri K, Grill JP, Schneider F (1996) Bifidobacteria strain behaviour toward cholesterol: coprecipitation with bile salts and assimilation. Curr Microbiol 33:187–193

Takano T (1998) Milk derived peptides and hypertension reduction. Int Dairy J 8:375–381

Taranto MP, Medici M, Perdigon G, Ruiz Holgado AP, Valdez GF (1998) Evidence for hypocholesterolemic effect of *Lactobacillus reuteri* in hypercholesterolemic mice. J Dairy Sci 81:2336–2340

Tuohy K, Probert HM, Smejkal CW, Gibson GR (2003) Using probiotics and prebiotics to improve gut health. Ther Focus 8:692–700

Walker DK, Gilliland SE (1993) Relationships among bile tolerance, bile salt deconjugation and assimilation of cholesterol by *Lactobacillus acidophilus*. J Dairy Sci 76:956–961

Weizman Z, Asli G, Alsheikh A (2005) Effect of a probiotic infant formula on infections in child care centers: comparison of two probiotic agents. Pediatrics 115(1):5–9

Wilson AA, Kwok LW, Hovav AH, Ohle SJ, Little FF, Fine A, Kotton DN (2008) Sustained expression of α_1-antitrypsin after transplantation of manipulated hematopoietic stem cells. Am J Respir Cell Mol Biol 39:133–141

Woo YP, Fung WY, Easa AM, Liong MT (2009) Probiotic-fermented soy foods: benefits and enhanced bioactivities. Acta Alimentaria Hung 38(3):381–391

Xu Xia, Duncan AM, Wangen KE, Kurzer MS (2000) Soy consumption alters endogenous estrogen metabolism in postmenopausal women. Cancer Epidemiol Biomark Prev 9:781–786

Yan F, Polk DB (2002) Prevention of cytokine-induced intestinal epithelial cell apoptosis by probiotic bacterium. J Biol Chem 277:50959–50965

Yasui H, Nagaoka N, Mike A, Hayakawa K, Ohwaki M (1992) Detection of *Bifidobacterium* strains that induce large quantities of IgA. Microb Ecol Health Dis 5:155–162

Yeo SK, Liong MT (2010) Angiotensin I-converting enzyme inhibitory activity and bioconversion of isoflavones by probiotics in soymilk supplemented with prebiotics. Int J Food Sci 61(2):161–181

Zhang M, Xiaomin H, Xiaobing F, Li D, Yang H (2008) Characterization and selection of *Lactobacillus* strains for their effect on bile tolerance, taurocholate deconjugation and cholesterol removal. World J Microbial Biotechnol 24:7–14

Carriers of Probiotic Microorganisms

Siok-Koon Yeo, Joo-Ann Ewe, Celestine Sau-Chan Tham, and Min-Tze Liong

Contents

1 Introduction 192
2 Dairy-Based Products as a Conventional Carrier of Probiotics 193
2.1 Challenges of Probiotics in Dairy-Based Products 193
2.2 Means to Overcome Limitations of Probiotics in Dairy-Based Products 196
3 Soy-Based Products as Carriers for Probiotics 199
3.1 Interaction of Probiotics and the Composition of Soy-Based Products 201
3.2 Challenges of Probiotics in Soy-Based Products 202
3.3 Means to Overcome Limitations of Probiotics in Soy-Based Products 203
4 Cereal-Based Products as Carriers for Probiotics 204
4.1 Interaction of Probiotics and the Composition of Cereal-Based Products 206
4.2 Challenges of Probiotics in Cereal-Based Products 207
4.3 Means to Overcome Limitation of Probiotics in Cereal-Based Products 208
5 Fruits and Vegetables as Carriers of Probiotics 208
5.1 Challenges of Probiotics in Fruit and Vegetable Juices 209
5.2 Means to Overcome Limitations of Probiotics in Fruit and Vegetable Juices 210
6 Meat and Meat Derivatives as Carriers of Probiotic Microorganisms 212
6.1 Interactions of Probiotics and the Composition of Meat and Meat Derivative 213
6.2 Challenges of Probiotics in Meat and Meat Derivatives 214
6.3 Means to Overcome Limitations of Probiotics in Meat and Meat Derivatives 214
7 Conclusions 215
References 215

Abstract There is a growing market potential for probiotic foods as an alternative to enhance human health. Milk-based products including milk beverage, yogurts, cheese and ice cream are conventionally used as the most suitable delivery vehicle for probiotics. Despite being an ideal substrate for probiotics, the growth of probiotics in these products is often inhibited due to excessive acidification, antagonistic effect of

S.-K. Yeo (✉) • J.-A. Ewe • C.S.-C. Tham • M.-T. Liong
School of Industrial Technology, Universiti Sains Malaysia, 11800 Penang, Malaysia
e-mail: siokkoon85@yahoo.com

M.-T. Liong (ed.), *Probiotics*, Microbiology Monographs 21,
DOI 10.1007/978-3-642-20838-6_8, © Springer-Verlag Berlin Heidelberg 2011

starter culture and the presence of oxygen during processing. Various means are evaluated to enhance the viability of probiotics, including supplementation with growth enhancer and protection by microencapsulation. The drawbacks of milk-based carrier associated with cholesterol contents and lactose intolerance have prompted the development of alternative carriers for probiotics. Currently, new foods such as soy-based products, cereal-based products, fruits, vegetables and meat products are developed as potential carriers. These nondairy-based products contain reasonable amounts of carbohydrates, fibers, proteins and vitamins that support the growth of probiotics. In addition, some components of these products are able to protect probiotics during transit through the harsh condition of gastrointestinal tract and during storage. However, growths in nondairy products such as sausage and fruit juices are inhibited by the presence of inhibitory substances such as nisin, organic acids and curing salts. Therefore, appropriate selection of culture used in these products is crucial in maintaining the viability of cells, without affecting the sensory and organoleptic property of the final products.

1 Introduction

Probiotics are live microorganisms that could confer health benefits on the host when consumed in sufficient amounts. The health benefits of probiotics are well documented, mainly involving the improvement of gastrointestinal health. With increasing health awareness among consumers, the scope of research on probiotics has widened to focus on their utilization to improve general health and as an alternative to the administration of drugs to treat a number of diseases such as hypertension, alleviation of postmenopausal disorders and prevention of cancers. Therefore, there has been a rise in the research and production of probiotic carriers.

Probiotics are normally added to foods as a part of the fermentation process. The probiotics involved in food fermentation are mainly *Lactobacillus* and *Bifidobacterium*, which are predominant inhabitants of the human gastrointestinal tract. The emphasis for prolonged survival of probiotics in the food matrix has resulted in the alteration of the functionality and efficacy of the food product, including the released of bioactive components. For instance, probiotics have been reported to release angiotensin-converting enzyme inhibitory peptides from the parent protein of dairy milk, cheese and soymilk upon fermentation, leading to an antihypertensive effect (Yeo and Liong 2010a). Probiotic dietary adjuncts could thus offer consumers both the benefits of beneficial microorganism, as well as the additional health promoting effects obtained via interaction of probiotics with foods.

In order to exert health benefits, probiotic bacteria must remain viable in the dietary adjunct with a minimum count of 10^6 CFU/g (Ewe et al. 2010). In addition to the necessity of having high viable counts in the products, survival rates during the passage through the harsh condition of gastrointestinal tract are also important. It has been suggested that the nature of food carrier could affect the stability of the probiotic microorganisms during gastrointestinal transit (Charalampopoulos et al. 2003).

To date, dairy-based products appear to be the main carriers for the delivery of probiotics to human. However, increasing demand for new flavor and taste has initiated a trend in using nondairy-based products as potential carriers of probiotic microorganisms. Numerous novel applications primarily emphasizing on soy, cereal, vegetables and fruits or meat fermentations have appeared in the market. In this chapter, a variety of products which serve as carriers for probiotics and efforts to improve growth and survival of probiotics in the products will be reviewed. In addition, several factors affecting the viability of probiotics in these products and various ways to overcome these challenges are also discussed.

2 Dairy-Based Products as a Conventional Carrier of Probiotics

Ever since the introduction of live lactic acid bacteria containing sour milk product by Elie Metchnikoff at the beginning of the twentieth century, milk-based products have been found to be an ideal carrier for delivering probiotics to the human gastrointestinal tract. Among various strains of lactic acid bacteria that have been characterized as probiotics, the genera of *Lactobacillus* and *Bifidobacterium* are the most employed type involved in the fermentation of dairy-based products.

Dairy-based products have been demonstrated as an excellent vehicle for the delivery of probiotics to humans. Probiotics proliferated well in dairy-based medium due to the lactose hydrolyzing enzyme and proteolytic system involved in casein utilization which provides probiotic cells with carbon source and essential amino acids for growth. Metabolism of these nutrients produced organic compounds which are essential for the development of flavor, preservation and appearance of the products.

To date, yogurt such as natural set, stirred and drink yogurt is the most established dairy-based medium to incorporate probiotic bacteria. Stirred and drink yogurts are well suited for the addition of probiotics after fermentation by starter culture and during the stirring of the product. On the other hand, probiotics were added together with *S. thermophilus* to perform fermentation in the production of probiotic-natural set yogurt. This is because fermentation has to take place in the final container and subsequent stirring would destroy the product's texture (Heller 2001). In addition to yogurt, there are also others probiotic dairy products available in the markets such as fermented milk, ice cream and cheese. The suitability of dairy products as a carrier for probiotics is presented in Table 1.

2.1 *Challenges of Probiotics in Dairy-Based Products*

The viability and stability of probiotics has been both a marketing and technological challenge for industrial producers. Probiotic food should contain suitable viable cells and be maintained at a minimum therapeutic level at the time of consumption

Table 1 Suitability of dairy-based medium as a carrier for probiotics

Dairy-based products	Probiotics	Viability in the products	References
Yogurt	*L. acidophilus, L. casei, B. bifidum*	>10^6 CFU/mL and remained high even after 30 days of storage at 4°C	Sendra et al. (2008)
Natural-set yogurt	*L. acidophilus, L. casei*	>10^6 CFU/mL over 28 days of storage at 4°C	Donkor et al. (2007a)
Whey drink	*L. acidophilus, Bifidobacterium, L. casei*	>10^7 CFU/mL over 28 days at 4°C	Drgalić et al. (2005)
Sweet-acidophilus milk	*L. gasseri*	>10^8 CFU/mL over 28 days of storage at 4°C	Usman and Hosono (1999)
Low-fat cheddar cheese	*L. casei*	10^7 CFU/g over 3 months of storage at 4°C	Sharp et al. (2008)
Whey cheese	*B. animalis, L. acidophilus, L. brevi, L. paracasei*	10^7 CFU/g over 28 days of storage at 7°C and provide protective effect in simulated gastrointestinal condition due to higher pH, dense matrix and fat content of cheese	Madudeira et al. (2005)
Ice cream	*L. johnsonii*	10^7 CFU/g over 8 months storage	Alamprese et al. (2002)

in order to exert beneficial effects in the host. Therefore, many studies have assessed the challenges faced to acquire better growth and survivability of probiotics during food manufacturing and storage.

The maintenance of the functionality of probiotics in a dairy-based matrix is often related to various intrinsic barriers existing during processing and storage. For technological purposes, starters are always added during the manufacturing process with the intention of acidification, flavor and texture development. However, antagonistic interaction between probiotics and starter cultures may result in growth inhibition by acid, peroxide, bacteriocins and other metabolites (Granato et al. 2010). Roy et al. (1995) reported that the addition of probiotics to the normal starters during the production of acidophilus milk and cheese reduced the growth of probiotics compared to when they were added alone in milk. This was because the typical starter culture in the fermented milk product could grow faster and lead to rapid acidification and subsequently reduced the fermentation time. Upon reaching the desired pH, fermentation will cease and result in lack of time for the extensive growth of probiotics (Vinderola et al. 2002a). In another study, Kheadr et al. (2004) reported that the growth of bifidobacteria in cheese was reduced in the presence of lactococci. This was due to the production of nisin by lactococci during fermentation which is inhibitory to bifidobacteria.

The presence of oxygen in yogurt and ice creams due to processing such as stirring and mixing of ingredients is inevitable. This oxygen content in the growth medium could pose a problem to probiotics. Probiotics such as *L. acidophilus* (microaerophilic) and bifidobacteria (obligate anaerobic) are strongly sensitive to oxygen. The viability of probiotic cells upon exposure to oxygen has been investigated by Ahn et al. (2001) and the authors found that the growth of *B. longum* was inhibited in the presence oxygen.

The presence of oxygen is harmful to most probiotics mainly because reaction of oxygen with water could form reactive oxygen species such as superoxide (O_2^-) and hydrogen anion (OH^-) that would attack proteins, lipids and nucleic acids and cause the death of probiotic cells (Ahn et al. 2001). Oxygen in the medium could also lead to the production of hydrogen peroxide either by the probiotic itself or by lactobacilli from the starter (Villegas and Gilliland 1998). Dave and Shah (1997) reported that *L. delbrueckii* created peroxide in the medium when oxygen is present. The production of this peroxide is detrimental and strongly affects the storage stability of probiotics in dairy-based products.

Being a fermented product such as yogurt and fermented milk, acidity (low pH) is always the limiting factor for the growth of probiotics during processing and storage. The organic acids and amino acids released as a result of the metabolic activity of yogurt cultures during storage may affect the viability of probiotics. The production of lactic acids by *L. delbrueckii* during refrigerated storage has been claimed to cause loss of viability of probiotic bacteria (Dave and Shah 1997). However, the sensitivity of probiotics towards acidic environment appears to be strain dependent. *L. casei* has been reported to have a better tolerance towards acids than *L. acidophilus* and *L. fermentum*, while *L. acidophilus* is more resistant to acid environments than bifidobacteria (Teh et al. 2009).

The addition of fruits that contribute to pH lowering of fermented dairy-based products may also affect the stability of probiotics during storage. In a study performed by Kailasapathy et al. (2008) which involved addition of mixed berry and passion fruit into yogurts, the low pH of the environment due to the presence of the fruit pulp substantially reduced the viability of *L. acidophilus*. In addition, increase in acidity caused by increasing percentage of fruit pulp also resulted in rapid viability loss of *B. animalis* (Kailasapathy et al. 2008).

The sensory properties of the food products is another important factor in the production of probiotics fermented dairy-based products. It is important that the incorporation of probiotics will not change the acceptability of the products. Nevertheless, technological options such as microencapsulation and increasing inoculum sizes could also cause alterations in the flavor and texture of the product during storage. Maragkoudasaki et al. (2006) demonstrated that increasing the inoculum of *Lactobacillus* in Greek yogurt via concentrated and encapsulated cells (10^{10}–10^{11} CFU/g) led to poor organoleptic properties. Therefore, adequate sensory evaluations are required to allow successful formulation of the food. The effects of probiotic addition to fermented milk-based products on the sensory attributes are presented in Table 2.

Table 2 Effect of addition of probiotic cultures to fermented milk-based products on their sensory properties

Fermented milk-based products	Additional supplements	Effect on sensory properties	References
Yogurt containing *L. acidophilus*, *B. bifidum*	Açai pulp	Increased preference with an increase in the proportion of açai pulp. High acceptance of flavor and color	Almeida et al. (2009)
Butterlike-fermented milk containing *B. animalis*	Flavored with fruit flavors (strawberry, vanilla, mint, *graviola*, *cupuaça*); added with sucrose and sucralose	Acceptability test results ranging from "like slightly" to "like moderately" for sucrose and sucralose-added samples	Antunes et al. (2009)
Ice cream containing *L. acidophilus*, *B. lactis*	Inulin, sugars	Increase in sugar concentration improved sensory properties. No effect on the addition of inulin	Akin et al. (2007)
Ice cream containing *B. longum*, *B. lactis*, *S. thermophilus*, *L. delbrueckii*	Acerola pulp	Low sensory acceptance of product stored for 15 weeks	Favaro-Trindade et al. (2006)
Synbiotic *petit-Suisse* cheeses containing *B. animalis* subsp. *lactis*, *L. acidophilus*	Inulin, oligofructose, and honey (oligosaccharides)	Highest acceptance for oligofructose and inulin containing cheese stored 28 days at 4°C	Cardarelli et al. (2008)

2.2 Means to Overcome Limitations of Probiotics in Dairy-Based Products

Although dairy-based products have been considered as an ideal vehicle for delivering probiotic bacteria to the human gastrointestinal tract, limitations including overacidification and antagonistic effect of starter culture have been reported to cause loss of viability of probiotic bacteria in the product. Therefore, much research has been conducted to overcome the limitations of probiotic bacteria in milk-based products.

One of the most constraining drawbacks associated with the use of probiotic cultures in fermented dairy products is the acidity of the products. The pH of yogurt may decline to a level below 4.0, which is inhibitive towards bifidobacteria (generally sensitive to pH values below 4.6). However, high bifidobacteria counts ($>10^6$ CFU/g) were observed in probiotic-buttermilk with a pH as low as 4.2 after storage for 28 days at 4°C, suggesting a selection of more acid-tolerant strains is possible and beneficial for high acid foods (Antunes et al. 2009). The application of ABT-yogurt starter cultures that consist of *L. acidophilus, B. bifidum* and *S. thermophilus* could also overcome the problem of overacidification and bring out the preferred flavors in fermented milk-based products (Lourens-Hattingh and Viljeon 2001).

The selection of compatible probiotics and starter cultures is crucial in the production of fermented milk-based products to prevent inhibition (Vinderola et al. 2002a). Inhibition by acid, peroxide, bacteriocins and other metabolites can be overcome when proper co-cultures are used. *L. acidophilus* and *B. bifidum* have been identified to have synergistic-growth promoting effects, while *L. acidophilus* has been found to release amino groups from milk proteins to support the growth of both *L. acidophilus* and *B. lactis* in milk (Moayednia et al. 2009).

In addition, proper selection of culture is also beneficial for improving the physical properties, organoleptic and acceptability of healthy (low-fat) dairy-based products. Costa et al. (2010) demonstrated that application of *Lactococcus lactis* in fat-reduced cheese substantially improved the cheese yield and both textural and cooking properties, without negatively affecting the flavor profiles of the cheese. This was due to the ability of the strains to produce exopolysaccharides which formed a porous network of the cheese (Fig. 1). These pores retained extra moisture in the cheese and diluted the casein matrix, leading to increased yield, improved texture and cooking properties of the cheese.

Supplementation with growth enhancer has also been demonstrated to increase the viability of probiotics in dairy-based products. Studies exhibiting the improvement of viability of probiotics during fermentation and storage in the fermented dairy-based products via the addition of growth supplements are presented in Table 3.

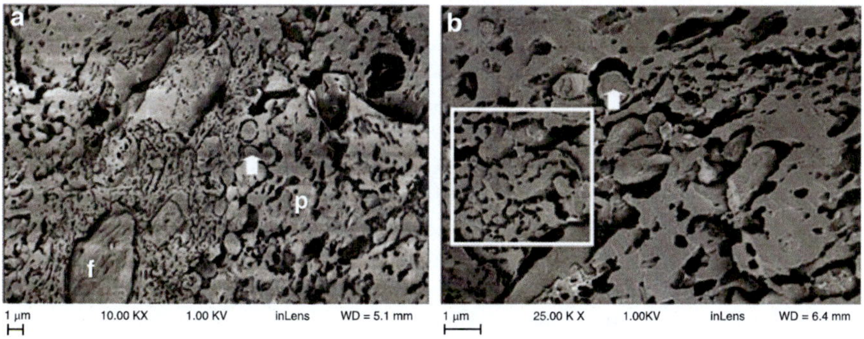

Fig. 1 Cryogenic-scanning electron micrographs of 1-day-old half-fat Cheddar cheeses manufactured using an exopolysaccharide (EPS)-producing culture (**a**, **b**). *Arrow* indicates bacterial cells, *square* indicates the network areas of protein–EPS interaction; f = fat globule; p = protein. Reprinted from Costa et al. (2010), with permission from Elsevier (License number: 2545681326740)

Table 3 Growth supplements added to improve growth or survival of probiotics in various dairy-based products

Dairy-based products	Growth supplements	Probiotics	References
Yogurt	Inulin	*L. acidophilus*, *L. bulgaricus*, *L. casei*, *Bifidobacterium*, *B. animalis* subsp. *lactis*	Donkor et al. (2007a) Vasiljevic et al. (2007)
	Mango/strawberry	*L. acidophilus*	Kailasapathy et al. (2008)
	Ascorbic acid	*S. thermophilus*, *L. delbrueckii* subsp. *bulgaricus*, *L. acidophilus*	Dave and Shah (1997)
	Oat β-glucan	*B. animalis* subsp. *lactis*	Vasiljevic et al. (2007)
	Polydextrose, maltodextrin, oligofructose	*L. acidophilus*, *L. rhamnosus*, *B. animalis* subsp. *lactis*	Oliveira et al. (2009)
	Sucrose, sucralose	*B. animalis* subsp. *lactis*	Antunes et al. (2009)
	Lemon/orange fibers	*L. acidophilus*, *L. casei*, *B. bifidum*	Sendra et al. (2008)
Fermented milk	Cysteine	*L. acidophilus*, *L. paracasei* subsp. *casei*, *B. bifidum*	Güler-Akln and Akln (2007)
	Inulin, sugar	*L. acidophilus*, *B. lactis*	Akin et al. (2007)
Ice cream	Sucrose, aspartame	*L. acidophilus*, *L. agilis*, *L. rhamnosus*	Başyğit et al. (2006)
Cheeses	Inulin + fructooligosaccharides	*L. acidophilus*	Rodrigues et al. (2011)
Petit-suisse cheese	Apple, pear	*L. acidophilus* *B. animalis* subsp. *lactis*	Kourkoutas et al. (2006)

In order to protect the viability of probiotic bacteria against undesirable growth conditions such as high acidity in yogurt, freezing injury and oxygen toxicity in ice cream, cell microencapsulation has offered some advantages for improving the viability of probiotic bacteria. Microencapsulation separates the cells from the adverse environment with protective coating, thus potentially reducing injury. Özer et al. (2008) reported that the viability of encapsulated probiotic bacteria remained above the therapeutic level while a drastic decline in the viability of nonencapsulated probiotic bacteria was detected during the scalding process (50–52°C for 1–2 min) involved in the manufacture of Kadar cheese. In addition, encapsulation with whey protein has also been reported to increase the survivability of probiotics in yogurt during storage and through simulated gastrointestinal

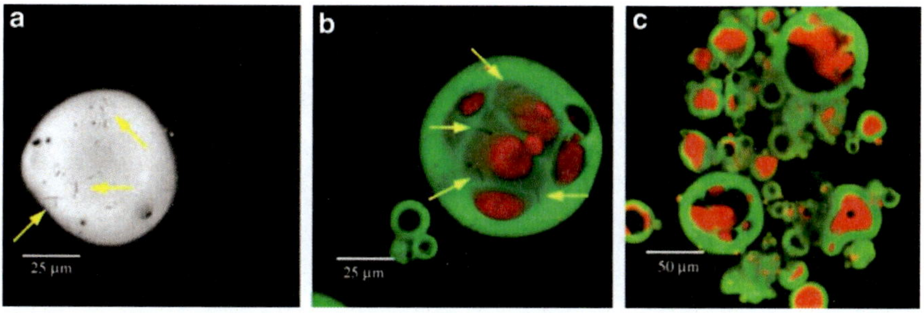

Fig. 2 Confocal laser-scanning micrographs of whey protein-based microcapsules produced with fresh culture and heat-denatured 10% w/w WPI solution (**a**), fresh culture, heat-denatured 10% w/w WPI solution and AMF containing 99.9% (w/w) lipids (o/w emulsion) (**b**), and micronized powder of freeze-dried culture heat-denatured 10% w/w WPI solution and AMF containing 99.9% (w/w) lipids (o/w emulsion) (**c**). Small dark rods indicated by *yellow arrows* represent the locations of the bacterial cells within the protein matrix. Reprinted from Picot and Lacroix (2004), with permission from Elsevier (License number: 2545660385556)

condition. Picot and Lacroix (2004) previously spray-dried whey protein-based microcapsules via three different methods (Fig. 2) using fresh culture or micronized powder of freeze-dried culture with or without anhydrous milk fat (AMF) containing lipids emulsion. Whey-based microcapsule produced using fresh culture without AMF was found to be the least destructive while milk fat was found to decrease the viability of probiotics.

The advancement of nanotechnology has also led to its increased applications for the protection of probiotic cells. Fung et al. (2010) has previously produced nanofibers from polyvinyl alcohol (PVA) and soluble dietary fiber fractions of agricultural wastes (agrowastes) such as okara, oil palm fronds and oil palm trunks, via the electrospinning method. These nanofibers were later used as nanoencapsulants for lactobacilli cells (Fig. 3a–c). Although cells were subjected to high voltage and shear stress during electrospinning, they remained viable upon nanoencapsulation (as proven by fluorescence microscopy; Fig. 3d) and upon 21 days of storage at 4°C. This method is relevant and suitable for applications in dairy foods, as two of the main viability challenges of probiotics in dairy products are oxygen and acidity. PVA is a biocompatible hydrophilic polymer which confers high oxygen barrier when dry, without affecting the bioactivity of the bacteria (Lopez-Rubio et al. 2009), while dietary fiber (fractions from agrowastes) are resistant towards gastrointestinal digestion and thus confer protection on acidity.

3 Soy-Based Products as Carriers for Probiotics

Soybean is a vegetarian diet well known for its inexpensive source of proteins and calories for human consumption. It is free of cholesterol and lactose and thus is a suitable diet for a lactose intolerance consumer. Soy also contains reasonable

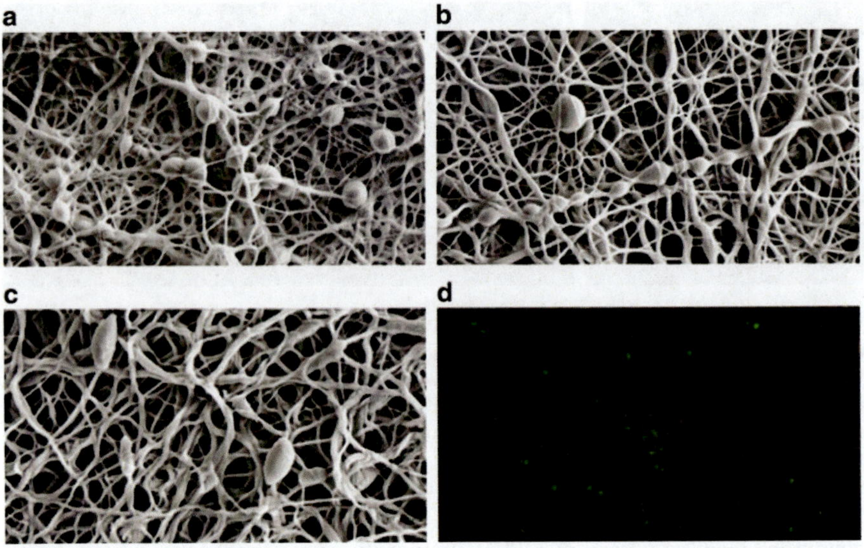

Fig. 3 SEM micrographs of *Lactobacillus acidophilus*-containing nanofibers from PVA:soluble dietary fiber of (**a**) okara, (**b**) oil palm trunks, and **c** oil plam fronds. Optical micrograph of the same nanoencapsulated cells with fluorescence illumination indicating live bacteria within the nanofibers (**d**). Micrographs courtesy of Fung WY, Universiti Sains Malaysia, Malaysia

amounts of bioactive ingredients such as isoflavones. There have been several promising evidences showing positive correlation between consumption of soy with various therapeutic effects including prevention of osteoporosis, breast cancer and cardiovascular diseases. Therefore, consumption of soy-based product has increased tremendously in the recent years.

Increasing public awareness on healthy lifestyle has prompted the demand for diet with versatile health-benefiting properties. Therefore, soy products with attractive nutritional properties and therapeutic significance have been extensively researched as a probiotic dietary adjunct. The formulation of soy products with probiotics offers consumers the benefits of both probiotics and soy.

Up to date, various studies have evaluated the efficiency of soy-based products as a potential carrier of probiotics (Table 4).

In addition to supporting the growth of probiotics, soy-based products also protected probiotic cells during transit through the harsh condition of gastrointestinal tract. Wang et al. (2009) demonstrated that the use of soymilk as a carrier for *L. casei* significantly improved the viability of this strain in simulated gastric transit (phosphate buffer containing pepsin; pH 2 and pH 2.5; 37°C). Additionally, de Boever et al. (2001) reported that soy products such as soygerm powder protected *L. reuteri* against bile salt toxicity in the small intestine. Soy protein has been found to have the ability to bind and aggregate bile acids (taurocholate and glycocholate), thus protecting the live bacterium from their toxicity.

Table 4 Suitability of soy-based medium as a carrier for probiotics

Soy-based products	Probiotics	Viability in the products	References
Soymilk	*Lactobacillus*, *Bifidobacterium*, *Streptococus thermophillus*	>10^6 CFU/mL after fermentation	Donkor et al. (2007b)
Soymilk	*L. acidophilus*, *L. casei*, *Bifidobacterium*	Ranged from 10^7 to 10^8 after fermentation	Yeo and Liong (2010b)
Soymilk	*L. acidophilus* and *L. gasseri*	>10^8 CFU/mL after fermentation	Ewe et al. (2010)
Soy cream cheese	*L. acidophilus*	>10^7 CFU/g over 20 days of storage at 4 and 25°C	Liong et al. (2009)
Tofufa, an Asian dessert make from soy	*L. bulgaricus*, *L. fermentum*	>10^6 CFU/g and maintained throughout 9 days of storage at 4 and 25°C	Ng et al. (2008)
Soy yoghurt	*L. acidophilus*, *B. lactis*, *L. paracasei*	>10^8 CFU/mL and remained consistent after 28 days of storage at 4°C	Donkor et al. (2005)

3.1 Interaction of Probiotics and the Composition of Soy-Based Products

Soy-based products could act as a good carrier for probiotics due to the natural pH of soy and the large pool of carbon and nitrogen sources. The presence of sucrose and simple sugars in soymilk supports the growth of probiotics, where sucrose, glucose and fructose are scarcely detected upon fermentation. In addition, the growth of probiotics in soy-based products could also be attributed to their ability to metabolize soy oligosaccharides such as raffinose and stachyose during fermentation. Past studies have reported that probiotics possess varying levels of α-galactosidase enzyme which enabled them to hydrolyze the α-galactosyl oligosaccharides into simpler sugar for growth (Yeo and Liong 2010b). The α-galactosidase enzyme acts upon the α-1,6 bonds of these oligosaccharides releasing galactose and sucrose, and an invertase will then hydrolyse the β-(1,2)-fructofuranoside bond of this disaccharide, releasing fructose and glucose (LeBlanc et al. 2004) in soy-based medium (Fig. 4). These simple sugars are easily metabolized by probiotics and lead to enhanced growth in the soy-based products.

Soy also contains reasonable amount of proteins including peptide and essential amino acid that could support the proliferation of probiotics. Probiotics such as *Lactobacillus* and *Bifidobacterium* possess a complex proteolytic system that enables them to hydrolyze soy proteins into peptides and amino acids for growth. Extracellular proteinases cleave proteins into oligopeptides, and further into simpler peptides and amino acids by intracellular peptidases such as endopeptidases, aminopeptidases, tripeptidases and dipeptidases.

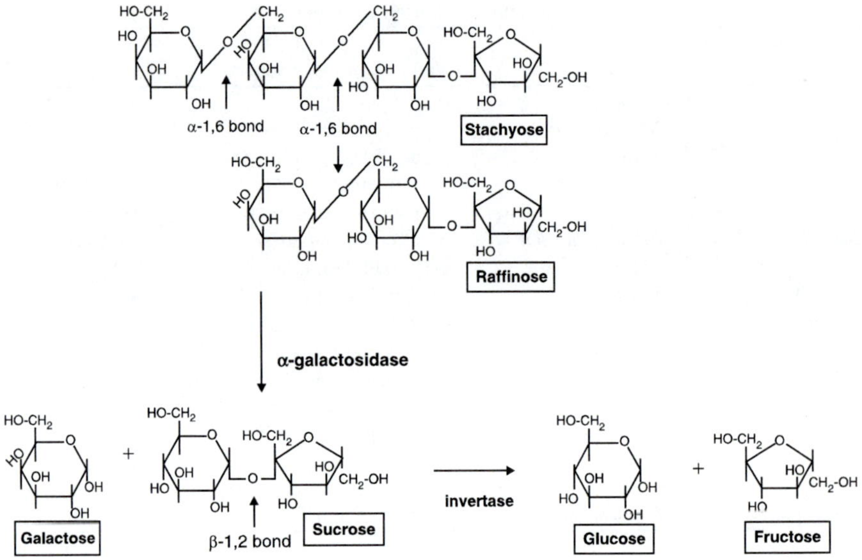

Fig. 4 Chemical structure of α-galactosyl oligosaccharides and the hydrolytic enzyme produced by probiotics

3.2 Challenges of Probiotics in Soy-Based Products

Despite well-documented evidences showing the excellent potential of soy as a carrier of probiotics, the production of probiotics–soy-based products would have to deal with several technological limitations. One of the main challenges faced by probiotic–soy-based products is the sensory properties of the products upon fermentation. Donkor et al. (2007b) demonstrated that probiotic–soy yogurts were less preferred compared to the soy yogurt without probiotics as determined by sensory evaluation. The authors postulated that the less preference to probiotic–soy yogurts was probably due to the high concentration of organic acids in probiotic–soy yogurts. Therefore, in order to increase the acceptability of consumer towards probiotics-fermented soy products, appropriate measures are taken to mask the excessive acidic taste of the products.

In addition to sensory properties, rheological and textural properties of the products are equally important in the production of probiotics food. Soymilk often has a lower buffering capacity compared to milk (Wang et al. 2009). Therefore, the reduction of pH upon fermentation with probiotics was more prevalent in soymilk and this could lead to increased viscosity, coagulation and alteration in the appearance of soymilk. At extreme pH away from the isoelectric point (4.6), weak gels were formed due to strong electrostatic repulsion (Liu 1997). The reduction of pH during fermentation also leads to lower water binding capacity of soymilk, which accelerates the coagulation and syneresis process of fermented soymilk (Chumchuere and Robinson 1999), leading to reduced acceptability of consumers.

The stability of probiotics during storage is essential in determining the shelf life of the products and this remained a challenge in the production of soy products as a carrier for probiotics. Past studies have indicated that there is a general reduction in the growth of probiotics in soy-based products over storage. The viability of *Lactobacillus* in soy products such as soy cheese reduced by at least 1 log cycle after storage at 10°C for 30 days (Liu et al. 2006). The reduction of growth were even more prevalent during storage at 25°C where *Bifidobacterium* reduced by approximately 2 log after 10 days of storage (Chou and Hou 2000). This was probably due to the accumulation of organic acids and reduced supply of carbohydrates in soymilk.

3.3 Means to Overcome Limitations of Probiotics in Soy-Based Products

Considering the importance of extending the shelf life of probiotics food with the required level of total viable probiotics, attempts are made to stimulate the growth of probiotics in soymilk. Viability of probiotics are enhanced via various factors, and the most common and inexpensive being the supplementation with various nutrients or growth enhancer and is presented in Table 5.

Another method to improve the potential of soy-based products as a carrier of probiotics involved the use of immobilized probiotic cells. Probiotics are commonly immobilized on matrices such as alginate, carrageenan and polyacrylates to increase their growth and storage stability. Chen (2008) reported that *L. acidophilus* immobilized within the wall material composed of κ-carrageenan and inulin was able to maintain viability above 5.0×10^8 CFU/g in soy energy bar stored at 4°C for 10 weeks. Other unconventional immobilizers were also shown to enhance the potential of soy as an efficient probiotic vehicle. Teh et al. (2009) demonstrated that lactobacilli cells immobilized on agrowastes powder obtained from rinds of durian,

Table 5 Means to enhance the viability of probiotics in soy-based products

Probiotics	Supplements Added	References
L. acidophilus, *L. gasseri*	1 mg/L B-vitamins	Ewe et al. (2010)
L. paracasei, *L. acidophilus*, *B. longum*	10% (w/v) sucrose, fructose and lactose	Wei et al. (2007)
B. infantis, *B. longum*	1% Bifitose (isomaltooligosccharide), glucose, lactose or galactose and 0.5%(w/v) yeast extract, peptone, tryptone, casitone or N-Z-Case plus	Chou and Hou (2000)
S. thermophilus, *L. delbrueckii* ssp. *bulgaricus*, *L. acidophilus*, *L. casei* and *B. animalis*	2% (w/v) inulin or 1% (w/v) each of raffinose and glucose	Donkor et al. (2007b)
L. acidophilus, *L. casei*, *Bifidobacterium*	1% (w/v) of fructooligosaccharides, mannitol, inulin, maltodextrin and pectin	Yeo and Liong (2010a)

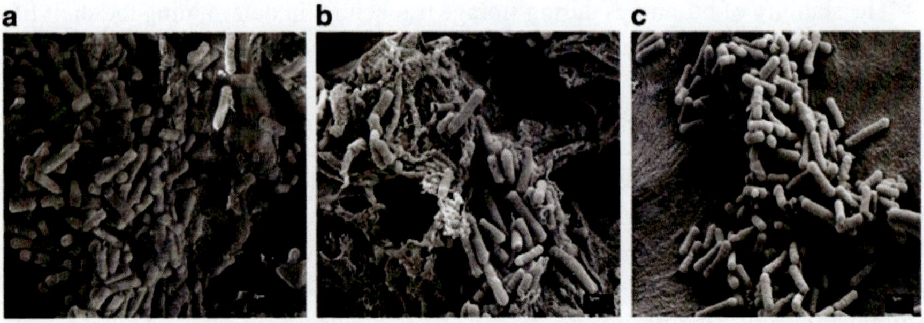

Fig. 5 Scanning electron micrographs of probiotic cells immobilized on (**a**) cempedak rind powder (**b**) durian rind powder and (**c**) mangosteen rind powder

mangosteen and cempedak survived longer and grew better in soymilk than free cells. Agrowastes provided good attachment support for cells (Fig. 5), leading to enhanced growth, utilization of substrates and production of metabolites.

Coagulation of soy protein due to decreased pH upon fermentation is one of the limiting factors for successful marketing of probiotics fermented soy products. Therefore, it is essential to overcome this problem in order to increase the acceptability of consumers towards fermented soy products. Curd formation and texture profile of fermented soy products could be improved or overcome by the addition of buffering agent which could stabilize the pH of the soy medium.

In addition, effective media buffering is also vital to prevent sublethal injury due to exposure to low pH during fermentation and storage (Heenan et al. 2002). Supplementation of calcium into soy products has been shown to increase the buffering capacity and stabilized the pH of soy products. Yazici et al. (1997) demonstrated that supplementation of calcium significantly slowed down the pH reduction of soy yogurt fermented by *Lactobacillus* and *Bifidobacterium* compared to the control yogurt without supplementation. The supplementation of calcium also improved the gel strength of soy yogurt compared to the control that exhibited a hard and brittle texture.

4 Cereal-Based Products as Carriers for Probiotics

Cereals such as rice, oats and wheat which are consumed as part of our daily diet are one of the most favored choices of probiotic delivery vehicle. Cereals are abundant in naturally occurring bioactive substances and fibers, which could act as fermentation substrates of indigenous gut microflora. The consumption of high-fiber food such as cereals has been associated with various health benefits including prevention and treatment of cardiovascular disease, regulation of body weight and promoting bowel regularities. Therefore, fermentation of cereal-based foods with probiotics could provide the beneficial effects of both the fiber and probiotics and thus could meet the demand of the consumers for food with versatile

Table 6 Suitability of cereal-based medium as a carrier for probiotics

Cereal-based products	Probiotics	Viability in the products	References
Rice medium	22 strains of bifidobacteria	10^7–10^8 CFU/mL upon fermentation	Lee et al. (1999)
Malt, barley and wheat extract	*L. fermentum*, *L. reuteri*, *L. plantarum*, *L. acidophilus*	10^7–10^{10} CFU/mL after fermentation	Charalampopoulos et al. (2002)
Oat-based beverage	*L. plantarum*	Max viability of 10^{10} CFU/mL after fermentation and no significant reduction in viability after 21 days (4°C)	Gupta et al. (2010)
Cereal extract	*L. plantarum*	>10^8 CFU/g even after 63 days (4°C)	Charalampopoulos and Pandiella (2010)
Wheat bran slurry	*L. plantarum*, *Pediococcus acidilactici*	10^7–10^8 CFU/mL after fermentation	Lamsal and Faubion (2009)

health-promoting properties. The efficiency of various cereal-based products in supporting the growth of probiotics has been reported (Table 6).

In addition, Charalampopoulos et al. (2003) demonstrated that cereal-based food including malt, barley and wheat extract provided the protective effect on probiotics against the low pH of gastric tract. These cereal extracts improved the viability of *L. plantarum* by approximately 3–4 log cycles in pH 2.5 for 4 h. The protective effect of the cereal extracts was mainly attributed to the presence of sugars such as maltose and glucose in the cereal extracts. Corcoron et al. (2005) also reported that glucose enhanced the survival of *L. rhamnosus* GG in simulated gastric juice at pH 2.0 for 90 min because the sugars play a significant role in pH homeostasis by activation of F_0F_1-ATPase. F_0F_1-ATPase is a multiple-subunit enzyme function to provide a mechanism for pH homeostasis in acidic conditions. In order to function efficiently, this enzyme requires sufficient ATP for exclusion of proton from the cell, thus maintaining the pH homeostasis and cell viability. The presence of glucose in acidic conditions aids the production of ATP by probiotics via glycolysis and subsequently enables optimal activity of F_0F_1-ATPase to maintain pH homeostasis and cell viability.

Cereal products also protected probiotics from destruction by bile salts. The cell membranes of probiotics which consist of lipids and fatty acids are generally very susceptible to destruction by bile. However, in the presence of cereal products, probiotics are found to exert higher resistance towards bile. The bile tolerance (2% bile for 4 h) of *L. acidophilus*, *L. plantarum* and *L. reuteri* was improved by at least 10^2 CFU/mL in the presence of malt, wheat and barley extract (Patel et al. 2004). The presence of fermentable carbon source such as resistant starch, β-glucan, stachyose, raffinose and fructooligosaccharides in cereals was reported to be the key factor in alleviating the detrimental effects of bile salts. Perrin et al. (2000) also reported that the presence of oligosaccharides led to improved resistance of *Bifidobacterium* strains against the bactericidal effects of bile salts. The oligosaccharides improved the resistance towards bile salts by an energy-dependant process. The metabolism of oligosaccharides by probiotics produced a high yield of energy required for the detoxification of bile salts and thus contributed towards increased bile salt resistance.

4.1 Interaction of Probiotics and the Composition of Cereal-Based Products

The ability of cereal-based products to support the growth of probiotics was mainly due to their high concentration of fibers. Cereals contain various fibers such as xylooligosaccharides, xylan and arabinoxylan which could act as a growth substrate for probiotics. The ability to grow in cereal products and utilize the different types of fibers in cereals varies largely with strains. Crittenden et al. (2001) evaluated the growth of 35 strains of probiotics on the dietary fiber of cereals and found that only *B. longum* strains grew well on the arabinoxylan. This was due to the ability of this strain to produce α-arabinofuranosidase, the key enzyme for utilization of arabinoxylan. Genome sequence of several *B. longum* has revealed that these strains exhibited genes for extracellular polymeric xylan-acting arabinofuranosidases AXH-m (GH family 51), which are able to cleave arabinose from mono-substituted β-D-xylopyranosyl backbone (Pastell et al. 2009). Therefore, the hydrolysis of the arabinosyl groups from polymeric arabinoxylan occurs extracellularly due to the production of the extracellular arabinofuranosidases by *B. longum*. Eventually, the released monosaccharide arabinoses were imported into the cell and readily utilized by *B. longum* for growth (Crittenden et al. 2001).

In addition, the starch and fiber component in cereals could act as a protectant or immobilizer to enhance the stability of probiotics during processing, storage and gastrointestinal environment. Wang et al. (1999) suggested that the starch granules of high amylose maize acted as an encapsulation material and provided surface of attachment for probiotics. Bosnea et al. (2009) reported that *L. casei* cells were immobilized on the surface and interior of wheat grain, which mainly consists of dietary fiber. The immobilized freeze dried cells remained stable even after storage for 12 months at −18°C (Fig. 6). Similarly, Guergoletto et al. (2010) reported that *L. casei* LC-1 could adhere to the matrices of oat bran fiber and such adherence enhanced the survivability of this strain during dehydration and storage. The adhered probiotics could better withstand the processing condition (dried at 45°C in a vacuum oven for 25 h) and viability was maintained above 10^7 CFU/g during storage for 28 days at room temperature and 10°C. The adhered cells were also protected against simulated gastrointestinal condition, mainly attributed to the production of exopolysaccharides which act as a stress protectant.

Generally, cereal-based products contain high concentration of starch. In order to utilize this carbohydrate source in cereals, α-amylase enzyme must be essentially present. Therefore, the ability of probiotics to proliferate well in cereal products

Fig. 6 Electron microscopy photographs showing *L. casei* cells immobilized on surface of wheat grains before freeze drying (**a**), freeze drying (**b**), and upon freeze-drying and storage for 12 months at −18°C (**c**). Reprinted from Bosnea et al. (2009), with permission from Elsevier (License number: 2545670301975)

depends primarily on the ability of the probiotics to exhibit amylolytic activity. Past studies have demonstrated that probiotics such as *Bifidobacterium* possessed amylase activity and were able to utilize the starch in rice medium for growth (Lee et al. 1999). Likewise, Calderon et al. (2003) reported that *L. fermentum* Ogi E 1 was able to produce α-amylase and consumed starch in cereals. The amylase of probiotics (*L. amylovorus* and *L. plantarum*) was found to exhibit a starch binding domain that can bind granular starch and increase the local concentration of substrate at the active site of the enzyme (Rodríguez-Sanoja et al. 2005). Upon binding at the active site, amylases disrupt the structure of the starch surface of raw and soluble starch by hydrolysis of the α-1,4 glycosidic linkages, thereby generating smaller dextrins and oligosaccharides which is easily utilized by probiotics.

4.2 *Challenges of Probiotics in Cereal-Based Products*

Although today's consumers are increasingly demanding for products with functional properties, the consumption of new products is strongly determined by their organoleptic qualities. Therefore, one of the primary challenges in developing a novel cereal-probiotic product is the need to ensure that fermentation of the food products by probiotics did not adversely affect its organoleptic properties. Nyanzi et al. (2010) reported that fermentation of maize beverage with *L. paracasei* contributed to sour, yeasty and off-flavor attributes as evaluated by ten trained sensory panels based on quantitative descriptive analysis. The off-flavor aroma is indeed a negative attribute and would affect the marketability of the product.

Cereal products such as energy bars and breakfast cereal usually exhibited a considerably long shelf-life. Therefore, the selection of suitable probiotics that survive in the cereal-based products for a prolonged period is indeed a challenge in developing cereal-based probiotic products. Previous studies have indicated that some strains of probiotics were unable to sustain the desired viability during storage. Gokavi et al. (2005) also demonstrated that the viability of *L. acidophilus* in oat-based beverage decreased significantly during storage at 4°C, and the viability was below the minimum requirement for probiotics food after 4 weeks of storage (10^5 CFU/mL).

4.3 *Means to Overcome Limitation of Probiotics in Cereal-Based Products*

Considering the importance of viable cells of probiotics in the food prior to consumption, various efforts have been attempted to increase or sustain the viability of probiotics during storage. The survivability of probiotics during storage could be increased by simultaneous inoculation of two or more strains in the cereal products. Past study has indicated that after 21 days of storage at 4–6°C, viable counts of *B. animalis* + *L. acidophilus* in water-based rice and maize puddings were 1–2 log higher than those obtained from both strains growing individually (Helland et al. 2004). This was probably due to the synergistic growth promoting effect between lactobacilli and bifidobacteria. In general, lactobacilli have a higher proteolytic activity leading to higher release of peptides and amino acids for the growth of bifidobacteria. The growth of lactobacilli will also reduce the redox potential of the growing medium, which is favorable for the growth of bifidobacteria (Helland et al. 2004). On the other hand, bifidobacteria produced more acetic acid which could possess a buffering effect and prevent a sharp decline in pH of the growing medium, thus preventing the growth inhibition of lactobacilli at low pH.

Past studies have reported that fermentation of cereal-based products with probiotics produced off flavor and thus could reduce the acceptability of consumers towards the fermented products. However, appropriate formulation of the cereal-based products could enhance the palatability and acceptability of the products. Sindhu and Khetarpaul (2004) demonstrated that fermentation of barley beverage formulated with milk coprecipitate, sprouted green gram paste and tomato pulp produced beverage with acceptable organoleptic properties and exhibited good acceptability even after 1 month storage at room temperature. This was probably due to the ability of the added ingredients to mask off flavor of the fermented products. Appropriate selection of the starter probiotics is also important in the development of probiotic–cereal products. Nyanzi et al. (2010) reported that different species of probiotics produced fermented-maize beverage with different acceptability in terms of flavor and aroma. The authors evaluated the sensory attributes of maize beverage fermented by four species of probiotics and found that the beverage fermented by *L. acidophilus* or *L. rhamnosus* were well accepted by 10 trained panels and 53 untrained panels.

5 Fruits and Vegetables as Carriers of Probiotics

Fruit and vegetable juices are often perceived as healthy foods and they have taste profiles that are appealing to all age groups. Fruits and vegetables contain beneficial compounds such as phytochemical and antioxidants that are important in maintaining human health. They are also rich in essential nutrients including sugars, mineral, vitamins and dietary fibers, making them ideal substrates for probiotics. Therefore,

Table 7 Growth of probiotics in fruits and vegetable juices

Fruits/vegetable juices	Probiotics	Viability in the products	References
Pomegranate juice	*L. plantarum*, *L. delbruekii*, *L. paracasei*, *L. acidophilus*	>10^8 CFU/mL after fermentation for 48 h	Mousavi et al. (2010)
Orange and pineapple juice	*L. casei*, *L. rhamnosus* GG, *L. paracasei*	>10^6 CFU/mL for 12 weeks of storage at 4°C	Sheehan et al. (2007)
Tomato juices	*L. acidophilus* LA39, *L. plantarum* C3, *L. casei* A4, *L. delbrueckii* D7	10^9 CFU/mL after fermentation and >10^6 CFU/mL after 4 weeks of storage at 4°C	Yoon et al. (2004)
Cabbage juice	*L. plantarum*	10^8 CFU/mL after fermentation and remained 10^7 CFU/mL after 4 weeks of storage at 4°C	Yoon et al. (2006)
Beet juice	*L. acidophilus*, *L. plantarum*, *L. casei*, *L. delbrueckii*	10^6–10^8 CFU/mL after 4 weeks of storage at 4°C	Yoon et al. (2005)
Carrot juice	*B. lactis* Bb-12, *B. bifidum* B7.1, B3.2	>10^8 CFU/mL after 6 h of incubation at 37°C	Kun et al. (2008)

there have been increasing interests in the application of vegetable and fruit juices as alternative carriers of probiotics. In the 1990s, industries have started to produce fruit drinks with added probiotics such as Gefilus®, Rela® and Biola® to meet consumer demands for nondairy-based probiotic products. The ability of fruits and vegetable juices to support the growth of probiotics is summarized in Table 7.

5.1 Challenges of Probiotics in Fruit and Vegetable Juices

Fruit and vegetable juices are highly perishable with limited shelf life. Therefore, in order to increase the shelf life of fruit and vegetable juices, preservatives such as nisin are added into the products. This preservative at certain concentrations are detrimental to the growth of some strains of probiotics. Vinderola et al. (2002b) reported that several strains of probiotics were inhibited by this preservative even at concentrations as low as 1 ppm. However, the sensitivity of probiotics towards this additive is highly strain dependent, where *L. acidophilus* and *S. thermophillus* were reportedly more resistant. Therefore, strain selection is crucial to prevent inferior viable counts of probiotics in juices containing preservatives.

Although processing and packaging may be oxygen free, bottles of fruit and vegetable juices shaken and opened by consumers could expose the products to excessive oxygen which may pose a threat to the viability of probiotics. Champagne et al. (2008) demonstrated that dissolved oxygen substantially increased by 15% after 1 week of storage at 7°C in opened bottles of apple–pear–raspberry juices. Additionally, agitation of bottles containing more than 250 mL of headspace could immediately increase the dissolved oxygen level to a maximum of 30%. Such high dissolved oxygen level is indeed unfavorable for probiotics especially *Bifidobacterium* and could prevent consumers from achieving the desirable viable counts at the time of consumption.

Fruit juices appear as a more complex system in the development of probiotic foods, typically due to the acidic pH (pH 2.5 and 3.7) of the products (Sheehan et al. 2007). Saarela et al. (2006) previously demonstrated that viability and stability of *B. animalis* subsp. *lactis* during storage was less prevalent in fruit juices (orange, grape, passion fruit) compared to milk. The authors suggested that this was probably due to the low pH of the juices (pH 3.7). Many probiotic strains are highly sensitive towards conditions of low pH, thus making them more prone to being killed in fruit juices.

5.2 *Means to Overcome Limitations of Probiotics in Fruit and Vegetable Juices*

The low pH of fruit juices always poses a problem in maintaining the survivability of probiotics. Thus, selection of strains that is more resistant to acidic environment is crucial in the development of a probiotic juice. The ability of various probiotics to tolerate the acidic environment in fruit juices was investigated by Sheehan et al. (2007). The authors found that *L. rhamnosus*, *L. casei* and *L. paracasei* displayed greatest robustness, where these strains were able to remain viable in orange juice (pH 3.65) at levels greater than 10^7 CFU/mL and in pineapple juice (pH 3.40) at levels above 10^6 CFU/mL upon storage at 4°C for 12 weeks.

Microencapsulation was shown to provide protection to acid-sensitive probiotics. The effect of microencapsulation on the viability of probiotic bacteria in orange and apple juices has been assessed by Ding and Shah (2008). Encapsulated probiotic bacteria was found to survive over 6 weeks of storage with viable counts of more than 10^5 CFU/mL while free probiotic cells lost their viability within 5 weeks. The authors claimed that the survivability of encapsulated cells in the juices was due to the protective layer of microcapsules which provides a more favorable anaerobic environment for the sensitive probiotic cells and also served as a physical barrier against the acidic environment of the juices.

In addition to microencapsulation, fruits such as apple and quince pieces have been proposed as immobilization supports, which could enhance the survivability of probiotics over extended storage periods compared to free cells (Kourkoutas et al. 2005). Vacuum impregnation into fruits or vegetables has been introduced to provide immobilization support to probiotic cells. Betoret et al. (2003) reported that vacuum impregnation of *L. casei* in dried apple could better sustain the viability of cells in juices and the viability maintained above 10^6 CFU/g after 2 months storage at 25°C. Similarly, Guergoletto et al. (2010) reported that *L. casei* LC-1 adhered to the matrices of fruit pieces such as apple and banana showed enhanced survivability during dehydration and storage. The adhered cells withstand better processing condition (dried at 45°C in a vacuum oven for 25 h; Fig. 7), and showed viability exceeding 10^7 CFU/g upon storage for 28 days at 25 and 10°C. Adhered cells were also protected against simulated gastrointestinal condition, mainly attributed to the production of exopolysaccharides which acted as a stress protectant.

Juices added with prebiotics can also improve the viability and stability of the probiotics. This was demonstrated by Vergara et al. (2009) who studied the effect of prebiotic in *L. johnsonii*-fermented cashew apple juice. The cashew apple juice was prefermented by *Leuconostoc mesenteroides*, which is a typical industrial dextran producer to produce prebiotic oligosaccharides that could promote the growth of probiotics in the cashew apple juice. The authors found that the growth of *L. johnsonii* in the fermented cashew apple juice was threefolds higher than the nonfermented juice.

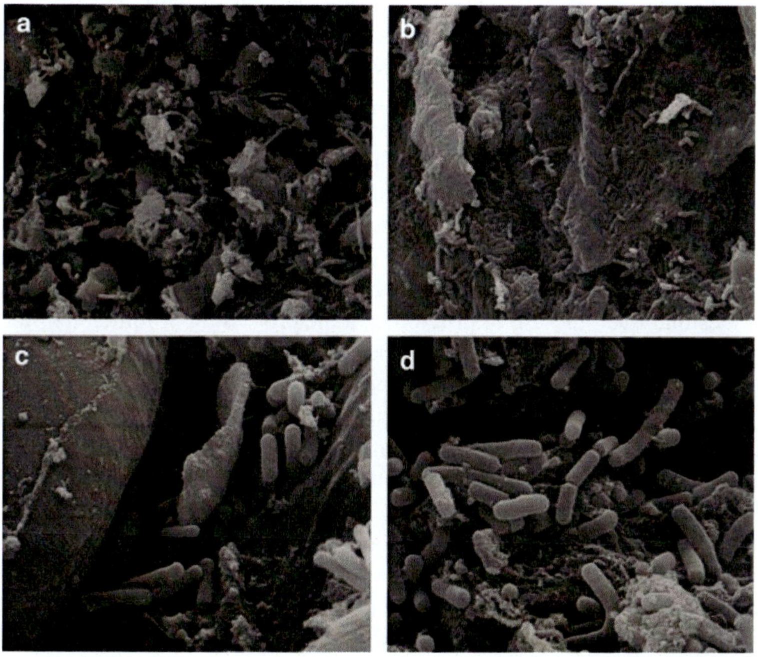

Fig. 7 Scanning electron micrograph of *L. casei* LC-1 cells immobilized on apple and banana pieces before (**a**, **c**) and after (**b**, **d**) drying at 45°C under vacuum. (**a**, **b**) Apple pieces, and (**c**, **d**) banana pieces. SEM courtesy of Dr Karla Bigetti Guergoletto, Universidade Estadual de Londrina (UEL), Campus Universitário, Brazil. Credits for Dr. Juca San Martin, Laboratorio de Microscopia Eletrônica e Microanálise (LMEM) – UEL

Probiotics incorporated food products often impart different taste and acceptance among consumers thus causing a problem in producing beverages that are widely accepted by all age groups. In a descriptive analytic study by Luckow and Delahunty (2004) using novel blackcurrant juices containing probiotics, the consumers claimed that the probiotic juices contained aromas ("perfumey", "dairy") and flavors ("sour", "savoury") that differed from the conventional blackcurrant juices. However, the addition of tropical fruit juices could successfully mask the off flavors caused by probiotics (Luckow et al. 2006). The authors reported that consumers showed a general liking for the sensory characteristics of the fruit juice that has undergone masking, with sensory descriptive analysis claiming that the masked juice contained low levels of "medicinal" odor and flavor.

6 Meat and Meat Derivatives as Carriers of Probiotic Microorganisms

Commercial application of probiotic microorganisms in meat and meat derivatives is still limited. The target products of meat derivatives with the inclusion of probiotic bacteria are mainly dry sausages which are processed by fermentation without heat treatment. Dry fermented sausages are meat products consisting of a mixture of lean meat and fat particles, salt, curing agents (nitrate and/or nitrite) and spices (mostly oregano and black pepper), which is stuffed into casings, subjected to fermentation followed by drying. The novel role of probiotics used in this food has been primarily as part of the fermentation process. It is believed that the meat matrix consisting of meat and fat are able to "encapsulate" the bacteria and protect them against the critical passage of gastrointestinal tract. Considering the ability of meat and meat derivatives to buffer the probiotic bacteria through the gastrointestinal tract, the use of probiotics in the processing of fermented sausages has since attracted much attention for health purposes.

Considering their mild heating and processing, dry fermented sausages have been viewed as potential carriers for probiotics. In order to produce fermented meat products with probiotics, cultures from dairy industry with well-documented health-promoting properties are normally used for the fermentation of dry sausages. Such strains are of intestinal origin and have been shown to survive the sausage manufacturing process. The first meat product to be inoculated with probiotics was demonstrated by Andersen (1998) who successfully fermented dry sausages using a mixture of traditional starter culture and probiotic *B. lactis* Bb-12. In another study, Erkkilä et al. (2001) showed that *L. rhamnosus* GG is suitable for use as starter cultures in dry sausages fermentation. The probiotic strain survived the processing of sausage and has been found to increase to 10^8 CFU/g after fermentation. A Turkish fermented sausage has also been manufactured using *L. acidophilus* and *L. casei* in the presence of traditional starter culture. A higher growth of probiotic bacteria was detected in the sausage after fermentation. Additionally, the probiotics studied were also able to survive to 10^9 CFU/g after fermentation and remained above 10^6 CFU/g after 8-month storage at 4°C.

6.1 *Interactions of Probiotics and the Composition of Meat and Meat Derivative*

Carbohydrates and proteins are the main sources for catabolism of probiotic bacteria during fermentation. Though glycogen serves as a source of glucose in meat, sugars such as glucose and lactose are usually included as fermentable substrates for the industrial manufacture of fermented meat products. Lactose is often added into fermented meat for lactose-fermenting *L. sakei*. Pentoses such as arabinose and xylose are alternative carbohydrates added for the fermentation of probiotics via the phosphoketolase pathway to produce lactic acid as the main end product accompanied by low concentration of acetic acid (Axelsson 1998).

Lactic acid bacteria have been found to possess weak proteolytic action on myofibrillar proteins (Sanz et al. 1999). Nevertheless, the requirements of probiotic bacteria towards amino acids are high in order to achieve their optimal growth in meat products. The proteolytic activities of lactic acid bacteria are often confined to the secondary hydrolysis of oligopeptides and small peptides, which proceeded after the degradation of myosin and actin into peptides by the endogenous enzymes of meat (Fig. 8) (Fadda et al. 1999). The oligopeptides generated are subsequently hydrolysed by their intracellular peptidases including endopeptidases, aminopeptidases, dipeptidases, tripeptidases and dipeptidyl peptidases (Kunji et al. 1996). Studies have shown that *Lactobacillus* exhibit proteolytic activity on porcine muscle myofibrillar and sarcoplasmic proteins (Fadda et al. 1999) followed by subsequent decomposition of peptides into amino acids for growth (Sanz et al. 1999).

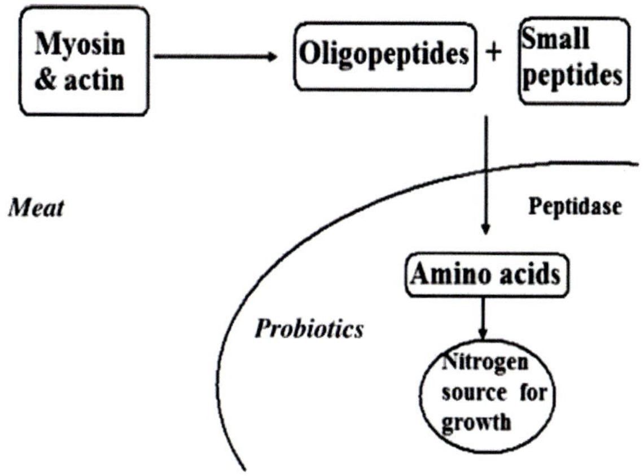

Fig. 8 Utilization of nitrogen sources by probiotics in meat and meat derivatives

6.2 Challenges of Probiotics in Meat and Meat Derivatives

The ability of probiotics to compete with the natural microbiota of the raw material could be affected by various growing conditions in the sausage. High salt concentrations, low water activity, low temperature, low pH and the presence of curing agents are the factors that might affect the survivability of probiotics in dry fermented sausages.

Salt acts as one of the first hurdles against pathogens during manufacturing of dry fermented sausages. The concentration of salt added in the batter is about 2.5–3% (w/w) initially and can reach 15% in the final product. The addition of salt subsequently resulted in lowering of water activity of the product. Addition of salt reduces the water activity to 0.94–0.98 while subsequent addition further decreases the water activity to less than 0.90. The low water activity of fermented meat is indeed unfavorable for the growth of probiotics.

Changes of pH also pose as a challenge for the survival of probiotics in fermented meat products. A reduction in pH from 5.6 to 4.9 after fermentation limits the survival of probiotics in the fermented meat over the entire fermentation and ripening process (Erkkilä et al. 2001). Curing agents such as sodium nitrite is added into the batter of meat for preservation purpose. The capability of probiotic bacteria to resist this compound is one of the challenges faced in meat fermentation. Arihara et al. (1998) previously found that such capability was strain dependent. *L. gasseri* was showed to grow better than *L. acidophilus, L. crispatus, L. amylovorus, L. gallinarum* and *L. johnsonii* in meat containing 2.5% NaCl and 5 ppm sodium nitrite.

6.3 Means to Overcome Limitations of Probiotics in Meat and Meat Derivatives

Probiotics are usually subjected to unfavorable conditions during fermentation in the meat environment. A promising strategy for the development of probiotic fermented meat product involves the use of bacteria with probiotic properties and is commonly associated with the meat environment. Lactobacilli and enterococci that exhibited probiotics properties have been isolated from natural sausage and recognized as potential cultures for fermenting sausages (Pennacchia et al. 2006). These probiotics strains are competitive against natural meat microflora to reach a high final population, and thus are preferred starter cultures for meat fermentation.

The sensitivity of probiotic bacteria to sodium chloride and sodium nitrite utilized for the preservation of meat can be reduced by physical treatment of the probiotic cells. Arihara and Itoh (2000) demonstrated that UV irradiation of *L. gasseri* generated several mutants resisting these compounds. The mutant strains showed satisfactory growth in meat containing 3.3% NaCl and 200 ppm sodium nitrite. The *L. gasseri* mutant also possessed indistinguishable biochemical characteristics from the original strain and hence could be utilized as a starter culture to develop probiotic meat products.

Poor survival of probiotic bacteria during meat fermentation can be overcome by microencapsulation. Muthukumarasamy and Holley (2006) demonstrated that microencapsulation of probiotics in alginate beads increases the survival of bacteria cell in sausages. The viable cell number of microencapsulated *L. reuteri* were 2 log cycle higher than unencapsulated bacteria after drying of sausages.

7 Conclusions

Increasing demand for probiotic foods has prompted the development of new probiotic carriers. Although the market for probiotic products shows a substantial increase in the recent years, scientific approaches are needed to further broaden the conventional use of dairy as probiotic vehicles. Studies have proven that it is possible to obtain probiotic foods from other matrices, such as cereals, soy, fruits, vegetables and also meat products. More research is needed to formulate these probiotic products into functional foods that could also meet the nutritional and palatability requirements of consumers. Future studies and challenges would involve new technological approaches to improve the organoleptic properties of these products and also their commercial feasibility.

Acknowledgement The authors would like to thank the USM RU grants for supporting this work (1001/PTEKIND/815056, 1001/PTEKIND/811141, 1001/PTEKIND/811089) and the USM fellowships for supporting this work.

References

Ahn JB, Hwang HJ, Park JH (2001) Physiological responses of oxygen-tolerant anaerobic *Bifidobacterium longum* under oxygen. J Ind Microbiol Biotechnol 11:443–451

Akin MB, Akin MS, Kirmaci Z (2007) Effects of inulin and sugar levels on the viability of yogurt and probiotic bacteria and the physical and sensory characteristics in probiotic ice cream. Food Chem 104:93–99

Alamprese C, Foschino R, Rossi M, Pompei C, Savani L (2002) Survival of *Lactobacillus johnsonii* La1 and the influence of its addition in retail-manufactured ice-cream produced with different sugar and fat concentration. Int Dairy J 12:201–208

Almeida MHB, Cruz AG, Faria JAF, Moura MRL, Carvalho LMJ, Freitas MCJ (2009) Effect of the açai pulp on the sensory attributes of probiotic yogurts. Int J Prob Preb 4:41–44

Andersen L (1998) Fermented dry sausages produced with the admixture of probiotic cultures. Proceedings of the 44th ICoMST, pp 826–827

Antunes AEC, Silva ÉRA, Dender AGFV, Marasca ETG, Moreno I, Faria EV, Padula M, Lerayer ALS (2009) Probiotic buttermilk-like fermented milk product development in a semiindustrial scale: physiological, microbiological and sensory acceptability. Int J Dairy Technol 62: 556–563

Arihara K, Itoh M (2000) UV-induced *Lactobacillus gasseri* mutants resisting sodium chloride and sodium nitrite for meat fermentation. Int J Food Microbiol 56:227–230

Arihara K, Ota H, Itoh M, Kondo Y, Sameshima T, Yamanaka H, Akimoto M, Kanai S, Miki T (1998) *Lactobacillus acidophilus* group lactic acid bacteria applied to meat fermentation. J Food Sci 63:544–547

Axelsson L (1998) Lactic acid bacteria: classification and physiology. In: Salminen S, von Wright A (eds) Lactic acid bacteria. Dekker, New York, pp 1–72

Başyğit G, Kuleaşan H, Karahan AG (2006) Viability of human-derived probiotic lactobacilli in ice-cream with sucrose and aspartame. J Ind Microbiol Technol 33:796–800

Betoret N, Puente L, Diaz MJ, Pagan MJ, Garcia MJ, Gras ML, Martinez-Monzo J, Fito P (2003) Development of probiotic-enriched dried fruits by vacuum impregnation. J Food Eng 56:273–277

Bosnea LA, Kourkoutas Y, Albantaki N, Tzia C, Koutinas AA, Kanellaki M (2009) Functionality of freeze-dried L. casei cells immobilized on wheat grains. LWT-Food Sci Tech 42:1696–1702

Calderon M, Loiseau G, Guyot JP (2003) Fermentation by Lactobacillus fermentum Ogi E1 of different combinations of carbohydrates occurring naturally in cereals: consequences on growth energetics and α-amylase production. Int J Food Microb 80:161–169

Cardarelli HR, Buriti FCA, Castro IA, Saad SMI (2008) Inulin and oligofructose improve sensory quality and increase the probiotic viable count in potentially synbiotic petit-suisse cheese. LWT Food Sci Technol 41:1037–1046

Champagne CP, Raymond Y, Gagnon R (2008) Viability of Lactobacillus rhamnosus 0011 in an apple-based fruit juice under simulated storage conditions at the consumer level. J Food Sci 73: M221–M226

Charalampopoulos D, Pandiella S (2010) Survival of human derived *Lactobacillus plantarum* in fermented cereal extracts during refrigerated storage. LWT-Food Sci Tech 43:431–435

Charalampopoulos D, Pandiella SS, Webb C (2002) Growth studies of potentially probiotic lactic acid bacteria in cereal-based substrates. J Appl Microb 92:851–859

Charalampopoulos D, Pandiella SS, Webb C (2003) Evaluation of the effect of malt, wheat and barley extracts on the viability of potentially probiotic lactic acid bacteria under acidic conditions. Int J Food Microb 82:133–41

Chen MO (2008) Development of a novel probiotic-fortified soy energy bar containing decreased α-galactosides. MSc thesis, University of Missouri

Chou CC, Hou JW (2000) Growth of bifidobacteria in soymilk and their survival in the fermented soymilk drink during storage. Int J Food Microb 56:113–121

Chumchuere S, Robinson RK (1999) Selection of starter cultures for the fermentation of soya milk. Food Microb 16:129–137

Corcoron BM, Stanton C, Fitzgerald GF, Rossi RP (2005) Survival of probiotics lactobacilli in acidic environment is enhanced in the presence of metabolizable sugars. App Environ Microb 7:3060–3067

Costa NE, Hannon JA, Guinee TP, Auty MAE, McSweeney PLH, Beresford TP (2010) Effect of exopolysaccharide produced by isogenic strains of Lactococcus lactis on half-fat cheddar cheese. J Dairy Sci 93:3469–3486

Crittenden R, Laitila A, Forssell P, Mättö J, Saarela M, Mattila-Sandholm T, Myllärinen P (2001) Adhesion of Bifidobacteria to granular starch and its implications in probiotic technologies. Appl Environ Microb 67:3469–3475

Dave RI, Shah NP (1997) Effectiveness of ascorbic acid as an oxygen scavenger in improving viability of probiotic bacteria in yoghurts made with commercial starter cultures. Int Dairy J 7:435–443

De Boever P, Wouters R, Verstraete W (2001) Combined use of Lactobacillus reuteri and soygerm powder as food supplement. Letters Appl Microb 33:420–424

Ding WK, Shah NP (2008) Survival of free and microencapsulated probiotic bacteria in orange and apple juices. Int Food Res J 15:219–232

Donkor ON, Henriksson A, Vasiljevic T, Shah NP (2005) Probiotic strains as starter cultures improve angiotensin-converting enzyme inhibitory activity in soy yoghurt. J Food Sci 70: M375–M381

Donkor ON, Henriksson A, Vasiljevic T, Shah NP (2007a) Rheological properties and sensory characteristics of set-type soy yogurt. J Agric Food Chem 55:9868–9876

Donkor ON, Nilmini SLI, Stolic P, Vasiljevic T, Shah NP (2007b) Survival and activity of selected probiotic organisms in set-type yogurt during cold storage. Int Dairy J 17:657–665

Drgalić I, Tratnik L, Božanić R (2005) Growth and survival of probiotic bacteria in reconstituted whey. Lait 85:171–179

Erkkilä S, Suihko ML, Eerola S, Petäjä E, Mattila-Sandholm T (2001) Dry sausages fermented by *Lactobacillus rhamnosus* strains. Int J Food Microbiol 64:205–210

Ewe JA, Wan Nadiah WA, Liong MT (2010) Viability and growth characteristics of *Lactobacillus* and *Bifidobacterium* in soymilk supplemented with B-vitamins. Int J Food Sci Nutr 61:87–107

Fadda S, Sanz Y, Vignolo G, Aristoy MC, Oliver G, Toldrá F (1999) Characterization of pork muscle protein hydrolysis caused by *Lactobacillus plantarum*. Appl Environ Microbiol 65:3540–3546

Favaro-Trindade CS, Bernardi S, Bodini RB, Balieiro JCC, Almeida E (2006) Sensory acceptability and stability of probiotic microorganisms and vitamin C in fermented acerola (*Malpighia emarginata* DC.) ice cream. J Food Sci 71:492–495

Fung WY, Yuen KH, Liong MT (2010) Characterization of fibrous residues from agrowastes and the production of nanofibers. J Agri Food Chem 58:8077–8084

Gokavi S, Zhang L, Huang MK, Zhao X, Guo M (2005) Oat-based symbiotic beverage fermented by *Lactobacillus plantarum, Lactobacillus paracasei ssp. casei, and Lactobacillus acidophilus.* J Food Sci 70:M216–M223

Granato D, Branco GF, Cruz AG, Faria JAF, Shah NP (2010) Probiotic dairy products as functional foods. Comp Rev Food Sci Food Safety 9:455–470

Güler-Akln MB, Akln MS (2007) Effects of cysteine and different incubation temperatures on the microflora, chemical composition and sensory characteristics of bio-yogurt made from goat's milk. Food Chem 100:788–793

Guergoletto KB, Magnani M, Martin JS, de Jesus Andrade CCT, Garcia S (2010) Survival of Lactobacillus casei (LC-1) adhered to prebiotic vegetal fibers. Innov Food Sci Emerg Tech 11:415–421

Gupta S, Cox S, Abu-Ghannam N (2010) Process optimization for the development of a functional beverage based on lactic acid fermentation of oats. Biochem Eng J 52:199–204

Heenan CN, Adams MC, Hosken RW, Fleet GH (2002) Growth medium for culturing probiotic bacteria for applications in vegetarian food products. Lebens-Wiss und-Tech 35:171–176

Helland MH, Wicklund T, Narvhus JA (2004) Growth and metabolism of selected strains of probiotic bacteria in milk- and water-based cereal puddings. Int Dairy J 14:957–965

Heller KJ (2001) Probiotic bacteria in fermented foods: product characteristics and starter organisms. Am J Clin Nutr 73:374S–379S

Kailasapathy K, Harmstorf I, Philips M (2008) Survival of *Lactobacillus acidophilus* and *Bifidobacterium animalis* spp. *lactis* in stirred fruit yogurts. LWT Food Sci Technol 41:1317–1322

Kheadr EE, Bernoussi NM, Lacroix C, Fliss I (2004) Comparison of the sensitivity of commercial strains and infant isolates of bifidobacteria to antibiotics and bacteriocins. Int Dairy J 14: 1041–1053

Kourkoutas Y, Bosnea L, Taboukos S, Baras C, Lambrou D, Kanellaki M (2006) Probiotic cheese production using *Lactobacillus casei* immobilized on fruit pieces. J Dairy Sci 89:1439–1451

Kourkoutas Y, Xolias V, Kallis M, Bezirtzoglou E, Kanellaki M (2005) Lactobacillus casei cell immobilization on fruit pieces for probiotic additive, fermented milk and lactic acid production. Process Biochem 40:411–416

Kun S, Rezessy-Szabo JM, Nguyen QD, Hoschke A (2008) Changes of microbial population and some components in carrot juice during fermentation with selected Bifidobacterium strains. Process Biochem 43:816–821

Kunji ERS, Mierau I, Hagting A, Poolman B, Konings WN (1996) The proteolytic systems of lactic acid bacteria. Antonie Leeuwenhoek 70:187–221

Lamsal BP, Faubion JM (2009) The beneficial use of cereal and cereal components in probiotic foods. Food Rev Int 25:103–114

LeBlanc JG, Garro MS, De Giori GS (2004) Effect of pH on *Lactobacillus fermentum* growth, raffinose removal, a-galactosidase activity and fermentation products. App Microb Biotech 65:119–123

Lee JH, Lee SK, Park KH, Hwang IK, Ji GE (1999) Fermentation of rice using amylolytic *Bifidobacterium*. Int J Food Microb 50:155–161

Liong MT, Easa AM, Lim PT, Kang JY (2009) Survival, growth characteristics and bioactive potential of *Lactobacillus acidophilus* in a soy-based cream cheese. J Sci Food Agric 89:1382–1391

Liu DM, Li L, Yang XQ, Liang SZ, Wang JS (2006) Survivability of *Lactobacillus rhamnosus* during the preparation of soy cheese. Food Tech Biotech 44:417–422

Liu KS (1997) Soybeans, chemistry, technology, and utilization. Chapman and Hall, New York

Lopez-Rubio A, Sanchez E, Sanz Y, Lagaron JM (2009) Encapsulation of living bifidobacteria in ultrathin PVOH electrospun fibers. Biomacromolecules 10:2823–2829

Lourens-Hattingh A, Viljeon BC (2001) Yogurt as probiotic carrier food. Int Dairy J 11:1–17

Luckow T, Delahunty C (2004) Consumer acceptance of orange juice containing functional ingredients. Food Res Int 37:805–804

Luckow T, Sheehan V, Fitzgerald G, Delahunty C (2006) Exposure, health information and flavour-masking strategies for improving the sensory quality of probiotic juice. Appetite 47:315–323

Madudeira AR, Pereira CI, Truszkowska K, Gomes AMP, Pintado ME, Malcata FX (2005) Survival of probiotic bacteria in a whey cheese vector submitted to environmental conditions prevailing in the gastrointestinal tract. Int Dairy J 15:921–927

Maragkoudasaki PA, Miarisa C, Rojeza P, Manalisb N, Magkanarib F, Kalantzopoulosa G, Tsakalidou E (2006) Production of traditional Greek yogurt using *Lactobacillus* strains with probiotic potential as starter adjuncts. Int Dairy J 16:52–60

Moayednia N, Ehsani MR, Jomeh ZE, Mazaheri AF (2009) Effect of refrigerated storage time on the viability of probiotic bacteria in fermented probiotic milk drinks. Int J Dairy Technol 62:204–208

Mousavi ZE, Mousavi SM, Rasavi SH, Emam-Djomeh Z, Kiani H (2010) Fermentation of pomegranate juice by probiotic lactic acid bacteria. World J Microb Biotech. doi:10.1007/s11274-010-0436-1

Muthukumarasamy P, Holley R (2006) Microbiological and sensory quality of dry fermented sausages containing alginate-microencapsulated *Lactobacillus reuteri*. Int J Food Microbiol 111:164–169

Ng KH, Lye HS, Easa AM, Liong MT (2008) Growth characteristics and bioactivity of probiotics in tofu-based medium during storage. Annals Microb 58:477–487

Nyanzi R, Jooste PJ, Abu JO, Beukes EM (2010) Consumer acceptability of a symbiotic version of the maize beverage mageu. Dev Southern Afr 27:447–463

Oliveira RPDS, Perego P, Converti A, Oliveira MND (2009) Effect of inulin on growth and acidification performance of different probiotic bacteria in co-cultures and mixed culture with *Streptococcus thermophilus*. J Food Eng 91:133–139

Özer B, Uzun YS, Kirmaci HA (2008) Effect of microencapsulation on viability of *Lactobacillus acidophilus* LA-5 and *Bifidobacterium bifidum* BB-12 during Kasar cheese ripening. Int J Dairy Technol 61:237–244

Pastell H, Westermann P, Meyer AS, Tuomainen P, Tenkanen M (2009) In vitro fermentation of arabinoxylan-derived carbohydrates by bifidobacteria and mixed fecal microbiota. J Agric Food Chem 57:8598–8606

Patel HM, Pandiella SS, Wang RH, Webb C (2004) Influence of malt, wheat, and barley extracts on the bile tolerance of selected strains of lactobacilli. Food Microb 21:83–89

Pennacchia C, Vaughan EE, Villani F (2006) Potential probiotic *Lactobacillus* strains from fermented sausages: further investigations on their probiotic properties. Meat Sci 73:90–101
Perrin S, Grill JP, Schneider F (2000) Effects of fructooligosaccharides and their monomeric components on bile salt resistance in three species of bifidobacteria. J Appl Microb 88:968–974
Picot A, Lacroix C (2004) Encapsulation of bifidobacteria in whey protein-based microcapsules and survival in simulated gastrointestinal conditions and in yogurt. Int Dairy J 15:505–515
Rodrigues D, Rocha-Santos TAP, Pereira CI, Gomes AM, Malcata FX, Freitas AC (2011) The potential effect of FOS and inulin upon probiotic bacterium performance in curdled milk matrices. Food Sci Tech 44:100–108
Rodríguez-Sanoja R, Ruiz B, Guyot JP, Sanchez S (2005) Starch-binding domain affects catalysis in two Lactobacillus α-amylases. Appl Environ Microb 71:297–302
Roy D, Desjardins M, Mondou F (1995) Selection of bifidobacteria for use under cheese making conditions. Milchwissenschaft 50:139–142
Saarela M, Virkajarvi I, Alakomi H, Sigvart-Mattila P, Matto J (2006) Stability and functionality of freeze-dried probiotic Bifidobacterium cells during storage in juice and milk. Int Dairy J 16:1477–1482
Sanz Y, Fadda S, Vignolo G, Aristoy MC, Oliver G, Toldrá F (1999) Hydrolysis of muscle myofibrillar proteins by *Lactobacillus curvatus* and *Lactobacillus sakei*. Int J Food Microbiol 53:115–125
Sendra E, Fayos P, Lario Y, Fernández-López J, Sayas-Barberá E, Pérez-Alvarez JA (2008) Incorporation of citrus fibers in fermented milk containing probiotic bacteria. Food Microbiol 25:13–21
Sharp MD, McMohan DJ, Broadbent JR (2008) Comparative evaluation of yogurt and low-fat Cheddar cheese as delivery media for probiotic *Lactobacillus casei*. J Food Sci 73: M375–M377
Sheehan VM, Ross P, Fitzgerald GF (2007) Assessing the acid tolerance and the technological robustness of probiotic cultures for fortification in fruit juices. Innov Food Sci Emerg Tech 8:279–284
Sindhu SC, Khetarpaul N (2004) Development, acceptability and nutritional evaluation of an indigenous food blend fermented with probiotic organisms. Nutr Food Sci 35:20–27
Teh SS, Ahmad R, Wan Abdullah WN, Liong MT (2009) Evaluation of agrowastes as immobilizers for probiotics in soy milk. J Agric Food Chem 57:10187–10198
Usman, Hosono A (1999) Viability of *Lactobacillus gasseri* and its cholesterol-binding and antimutagenic activities during subsequent refrigerated storage in nonfermented milk. J Dairy Sci 82:2536–2542
Vasiljevic T, Kealy T, Mishra VK (2007) Effects of β-glucan addition to a probiotic containing yogurt. Food Chem Toxicol 72:C405–C411
Vergara CMAC, Honorato TL, Maia GA, Rodrigues S (2010) Prebiotic effect of fermented cashew apple (Anacardium occidentale L) juice. Food Sci Tech 43:1–5
Villegas E, Gilliland SE (1998) Hydrogen peroxide production by *Lactobacillus delbrueckii* subsp. *lactis* at 5°C. J Food Sci 63:1070–1075
Vinderola CG, Costa GA, Regenhardt S, Reinheimer JA (2002a) Influence of compounds associated with fermented dairy products on the growth of lactic acid starter and probiotic bacteria. Int Dairy J 12:579–589
Vinderola CG, Mocchiutti P, Reinheimer JA (2002b) Interactions among lactic acid starter and probiotic bacteria used for fermented dairy products. J Dairy Sci 85:721–729
Wang J, Guo Z, Zhang Q, Yan L, Chen W, Liu XM, Zhang HP (2009) Fermentation characteristics and transit tolerance of probiotic *Lactobacillus casei* Zhang in soymilk and bovine milk during storage. J Dairy Sci 92:2468–2476
Wang X, Brown IL, Evans AJ, Conway PL (1999) The protective effects of high amylose maize (amylomaize) starch granules on the survival of *Bifidobacterium* spp. in the mouse intestinal tract. J Appl Microb 87:631–639

Wei QK, Chen TR, Chen JT (2007) Using of *Lactobacillus* and *Bifidobacterium* to product the isoflavone aglycones in fermented soymilk. Int J Food Microb 117:120–124

Yazici F, Alvarez VB, Hansen PMT (1997) Fermentation and properties of calcium-fortified soy milk yogurt. J Food Sci 62:457–461

Yeo SK, Liong MT (2010a) Effect of prebiotics on viability and growth characteristics of probiotics in soymilk. J Sci Food Agric 90:267–275

Yeo SK, Liong MT (2010b) ACE inhibitory activity and bioconversion of isoflavones by lactobacilli and bifidobacteria in soymilk supplemented with prebiotics. Int J Food Sci Nutr 61:161–181

Yoon KY, Woodams EE, Hang YD (2004) Probiotication of tomato juice by lactic acid bacteria. J Microb 42:315–318

Yoon KY, Woodams EE, Hang YD (2005) Fermentation of beet juice by beneficial lactic acid bacteria. Lebensm-Wiss U-Tech 38:73–75

Yoon KY, Woodams EE, Hang YD (2006) Production of probiotic cabbage juice by lactic acid bacteria. Bioresour Tech 97:1427–1430

Use of Probiotics for Dermal Applications

Benedetta Cinque, Cristina La Torre, Esterina Melchiorre, Giuseppe Marchesani, Giovanni Zoccali, Paola Palumbo, Luisa Di Marzio, Alessandra Masci, Luciana Mosca, Paola Mastromarino, Maurizio Giuliani, and Maria Grazia Cifone

Contents

1 Introduction 222
2 Probiotics and Skin Microflora 223
2.1 Production of Acids by Probiotics 223
2.2 Antimicrobial Substances 223
2.3 Beta Defensins 225
2.4 Probiotics and Wound Protection 225
3 Probiotics and Disturbed Skin Barrier 227
3.1 Cutaneous pH 227
3.2 Ceramides 228
3.3 Hyaluronic Acid 230
3.4 Collagen 231

B. Cinque • C. La Torre • E. Melchiorre • G. Marchesani • G. Zoccali • P. Palumbo • M. Giuliani • M.G. Cifone (✉)
Department of Health's Sciences, University of L'Aquila, 67100 Coppito, L'Aquila, Italy
e-mail: benedetta.cinque@univaq.it; crix_latorre@libero.it; esterina.melchiorre@univaq.it; giuseppe.marchesani@univaq.it; zoccali.giovanni@virgilio.it; paola.palumbo@univaq.it; maurizio.giuliani@univaq.it; cifone@univaq.it

L. Di Marzio
Department of Drug Sciences, University of Chieti-Pescara Gabriele D'Annunzio, Via dei Vestini, 66013 Chieti, Italy
e-mail: l.dimarzio@unich.it

A. Masci • L. Mosca
Department of Biochemical Sciences, Sapienza University, p.le Aldo Moro, 5, 00185 Rome, Italy
e-mail: alexmasci@katamail.com; luciana.mosca@uniroma1.it

P. Mastromarino
Department of Public Health and Infectious Diseases, Section of Microbiology, Sapienza University, p.le Aldo Moro, 5, 00185 Rome, Italy
e-mail: paola.mastromarino@uniroma1.it

M.-T. Liong (ed.), *Probiotics*, Microbiology Monographs 21,
DOI 10.1007/978-3-642-20838-6_9,

4 Probiotics and Skin Inflammatory/Immune System 231
4.1 Barrier Function and Skin Reactivity 231
4.2 Environmental Stress 232
4.3 Protection Against NO 234
4.4 Anti-inflammatory Potential 235
5 Conclusions 237
References 237

Abstract The concept of probiotic bacteria is considerably evolving. Clinical and experimental researches extensively document that beyond probiotic capacity to influence positively the intestinal functions, they can exert their benefits at the skin level thanks to their peculiar properties. Indeed, scientific and evidence-based reports strengthen the assumption that certain probiotics can contribute to modulate cutaneous microflora, lipid barrier, and skin immune system, leading to the preservation of the skin homeostasis. In this chapter, the most relevant evidences available from scientific literature as well as registered patents have been summarized in relation to actual or potential topical applications of probiotics in the field of dermatology. Altogether the evidences reported in this review afford the possibility of designing new strategies based on a topical approach for the prevention and treatment of cutaneous disorders.

1 Introduction

Most often in foods and oral formulations, the probiotics are mainly used as a means for restoring microbial balance, particularly in the gastrointestinal tract (Williams 2010). This approach appears particularly significant since the intestinal microbiota is involved in physiological balance and in the intestinal development and maturation of the host immune system (Oelschlaeger 2010). Thus, in the last years an increasing interest has been focused on the possible use of ingested probiotics for treating inflammatory and allergic conditions being specific strains able to modulate the immune system at the local and systemic levels (Caramia et al. 2008). Arck et al. (2010) have recently hypothesized a new, unifying model i.e., gut–brain–skin axis, suggesting that modulation of the microbiome by deployment of probiotics can exert profound beneficial effects on skin homeostasis, skin inflammation, hair growth, and peripheral tissue responses to stress. On the other hand, new insights could now fundamentally change the impact of probiotics on dermatology. Indeed, an emerging approach to help preventing and treating skin conditions, including the external signs of aging, acne, rosacea, yeast and bacterial infections, psoriasis, and dermatitis, is represented by topical probiotics, as shown by the growing marketplace for topical probiotic formulations available for skin care and antiaging benefits. In this chapter, the key evidences available from scientific literature as well as registered patents will be summarized in relation to actual or potential topical applications of probiotics in the field of dermatology.

2 Probiotics and Skin Microflora

The skin is able to act as a physical barrier exerting several functions such as fluid homeostasis, thermoregulation, immune responses, neurosensory functions, metabolic functions, and primary protection against infection. The skin microflora plays a significant role in competitive exclusion of pathogens that are aggressive and provoke infection in the skin and in the processing of skin proteins, free fatty acids (FFAs), and sebum. Interestingly the resident microbiota may be regarded as "beneficial" to the normal, healthy host, but may become dangerous to the host with disturbed skin integrity. Microorganisms may have a role even in atopic dermatitis (AD), eczema, rosacea, psoriasis, and acne. Even if there are only very few studies pursuing on probiotic approach for microflora-related skin disorders, it is intriguing to suppose that topical probiotic application can be beneficial either for preventing or for treating altered microflora-associated skin diseases (Krutmann 2009; Simmering and Breves 2009).

2.1 *Production of Acids by Probiotics*

Preservation of the resident microflora is thought to be an effective way to achieve maintenance of healthy normal skin functions; however, the number of microbial species colonizing human skin is limited depending on the hard physical and biochemical factors. Probiotic microorganisms use different mechanisms, such as by lowering pH, to preserve skin health and to inhibit the growth of pathogens. The acidic skin environment is indeed very important as it discourages bacterial colonization and provides a moisture barrier through absorption or moisture by aminoacids, salts, and other substances in the acid mantle (Lambers et al. 2006; Mauro 2006). An interesting property of probiotics is the fermentative metabolism that involves the production of acid molecules (i.e., lactic acid), thus acidifying the surrounding environment (Krutmann 2009).

2.2 *Antimicrobial Substances*

The potential topical use of probiotic strains capable of producing potent antimicrobial toxins (i.e., bacteriocins, bacteriocin-like substances, organic acids, and H_2O_2) has received increasing attention to successfully prevent pathogen adhesion and outcompete undesired species (Oh et al. 2006; Gillor et al. 2008). Topical compositions containing probiotic bacteria, spores, and extracellular products and uses thereof represented the basis of the invention of Farmer (2005) suitable for topical application to the skin which can be utilized to inhibit the growth of bacteria, yeasts, fungi, viruses, and combinations thereof. The invention also disclosed methods of treatment and therapeutic systems for inhibiting the growth of pathogens and combinations thereof, by topical application of therapeutic compositions which were comprised, in part, of isolated *Bacillus* species, spores, or an extracellular product of *Bacillus coagulans* comprising a supernatant or filtrate of a culture of *Bacillus coagulans* strain. A method and composition were also provided for application of probiotic microorganisms to a surface to prevent or inhibit contamination by pathogenic microorganisms (Spigelman and Ross 2008). The probiotic microorganisms may be bacteria, yeast, or mold. Suitable probiotics should be selected according to one or more particular properties, being the preferred properties of their competitive exclusion of pathogenic organisms from the surface to which they are applied, adherence to human tissue, sensitivity to antibiotics, antimicrobial activity, acid tolerance, and a high oxygen tolerance. In particular, the method consists of different application modalities (i.e., lotions, spraying, wipe paper) of one or more probiotic microorganisms to a wide variety of surfaces, such as human skin and hospital equipment and fixtures, in an amount effective to, at least partly, prevent their contamination, colonization, growth, and cross contamination by the pathogenic bacteria. The method relies on the probiotic ability to form isolated colonies producing a protective layer that can inhibit and exclude pathogenic bacteria generally unable to grow on top of other bacteria. The probiotic application is recommended for a sufficient time depending upon such factors as the therapeutically effective amount, the type, the mode of probiotic application, or the degree of contamination of the biological or nonbiological surface. Accordingly, the method propones the use of a single or a plurality of different probiotic microorganisms, applying multiple bacteria serially, in layers to fight several or single resistant types of pathogenic organisms.

In regards to the potential use of bacteriocin-producing strains as probiotic and bioprotective agents, a number of bacteriocins produced by various lactic acid bacteria species including *Lactobacillus, Lactococcus, Pediococcus, Propionibacterium, Leuconostoc, and Carnobacterium* have been reported (Klaenhammer 1993; Oh et al. 2006). Of interest, Oh et al. (2006) reported the efficacy of the bacteriocin from *Lactococcus* sp. HY 449 in controlling skin-inflammation and acnes by clinical skin irritation test. This study demonstrated that this bacteriocin was able to inhibit the growth of skin inflammatory bacteria such as *Staphylococcus epidermidis Staphylococcus aureus, Streptococcus pyogenes,* and *Propionibacterium acnes.* The experimental data highlighted that the inhibitory effect of the bacteriocin employed in this study was due to bacteriolytic action on the cell wall and cell membranes of *P. acnes*. Thanks to its antimicrobic properties, *Lactococcus* bacteriocin could be used in cosmetic application for a variety of purposes.

The invention of Teodorescu (1999) disclosed the use of eubiotic product, consisting of a mixture of three *Lactobacillus acidophilus* strains, LD-11, LR-13, and LV-17, associated in equal parts, for the maintenance and treatment of tegument. In comparison with the common strains, LD-11, LR-13, and LV-17 can also ferment raffinose, trehalose, and dextrin, respectively. The most important aspect of this finding lies on an optimum association of three *L. acidophilus* anallergical strains for realizing a natural eubiotic product, capable of maintaining the skin pH at physiological values, to destroy the pathogenic microflora and to be resistant in cosmetic composition.

2.3 Beta Defensins

Sullivan et al. (2009) claimed that extracts of *Lactobacillus* could stimulate the production of beta-defensins in skin cells, which may be useful in the reduction or prevention of growth of microbial populations on the skin, in a dose-dependent manner. Effective amounts of *Lactobacillus* extracts are applied to an open cut or wound on the skin that may have been in contact with dirt or undesirable microbes; or on a chronic basis, applied to clean skin to maintain a healthy level of skin flora. According to the authors, the extracts could also be useful in the treatment of acne. Indeed, topical compositions containing *L. plantarum* extract are shown to reduce the incidence of both inflamed and noninflamed acne lesions when used regularly over a period of 2 months. The extracts had further been proposed as a preservative in cosmetic of pharmaceutical products, in particular the *L. plantarum,* which possesses a broad spectrum of activity against both Gram-positive and Gram-negative bacteria. *Acne vulgaris* is multifactorial condition and is characterized by hypercolonization with *Propionibacterium acnes*, inflammation, and immune responses. The synbiotic ability of probiotic bacteria and Konjac glucomannan hydrolysates to inhibit the growth of *Propionibacterium acnes* in an in vivo study has been recently reported suggesting that the development of a new alternative involving probiotic therapy for reducing acne episodes in vivo could be encouraging (Al-Ghazzewi and Tester 2009).

2.4 Probiotics and Wound Protection

Chronic wounds are, by definition, the ones that remain in a chronic inflammatory state and therefore fail to follow the normal patterns of the healing process; factors that may impede healing must be identified and, if possible, corrected, for healing to occur. Chronic wounds, and burns in particular, are rarely, if ever, sterile. The burn wound surface is sterile immediately following injury, although it is repopulated within 48 h with Gram-positive organisms from hair follicles, skin appendages, and environment; successively, more virulent Gram-negative organisms replace the Gram-positive organisms after 5–7 days. Burns produce disruption of the mechanical integrity of the skin and generalized immune suppression that allows microorganisms to multiply freely. Currently, the common pathogens isolated from burn are *Staphylococcus aureus*, *Pseudomonas aeruginosa*, *Streptococcus pyogenes*, and various coliforms. Other streptococci, anaerobic organisms, and fungi can also cause infections. Bacterial infections are generally treated by administration of antibiotics; however, this is not always efficacious. Contributing to the lack of successful antibiotic treatment is the ability of colonizing bacteria to establish themselves and proliferate in a biofilm, a characteristic architecture of microcolonies embedded in a self-made matrix of biopolymers that offers structural stability and protection. In this composite state, the bacteria resist the action of a variety of antimicrobial measures, and moreover they are extremely resistant to antibiotics, antiseptics, and to the host immune response, so new therapeutic modalities may be required. An effective approach to prevent or contrast infection could be bacteriotherapy, i.e., the use of probiotics to displace pathogenic organisms. Valdèz et al. (2005) suggested that *L. plantarum* and/or its products are potential therapeutic agents in the local treatment of *P. aeruginosa* burn infections. The authors showed that the in vitro treatment with *L. plantarum* was able to inhibit the production of the *P. aeruginosa* quorum-sensing signal molecules, acyl-homoserine-lactones, and two virulence factors controlled by these signal molecules elastase and biofilm. On the other hand, the subcutaneous injection of *L. plantarum* on a burned-mouse model with *P. aeruginosa* burn infection, induced inhibition of *P. aeruginosa* colonization as observed in skin, liver, and spleen samples taken after 5, 10, and 15 days upon infection, thus leading to a significant improvement in tissue repair. On the basis of these encouraging findings, a preliminary study (Peral et al. 2009a) was carried out to determinate the effect of topical *L. plantarum* treatment on infected and non-infected second-degree burn patients and on infected third-degree burn patients. As a result, in non-infected third-degree burns, the ability of *L. plantarum* to prevent infection, to promote granulation tissue, and to heal wounds was comparable to the silver sulphadiazine cream one; in infected second-degree burns, *L. plantarum* treatment was as effective as the silver sulphadiazine one with reference to decrease in the bacterial load, promotion of the appearance of granulation tissue, and wound healing. In infected third-degree burns, treatment with *L. plantarum* would show great efficacy. A further study of the same group showed the efficacy of *L. plantarum* bacteriotherapy on the chronic infected leg ulcers of diabetic and nondiabetic patients (Peral et al. 2009b).

The antimicrobial efficacy of nitric oxide (NO) is well known. Of note, a recent report (Jones et al. 2010) showed that the NO-producing probiotic patch device containing lyophilized alginate-immobilized *L. fermentum*, glucose, and nitrite salts can produce sufficient levels of gaseous NO over a therapeutically relevant duration, to kill common bacterial and fungal pathogens existed in the wounds of humans.

A recent invention of Hansen and Jespersen (2010) was directed to a wound or tissue dressing comprising bacteria having the property of producing lactic acid by fermentation of the sugars, to use in healing wounds or in accelerating the wound healing. Particularly, preferred species of lactic acid bacteria including *L. sporogenes*, *L. acidophilus*, *L. plantarum*, *L. casei*, *L. brevis*, *L. delbruckii*, and *L. lactis* are present in the dressing, capable of (1) lowering the pH in an open wound environment, (2) securing an intraspecies competitive exclusion thus preventing growth of undesirable bacterial species, (3) exerting an immunomodulatory effect by inducing "wound healing-promoting substances" (i.e., cytokines, growth factors), and (4) producing certain bacteriocins such as toxins that can sustain a wound-healing process.

Actual management of chronic wound and burn patients is enormously expensive. Moreover, traditional treatments and caring are often ineffective and fail to eradicate bacteria, especially in biofilm form. Bacteriotherapy with specific probiotic strains could represent a new therapeutic modality thanks to their efficacy, innocuousness, easy access to application, and low costs, even if this issue deserves further investigation for possible use in topical wound treatments.

3 Probiotics and Disturbed Skin Barrier

Cutaneous pH plays an important role in the barrier function; in particular, human normal skin pH is acidic and varies from an acidic pH of 3.0 to an almost neutral pH of 6.5. Acidic skin surface has been attributed to microbial factors, esocrine gland presence, generation of urocanic acid by histidase-catalyzed deimination of histidin, secretory phospholipase A_2-mediated generation of FFAs from phospholipids, and a nonenergy-dependent sodium–proton exchanger (Feingold 2007; Cinque et al. 2010). Cutaneous pH can control bacterial populations on skin surface affecting resident microbiota and can regulate epidermal permeability barrier homeostasis and stratum corneum (SC) integrity (Feingold 2007; Cinque et al. 2010).

3.1 Cutaneous pH

The two key cutaneous lipid-processing enzymes, β-glucocerebrosidase and acidic sphingomyelinase (aSMase), which generate a family of ceramides from glucosylceramide and sphingomyelin (SM), respectively, exhibit low pH optima (Feingold 2007; Cinque et al. 2010). An acidic pH directly impacts lipid–lipid interactions in the SC extracellular lamellar bilayers confirming the link between SC pH and barrier homeostasis and SC integrity/cohesion. Elevations of SC pH are accompanied by perturbed cutaneous permeability barrier homeostasis, and by an increase in serine protease activity that mediate degradation of corneodesmosome, resulting in abnormality of SC integrity/cohesion. In conclusion, an acute increase in SC pH reduces the activity of certain key lipid-processing enzymes in the SC, resulting in abnormal lipid processing and the formation of defective lamellar membranes. Moreover, an elevation in SC pH is associated with several cutaneous disorders, such as acute eczema, atopic dermatitis, and seborrheic dermatitis. In these diseases the increased pH could adversely affect cutaneous functions exacerbating these conditions further with more severe clinical manifestations (Feingold 2007; Cinque et al. 2010). Thus, in order to restore the altered cutaneous functions because of increased pH, alternative strategies could be pursued in order to modulate SC pH. As stated above, an interesting approach could envisage the use of probiotic as lactic acid bacteria with fermentative metabolism that are able to produce lactic acid and obtain energy from the fermentation of lactose, glucose, and other sugars to lactate via homofermentative metabolism (Teodorescu 1999; Farmer 2005; Chiba 2007)

Of interest, Yadav et al. (2007) reported that the lipolysis of milk fat by probiotic lactobacilli increases the production of FFAs and produces conjugated linoleic acid by using internal linoleic acid. This acid-producing mechanism inhibits growth of other organisms and favors the growth of lactobacilli that thrive in low pH environments conferring a further health benefit to the host.

3.2 Ceramides

The SC lipids are secreted from keratinocytes in lamellar bodies containing lipid precursors (i.e., glucosylceramides, phospholipids, and cholesterol sulfate) and enzymes (β-glucocerebrosidase, acidic SMase, secretory phospholipase A_2, and steroid sulfatase) that generate ceramides, FFAs, and cholesterol. Within the epidermal membrane structure, the ceramides are the dominant lipid class by weight (~50%) and play an essential role in maintaining and structuring the lipid barrier of the skin. Therefore, a decrease of ceramide in SC will cause water loss and barrier dysfunction in the epidermis, including a loss of protection against antigens, including bacterial, and can result in a skin abnormality such as AD (Feingold 2007; Mizutani et al. 2009; Cinque et al. 2010). The rate-limiting enzyme for the synthesis of ceramides is serine palmitoyl-CoA transferase since its inhibition leads to delayed barrier repair. Ceramides result also from hydrolysis of cerebrosides and SM through β-glucocerebrosidase and SMase, respectively. The SC ceramide levels are thus regulated by the balance among these ceramide-generating enzymes and the degradative-enzyme ceramidase. Both SM and SMase are present in the epidermis and are originally contained in lamellar bodies; structure contains a mixture of ceramides, cholesterol, and FFA. They play a role in the formation of the lipid component of the skin barrier and in the maintenance of the SC stability. To effectively maintain the permeability barrier homeostasis, all three main lipidic components of the stratum corneum, cholesterol, FFA, and ceramides must be present. The absence or decrease of even one of these components in perturbed skin can delay barrier recovery (Feingold 2007; Mizutani et al. 2009; Cinque et al. 2010). A previous study of our group (Di Marzio et al. 1999) reported that high levels of neutral SMase was detected in sonicated *Streptococcus salivarium* ssp. *thermophilus* which was able to induce generation of relevant ceramide levels in keratinocytes in vitro. Both hydroxyceramide and nonhydroxyceramide levels strongly and gradually increased in the presence of sonicated *S. thermophilus* in a time-dependent manner due to SM hydrolysis in keratinocytes which in turn could be correlated with the high levels of a neutral SMase activity of *S. thermophilus*.

These results were confirmed by in vivo studies in healthy and young volunteers. A relevant increase in SC ceramide levels was indeed observed in all the analyzed subjects after topical application of experimental cream containing lysed *S. thermophilus* (Di Marzio et al. 1999). The presence of high levels of neutral SMase activity in the volunteers was responsible for the observed increase of SC ceramide levels, thus leading to an improvement in barrier function and maintenance of SC flexibility. The use of SMase obtained from selected strains of lactic acid bacteria to increase the levels of skin ceramides, and dermatological and cosmetic compositions suitable for topical application containing same, represented the basis of the invention of Cavaliere and De Simone (2001). According to this patent the bacterial SMase could also be used as a cutaneous permeation or absorption enhancer, either alone or in admixture with other enhancers, for preparing pharmaceuticals or cosmetic compositions suitable for transdermal administration. Thus, a method claimed to prevent or treat all conditions associated with abnormal ceramide levels (deficiencies), including skin aging, atopic eczema, dermatosis or dermatitis, atopic dermatitis, psoriasis, ichtyosis, Fabry's disease, Gaucher's disease, Tay-Sachs disease, or Sjogren-Larsson's syndrome comprising topical applying on the affected skin with a pharmaceutical composition which contain neutral SMase obtained from sonicated lactic acid bacteria. To test the possibility that the use of the *S. thermophilus*-containing cream could improve SM dysmetabolism in AD patients, our group previously conducted a study to determine the effect of sonicated *S. thermophilus* on the level of ceramides in vivo in the skin of AD patients (Di Marzio et al. 2003). A 2-week application of the cream containing a sonicated *S. thermophilus* in the forearm skin of the patients led to a significant and relevant increase of skin ceramide amounts, which could result from the SM hydrolysis through the bacterial SMase. Of note, the topical treatment consequently led to a reduction of the AD-associated signs and symptoms such as erythema, scaling, and pruritus in all patients.

Some reports in the last decade have described the changes in total SC with increasing age, in particular the reduced corneum lipid levels (Cinque et al. 2010). Indeed, all major lipid species in the stratum corneum of aged human and mice skin decreased by approximately 30%. In addition, the neutralization of the normally acidic SC has deleterious effect in permeability barrier homeostasis and SC integrity/cohesion. Considering the role of ceramides in SC, our group investigated the short-term topical application of a probiotic formulation on healthy skin of old Caucasian women (Di Marzio et al. 2008). Increase in skin ceramide levels in aged subjects following a short-term topical application of bacterial SMase from *S. thermophilus* has been reported. Consequently, the skin hydration and ceramide levels as markers of functional skin were determined. The results of this study showed hydration effects increase in the skin of subjects treated with *S. thermophilus*-containing cream when compared with controls. The hydration skin increase could be attributed to the enhanced SC ceramide levels probably due to the SMase presence in *S. thermophilus*. These findings suggested that topical application of a sonicated *S. thermophilus* preparation may contribute to the improvement of lipid barrier and a more effective resistance against aging-associated skin xerosis.

The invention of Gueniche et al. (2006a, b) related to the use of an effective amount of at least one microorganism belonging to the species *L. paracasei* or *casei*, a fraction thereof or a metabolite thereof, in combination with an effective amount of at least one microorganism belonging to the species *Bifidobacterium lactis*, a fraction thereof or a metabolite thereof, for producing a dermatological composition intended for treating and/or preventing reactive, irritable and/or intolerant, acquired dry skin and/or constitutional dry skin.

The cosmetic and/or dermatologic use of hesperidin (a flavonoid) in combination with probiotic microorganism for preventing a reduction in and/or for reinforcing the barrier function of the skin associated with aging and/or photoaging was also proposed in a recent patent of Gueniche and Castiel (2009). The authors also presented an invention related to the cosmetic use of an effective amount of at least one microorganism, especially a probiotic microorganism, or a fraction thereof, as an agent for preventing the appearance and/or for treating the manifestation of sensations of discomfort and/or cutaneous signs associated with a surface skin treatment or an invasive treatment for esthetic purposes (Castiel and Gueniche 2009).

Epidermal keratinocyte differentiation is another essential event that coordinates epidermal turnover and construction of skin barrier function. In this context, Baba et al. (2006) reported that *L. helveticus*-fermented milk was able to promote differentiation of cultured normal human epidermal keratinocytes by enhancing production of the differentiation-related element profilaggrin, a precursor of a natural moisturizing factor that controls normal epidermal hydration and flexibility.

3.3 Hyaluronic Acid

Fermentation technology using lactic acid bacteria was also proposed to produce many other beauty factors, including hyaluronic acid (HA) and collagen, that provide beneficial activities in maintaining skin health and preventing skin aging (Chong et al. 2005; Chiba 2007). In the skin, HA represents a predominant voluminous molecule of extracellular matrix (ECM). It is synthesized by keratinocytes and fibroblasts and performs important biological role in the skin by having specific rheological characteristics and good water-holding properties. HA has several applications in medicine and cosmetics, including skin moisturizers. The ability of *Bifidobacterium*-fermented soy milk extract (BE) to significantly enhance HA synthesis in vitro and in vivo has also been performed (Miyazaki et al. 2003, 2004). The authors established that topical application of BE was able to ameliorate the elasticity and viscoelasticity of mouse skin, to increase HA level and thus preventing the age-dependent loss of cutaneous HA. The topical application of a gel formula containing BE on human skin significantly enhanced skin elasticity suggesting BE as a new cosmetic ingredient to improve skin elasticity through augment of

HA production. Recently, Izawa et al. (2010) described the optimized fermentation conditions in a skimmed milk-based medium which had significantly (about 20-fold) increased HA yield from *S. thermophilus* YIT 2084. Because of the safety of this bacterium, the fermentation technique would have an impact on the application of the bacterial HA.

3.4 Collagen

Collagen is a major constituent of the human skin and accounts for a high proportion of the skin's elasticity and physical properties. It is well documented that exposure to sunlight damages the skin structures. In response to this damage, the skin repairs itself through the rapid production of collagen and other associated dermal components such as polysaccharides. A recent invention of Lieurey and Watkins (2009) reported that the use of a fermented milk product comprising nonhydrolysed and casein-free whey proteins improved skin firmness when topically applied to skin. The milk fermented with classic lactic acid bacteria (*S. thermophilus* and *L. bulgaricus*) could improve the structuring of skin's collagen without promoting collagen synthesis. The inventors suggested the topical use of the fermented milk on damaged skin to improve the collagen natural generation as part of the repair process, thus preventing a skin condition termed elastosis.

4 Probiotics and Skin Inflammatory/Immune System

Skin consists of stratified epithelium with various cell types, including keratinocytes that have been specialized to act as the outpost of the innate defense system and, in a lower proportion, dendritic cells, melanocytes, and Langerhans cells. Each of these cell types contributes to skin protection. Moreover, the underlying dermal compartment harbors leukocytes, mastocytes, and macrophages that are key actors of cell defense. Probiotics actions on the skin can be mediated by modulation of host's immune response including innate as well as adaptive.

4.1 Barrier Function and Skin Reactivity

Some probiotic strains display potent immune-modulatory properties at the skin level. Recently, the ability of *L. paracasei* CNCM-I 2116 (ST11) to modulate reactive skin-associated inflammatory mechanisms has been evaluated (Gueniche et al. 2010a). The authors showed that ST11 was able to abrogate vasodilation, edema, mast cell degranulation, and TNF-alpha release which induced by substance P, compared to control. Moreover, using ex vivo skin organ culture, the authors showed that ST11-conditioned medium induced a significantly faster barrier function recovery after sodium lauryl sulphate disruption, compared to control. These results support a beneficial role of ST11 on key biological processes associated with barrier function and skin reactivity. Gueniche et al. (2010b) performed an in vitro and a clinical trial using *B. longum* sp. extract proved that these nonreplicating bacteria forms applied to the skin were able to improve sensitive skin in various parameters associated with inflammation such as decrease in vasodilation, edema, mast cell degranulation, and TNF-alpha release. These findings suggest that new approaches, based on a bacteria lysate, could be developed for the treatment and/or prevention of symptoms related to reactive skin. A recent invention of Gueniche (2010) disclosed methods directed to the cosmetic use of an effective amount of at least one probiotic microorganism especially from the genus *Lactobacillus* and/or *Bifidobacterium*, or a fraction thereof and/or a metabolite thereof, as an active agent for limiting, preventing or treating skin irritation, and/or irritative skin disorders.

The innate immune system uses pattern recognition receptors such as Toll-like receptors (TLRs) to recognize microorganisms or their products on the cell membranes (Guan and Mariuzza 2007). Keratinocytes were found to express TLR2, which is specifically involved in the recognition of peptidoglycan (PGN), a mesh-like layer outside the plasma membrane of bacteria forming the cell wall that protects bacteria from environmental stress. It has been recently demonstrated that TLR2 can differentially recognize PGN from Gram-positive and Gram-negative bacteria (Asong et al. 2009). DAP-containing muropeptides were bound with high affinity to TLR2, whereas only a restricted number of Gram-positive lysine-containing muropeptides, derived from PGN remodeled by bacterial autolysins, were recognized. The difference in recognition of the two classes of muropeptides is proposed to be a strategy by the host to differentially respond to Gram-negative and Gram-positive bacteria, which produce vastly different quantities of PGN. Lactobacilli were demonstrated to stimulate innate immune response via TLR2 and nucleotide-binding oligomerization domain-2 thus modulating dendritic cell function (Zeuthen et al. 2008). They were also demonstrated to induce the production of interleukin-12 and other regulatory factors by macrophages (Sun et al. 2005; Shida et al. 2006; Bleau et al. 2007). Most notably, *L. casei* strain Shirota, both as intact cells or as cell wall-derived polysaccharide–peptidoglycan complex (PSPG), inhibited IL-6 production in lipopolysaccharide (LPS)-stimulated lamina propria mononuclear cells isolated from murine models of inflammatory bowel disease and induced an improvement of disease conditions in mice (Matsumoto et al. 2009). In the light of these findings, it is reasonable to hypothesize a possible role of *Lactobacillus* surface molecules in modulating inflammatory response in skin.

4.2 Environmental Stress

Skin is the largest human organ and is directly exposed to environmental stress which may cause oxidative stress, especially UV irradiation. Indeed, UVB (290–320 nm) and UVA (320–400 nm) light can both induce the generation of reactive oxygen and nitrogen species (ROS/RNS) in human skin (Xu and Fisher 2005). When the generation of ROS/RNS exceeds the skin antioxidant defenses, the consequence is epidermal oxidative stress. Excess production of ROS/RNS may affect the cell function leading to apoptotic or necrotic cell death. Oxidative stress is thought to play a central role in initiating and driving the signaling events that lead to cellular response following skin UV irradiation. Increased ROS/RNS production induced by UV light alters gene and protein structure and function, leading to skin damage through the activation of multiple cytokine and growth factor cell surface receptors (Rittié and Fisher 2002). Many of the molecular alterations observed following UV irradiation of skin also occur during aging, a condition which is known to be associated with oxidative stress. Indeed, it is well known that endogenous antioxidants are decreased in skin and blood during UV exposure and in senescence (Rittié and Fisher 2002). Hence, a treatment aimed at counteracting oxidative stress may be helpful in the prevention of damages caused by UVB and UVA light or skin aging. Furthermore, epidemiological studies clearly indicate UV light exposure as the major cause of skin cancer (Armstrong and Kricker 2001); hence it is recommended to adopt preventive measures by sunscreens via the topical route, in addition to antioxidants via the systemic route. A number of reports indicate that food supplementation with antioxidant molecules (such as vitamins C and E, carotenoids, flavonoid, polyphenols, thiol compounds, and selenium) is able to counteract skin damage by UVA and UVB (Greul et al. 2002).

Many evidences indicate that probiotics may be helpful as antioxidant agents, both in vitro and in vivo. A number of probiotic strains were demonstrated to possess antioxidative action in vitro. Lin and Yen (1999) and Lin and Chang (2000) demonstrated that various *Lactobacillus* and *Bifidobacterium* strains were able to exert antioxidant action in vitro. Both intact cells and cell-free extracts were able to inhibit ascorbate autoxidation, to exert metal-chelating ability, to scavenge superoxide anion and other ROS, and to inhibit lipid peroxidation. Probiotics' ability to act as antioxidant can be attributed to the presence of antioxidant enzymes such as superoxide dismutase (Shen et al. 2010) to the release of antioxidant compounds such as glutathione (Peran et al. 2006) and to the production of extracellular polysaccharide (EPS) biomolecules that probiotic bacteria release into the surroundings to protect themselves under starvation conditions and also from extreme pH and temperature conditions (Kodali and Sen 2008). Ingestion of probiotics may also exert systemic protection from oxidative stress. Lactic acid bacteria extracts or fermented milks were found to decrease human low-density lipoprotein oxidation and to prolong the resistance of the lipoprotein fraction to oxidation (Terahara et al. 2001; Kullisaar et al. 2003; Gueniche et al. 2006b, 2008; Peguet-Navarro et al. 2008; Bouilly-Gauthier et al. 2010). Thus, probiotics may represent a useful therapeutic tool for the prevention of epidermal oxidative stress either via the topical route or via ingestion. On the basis of these premises, our group has recently performed a number of experiments in order to study the antioxidant activity of a specific strain of lactic acid bacteria, the *S. thermophilus* S244 (provided by VSL Pharmaceuticals, Inc., Gaithersburg, MD, USA) (manuscript in preparation). Extracts of this bacterium were found to be good free radical scavengers, as compared to Trolox used as reference antioxidant compound. The oxygen radical absorbance capacity of the bacterial lysate, measured as the area under the curve, revealed that the bacterial extract efficiently inhibited the free radical-dependent loss of phycoerythrin-E fluorescence in a dose-dependent fashion (not shown). Encouraged by these results we assessed the photoprotective activity of the bacterial lysate. When human HaCaT keratinocytes were irradiated with UVB light, a 50% reduction in cells viability was observed. However, when the cells were treated with UV in the presence of the bacterial extract, a dose-dependent protection of cell viability was observed (not shown). In the light of our data on the antioxidant activity of bacterial extracts, we propose that the photoprotective effect may be at least in part due to the good free radical-scavenging properties of the extract.

4.3 Protection Against NO

The NO pathway has been shown in several cell types that reside in the skin, including keratinocytes, melanocytes, Langerhans cells, fibroblasts, and endothelial cells (Bruch-Gerharz et al. 1998). Convincing evidence suggests that NO synthesis in these cells can be modulated by calcium-mobilizing agonists as well as diverse inflammatory and immune stimuli, and thereby contributes to the pathogenesis of several human skin diseases. Characterization of these intrinsic and extrinsic regulatory stimuli of NO synthesis has afforded substantial insights into the role of NO in inflammatory, hyperproliferative, and autoimmune skin diseases, as well as skin cancer, and may ultimately form the basis for future therapeutic intervention. NO is synthesized from arginine and oxygen by various nitric oxide synthase (NOS) enzymes. NOS is a group of enzymes responsible for the synthesis of NO from the terminal nitrogen atom of L-arginine in the presence of oxygen and some cofactors. The presence of arginine deiminase in *L. brevis*, previously reported by our group (Di Marzio et al. 2001), was further characterized through activity and expression studies. *L. brevis* arginine deiminase, being able to metabolize arginine to citrulline and ammonia, allows to inhibit NO generation by competing with NOS for the same substrate, arginine. Considering the role of NO in inflammatory conditions our group had also analyzed the ability of *L. brevis* to inhibit NOS activity as well other inflammatory parameters including IFN-γ and PGE_2 release and MMP expression in murine macrophages activated by LPS (Della Riccia et al. 2007). The results suggested that the presence of *L. brevis* extracts in cell culture strongly inhibited inducible NOS activity, IFN-γ/PGE_2 production, and MMP activity in LPS-activated macrophages. These effects could be attributed to *L. brevis*' ability to prevent inducible NOS activity responsible for NO, a key inflammatory molecule. The invention of De Simone (2003) disclosed the use of bacteria endowed with arginine deiminase to induce apoptosis and/or reduce an inflammatory reaction, and pharmaceutical compositions containing such bacteria, including creams and ointments. The inventor also included a strain of *L. brevis* referred to as CD2 highly endowed with arginine deiminase.

4.4 Anti-inflammatory Potential

In the context of the anti-inflammatory potential of probiotics, the in vitro and in vivo effects of supernatants from *L. acidophilus* (ATCC strains 4356 and 43121) on tissue repair and angiogenesis were investigated by Halper et al. (2003). The authors suggested that *Lactobacillus* supernatant promoted proinflammatory processes including chemoattraction of polymorphonuclear cells, macrophages, and angiogenesis in addition to previously described stimulation of production of TNFα and other cytokines including interleukins and interferons.

Trautmann et al. (2001) demonstrated that in atopic dermatitis and allergic contact dermatitis, skin-activated T cells stimulated Fas-induced keratinocyte apoptosis. In particular, diseased skin-infiltrating T cells produce IFN-γ that increasing Fas receptor number on keratinocyte membrane renders them susceptible to apoptosis by Fas ligand expressed on or released by T cell surface. The knowledge of these mechanisms provided a useful molecular model to focus on innovative therapeutic applications. In order to examine the role of probiotics on inflammatory skin disease, our group investigated the effect of a selected extract from *B. infantis* on human keratinocyte cell line (HaCaT) abnormal apoptosis induced by activated T-lymphocyte (Cinque et al. 2006). In particular, the probiotic effect on inflammatory skin disease was investigated in the experimental model of AD as proposed by Trautmann (2001). In this in vitro model of atopic AD, the ability of the probiotic extract to protect HaCaT from apoptosis induced by soluble factors (IFN-γ and CD95 ligand) released by human T-lymphocytes in vitro, activated with anti-CD3/CD28 mAbs or phytohemoagglutinin, has been evaluated. The obtained results highlighted the bacterial extracts' ability to totally prevent T-lymphocyte-induced HaCaT cell apoptosis in vitro. The mechanism underlying this inhibitory effect has been suggested to depend on the ability of the bacterial extracts to significantly reduce anti-CD3/CD28 mAbs and mitogen-induced T-cell proliferation, IFN-γ generation, and CD95 ligand release. These results may represent an experimental basis for a potential therapeutic approach mainly targeting the skin disorders-associated immune abnormalities.

Cutaneous immune responses must be tightly controlled to prevent unnecessary inflammation in response to innocuous antigens, while maintaining the ability to combat skin-tropic pathogens. Regulatory T cells (Treg) play an important protective role against autoimmune response to self-antigens to maintain self-tolerance, functioning by suppressing the activation, cytokine production, and proliferation of other T cells. Tregs are CD4- and CD25-positive cells but the most specific marker for these cells is FoxP3 (forkhead box P3), which is localized intracellularly. Dysregulation in Treg cell frequency or functions may lead to the development of autoimmune disease. Treg modulation is considered to be a promising therapeutical approach to treat some selected disorders, such as allergies, and to prevent allograft rejection (Clark 2010). In normal human skin these cells represent between 5 and 10% of the T cells resident and proliferating in inflamed skin serving as a brake for cutaneous inflammation. Indeed their number increases in the skin lesions of contact dermatitis and in DTH reactions (Teraki and Shiohara 2003; Vukmanovic-Stejic et al. 2008). Of note, in a recent study, de Roock et al. (2010) examined the ability of specific probiotic strains to induce Foxp3-positive Tregs from human peripheral blood mononuclear cells (PBMC) in vitro. The authors highlighted a different capacity of tested probiotic strains to stimulate Treg population demonstrating that *L. acidophilus* proved to be the most potent Treg inducing bacterium. Moreover *L. acidophilus*-induced Foxp3 cells were also able to diminish effector T cell proliferation. All examined probiotic strains were tested for the ability to induce cytokine secretion, and in particular the results showed that all bacteria induced a comparable IFN-γ release while IL-17 and IL-13 were produced in low levels, and IL-4 did not detect. Induction of regulatory T cells is an attractive target in the therapy of disorders where an abnormal immune response is present including autoimmune diseases, asthma, and allergy. For this purpose there is a need to increase understanding about specific role of each single bacterium in modulating the Treg function in distinct compartments, both in the intestine and locally in inflamed tissue. Volz and Biedermann (2009) sustained the topical application of probiotics for prophylaxis and therapy of overwhelming cutaneous

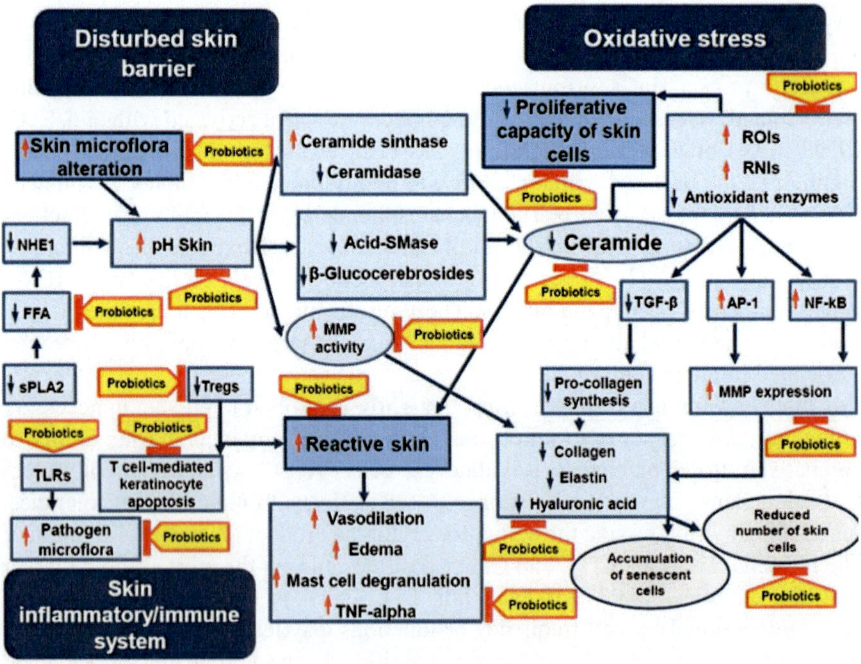

Fig. 1 Comprehensive model that summarizes the main actions carried out by probiotics in different skin conditions associated with altered microflora, abnormal oxidative stress, disturbed skin barrier, and/or inflammatory/immune skin reactions. It is important to note that, for simplicity, the generic term "probiotics" is indicated but it must be implied that the claimed effect should be attributed to specific species or specific strains of probiotics as indicated in the text

proinflammatory immune reactions, considering it very promising also on the basis of the ability of specific probiotic strains to trigger the production of tolerogenic IL-10 by activating anti-inflammatory Treg.

5 Conclusions

Topical probiotic formulations are becoming increasingly available for healthy skin care, prevention and treatment of skin diseases, and antiaging benefits, thus representing an emerging area for skin health. The potential benefits of skin probiotics could strongly depend on how each species or strain is selected as the specific mechanisms underlying a specific effect on the healthy or disturbed skin. It appears, therefore, particularly important to stress that it is not possible to generalize the effects that each of them, any association or combination thereof or extracts thereof, has on the skin. A comprehensive model that summarizes the main actions carried out by probiotics in different skin conditions associated with altered microflora, abnormal oxidative stress, disturbed skin barrier, and/or inflammatory/immune skin reactions is shown in Fig. 1.

References

Al-Ghazzewi FH, Tester RF (2009) Effect of konjac glucomannan hydrolysates and probiotics on the growth of the skin bacterium *Propionibacterium acnes* in vitro. Int J Cosmet Sci 32:139–142

Arck P, Handjiski B, Hagen E, Pincus M, Bruenahl C, Bienenstock J, Paus R (2010) Is there a 'gut-brain-skin axis'? Exp Dermatol 19:401–405

Armstrong BK, Kricker A (2001) The epidemiology of UV induced skin cancer. J Photochem Photobiol B 63:8–18

Asong J, Wolfert MA, Maiti KK, Miller D, Boons GJ (2009) Binding and cellular activation studies reveal that Toll-like receptor 2 can differentially recognize peptidoglycan from gram-positive and gram-negative bacteria. J Biol Chem 284:8643–8653

Baba H, Masuyama A, Takano T (2006) Effects of *Lactobacillus helveticus*-fermented milk on the differentiation of cultured normal human epidermal keratinocytes. J Dairy Sci 89:2072–2075

Bleau C, Savard R, Lamontagne L (2007) Murine immunomodulation of IL-10 and IL-12 induced by new isolates from avian type 2 *Lactobacillus acidophilus*. Can J Microbiol 53:944–956

Bouilly-Gauthier D, Jeannes C, Maubert Y, Duteil L, Queille-Roussel C, Piccardi N, Montastier C, Manissier P, Piérard G, Ortonne JP (2010) Clinical evidence of benefits of a dietary supplement containing probiotic and carotenoids on ultraviolet-induced skin damage. Br J Dermatol 163:534–543

Bruch-Gerharz D, Ruzicka T, Kolb-Bachofen V (1998) Nitric oxide in human skin: current status and future prospects. Investig Dermatol 110:1–7

Caramia G, Atzei A, Fanos V (2008) Probiotics and the skin. Clin Dermatol 26:4–11

Castiel I, Gueniche A (2009) Cosmetic use of microorganisms. WO Patent 031099 A2, 12 Mar 2009

Cavaliere widow Vesely RMA, De Simone C (2001) Sphinogomyelinase composition and use thereof. US patent 6,258,355 B1, 10 Jul 2001

Chiba K (2007) Development of functional cosmetic ingredients using lactic acid bacteria in Japan. Jpn J Lactic Acid Bact 18:105–112

Chong BF, Blank LM, Mclaughlin R, Nielsen LK (2005) Microbial hyaluronic acid production. Appl Microbiol Biotechnol 66:341–351

Cinque B, Di Marzio L, Della Riccia DN, Bizzini F, Giuliani M, Fanini D, De Simone C, Cifone MG (2006) Effect of *Bifidobacterium infantis* on Interferon-gamma-induced keratinocyte apoptosis: a potential therapeutic approach to skin immune abnormalities. Int J Immunopathol Pharmacol 19:775–786

Cinque B, Palumbo P, La Torre C, Melchiorre E, Corridoni D, Miconi G, Di Marzio L, Cifone MG, Giuliani M (2010) Probiotics in aging skin. In: Textbook of aging skin. Springer, Berlin, pp 811–819

Clark RA (2010) Skin resident T cells: the ups and downs of on site immunity. Invest Dermatol 130:362–370

de Roock S, van Elk M, van Dijk ME, Timmerman HM, Rijkers GT, Prakken BJ, Hoekstra MO, de Kleer IM (2010) Lactic acid bacteria differ in their ability to induce functional regulatory T cells in humans. Clin Exp Allergy 40:103–110

De Simone C (2003) Use of bacteria endowed with arginine deiminase to induce apoptosis and/or reduce an inflammatory reaction and pharmaceutical or dietetic compositions containing such bacteria. US Patent 0215429 A1, 20 Nov 2003

Della Riccia DN, Bizzini F, Perilli MG, Polimeni A, Trinchieri V, Amicosante G, Cifone MG (2007) Anti-inflammatory effects of *Lactobacillus brevis* (CD2) on periodontal disease. Oral Dis 13:376–385

Di Marzio L, Cinque B, De Simone C, Cifone MG (1999) Effect of lactic acid bacterium *Streptococcus thermophilus* on ceramide levels in human keratinocytes in vitro and stratum corneum in vivo. J Invest Dermatol 113:98–106

Di Marzio L, Russo FP, D'Alò S, Biordi L, Ulisse S, Amicosante G, De Simone C, Cifone MG (2001) Apoptotic effects of selected strains of lactic acid bacteria on a human T leukemia cell line are associated with bacterial arginine deiminase and/or sphingomyelinase activities. Nutr Cancer 40:185–196

Di Marzio L, Centi C, Cinque B, Masci S, Giuliani M, Arcieri A, Zicari L, De Simone C, Cifone MG (2003) Effect of the lactic acid bacterium *Streptococcus thermophilus* on stratum corneum ceramide levels and signs and symptoms of atopic dermatitis patients. Exp Dermatol 12:615–620

Di Marzio L, Cinque B, Cupelli F, De Simone C, Cifone MG, Giuliani M (2008) Increase of skin-ceramide levels in aged subjects following a short-term topical application of bacterial sphingomyelinase from *Streptococcus thermophilus*. Int J Immunopathol Pharmacol 21 (1):137–143

Farmer S (2005) Topical compositions containing probiotic bacillus bacteria, spores, and extracellular products and uses thereof. US Patent 6905692, 14 June 2005

Feingold KR (2007) Thematic review series: skin lipids. The role of epidermal lipids in cutaneous permeability barrier homeostasis. J Lipid Res 48:2531–2546

Gillor O, Etzion A, Riley MA (2008) The dual role of bacteriocins as anti- and probiotics. Appl Microbiol Biotechnol 81:591–606

Greul AK, Grundmann JU, Heinrich F, Pfitzner I, Bernhardt J, Ambach A, Biesalski HK, Gollnick H (2002) Photoprotection of UV-irradiated human skin: an antioxidative combination of vitamins E and C, carotenoids, selenium and proanthocyanidins. Skin Pharmacol Appl Skin Physiol 15:307–315

Guan R, Mariuzza RA (2007) Peptidoglycan recognition proteins of the innate immune system. Trends Microbiol 15:127–134

Gueniche A (2010) Use of probiotic microorganisms to limit skin irritation. US Patent 0226892 A1, 9 Sept 2010

Gueniche A, Castiel I (2009) Use of combination of hesperidin and of a microorganism for influencing the barrier function of the skin. US Patent 00681161 A1, 12 Mar 2009

Gueniche A, Benyacoub J, Buetler TM, Smola H, Blum S (2006a) Supplementation with oral probiotic bacteria maintains cutaneous immune homeostasis after UV exposure. Eur J Dermatol 16:511–517

Gueniche A, Breton L, Ballevre O, Blum-Sperise S, Bureau-Franz I, Benyacoub J (2006b) Cosmetic and/or dermatological composition for sensitive skin. US Patent 0171936 A1, 3 Aug 2006

Gueniche A, Buetler T, Benyacoub J, Blum S (2008) *Lactobacillus johnsonii* provides a dose-dependent protection against UVR-induced immunosuppression. Eur J Dermatol 18:476–477

Gueniche A, Bastien P, Ovigne JM, Kermici M, Courchay G, Chevalier V, Breton L, Castiel-Higounenc I (2010b) *Bifidobacterium longum* lysate, a new ingredient for reactive skin. Exp Dermatol 19:e1–e8

Gueniche A, Benyacoub J, Philippe D, Bastien P, Kusy N, Breton L, Blum S, Castiel-Higounenc I (2010a) *Lactobacillus paracasei* CNCM I-2116 (ST11) inhibits substance P-induced skin inflammation and accelerates skin barrier function recovery in vitro. Eur J Dermatol 20(6):731–737

Halper J, Leshin LS, Lewis SJ, Li WI (2003) Wound healing and angiogenic properties of supernatants from *Lactobacillus* cultures. Exp Biol Med 228:1329–1337

Hansen JE, Jespersen LK (2010) Wound or tissue dressing comprising lactic acid bacteria. US Patent 0143447 A1, 10 June 2010

Izawa N, Hanamizu T, Sone T, Chiba K (2010) Effects of fermentation conditions and soybean peptide supplementation on hyaluronic acid production by *Streptococcus thermophilus* strain YIT 2084 in milk. J Biosci Bioengn 109:356–360

Jones ML, Ganopolsky JG, Labbé A, Prakash S (2010) A novel nitric oxide producing probiotic patch and its antimicrobial efficacy: preparation and in vitro analysis. Appl Microbiol Biotechnol 87:509–516

Klaenhammer TR (1993) Genetics of bacteriocins produced by lactic acid bacteria. FEMS Microbiol Rev 2:39–85

Kodali VP, Sen R (2008) Antioxidant and free radical scavenging activities of an exopolysaccharide from a probiotic bacterium. Biotechnol J 3:245–251

Krutmann J (2009) Pre- and probiotics for human skin. J Dermatol Sci 54:1–5

Kullisaar T, Songisepp E, Mikelsaar M, Zilmer K, Vihalemm T, Zilmer M (2003) Antioxidative probiotic fermented goats' milk decreases oxidative stress-mediated atherogenicity in human subjects. Br J Nutr 90:449–456

Lambers H, Piessens S, Bloem A, Pronk H, Finkel P (2006) Natural skin surface pH is on average below 5, which is beneficial for its resident flora. Int J Cosmet Sci 28:359–370

Lieurey S, Watkins K (2009) Use of fermented milk product for skin treatment. US Patent 0252812 A1, 8 Oct 2008

Lin MY, Chang FJ (2000) Antioxidative effect of intestinal bacteria *Bifidobacterium longum* ATCC 15708 and *Lactobacillus acidophilus* ATCC 4356. Dig Dis Sci 45:1617–1622

Lin MY, Yen CL (1999) Antioxidative ability of lactic acid bacteria. J Agric Food Chem 47:1460–1466

Matsumoto S, Hara T, Nagaoka M, Mike A, Mitsuyama K, Sako T, Yamamoto M, Kado S, Takada T (2009) A component of polysaccharide peptidoglycan complex on *Lactobacillus* induced an improvement of murine model of inflammatory bowel disease and colitis-associated cancer. Immunology 128:e170–e180

Mauro T (2006) SC pH: measurement, origins, and functions. In: Elias P, Feingold K (eds) Skin barrier. Taylor & Francis, New York, pp 223–229

Miyazaki K, Hanamizu T, Iizuka R, Chiba K (2003) *Bifidobacterium*-fermented soy milk extract stimulates hyaluronic acid production in human skin cells and hairless mouse skin. Skin Pharmacol Appl Skin Physiol 16:108–116

Miyazaki K, Hanamizu T, Sone T, Chiba K, Kinoshita T, Yoshikawa S (2004) Topical application of *Bifidobacterium*-fermented soy milk extract containing genistein and daidzein improves rheological and physiological properties of skin. J Cosmet Sci 55:473–479

Mizutani Y, Mitsutake S, Tsuji K, Kihara A, Igarashi Y (2009) Ceramide biosynthesis in keratinocyte and its role in skin function. Biochimie 91:784–790

Oelschlaeger TA (2010) Mechanisms of probiotic actions – a review. Int J Med Microbiol 300:57–62

Oh S, Kim SH, Ko Y, Sim JH, Kim KS, Lee SH, Park S, Kim YJ (2006) Effect of bacteriocin produced by *Lactococcus* sp. HY 449 on skin-inflammatory bacteria. Food Chem Toxicol 44:552–559

Peguet-Navarro J, Dezutter-Dambuyant C, Buetler T, Leclaire J, Smola H, Blum S, Bastien P, Breton L, Gueniche A (2008) Supplementation with oral probiotic bacteria protects human cutaneous immune homeostasis after UV exposure-double blind, randomized, placebo controlled clinical trial. Eur J Dermatol 18:504–511

Peral MC, Martinez MA, Valdez JC (2009a) Bacteriotherapy with *Lactobacillus plantarum* in burns. Int Wound J 6:73–81

Peral MC, Rachid MM, Gobbato NM, Huaman Martinez MA, Valdez JC (2009b) Interleukin-8 production by polymorphonuclear leukocytes from patients with chronic infected leg ulcers treated with *Lactobacillus plantarum*. Clin Microbiol Infect 16:281–286

Peran L, Camuesco D, Comalada M, Nieto A, Concha A, Adrio JL, Olivares M, XausJ ZA, Galvez J (2006) *Lactobacillus fermentum*, a probiotic capable to release glutathione, prevents colonic inflammation in the TNBS model of rat colitis. Int J Colorectal Dis 21:737–746

Rittié L, Fisher GJ (2002) UV-light-induced signal cascades and skin aging. Ageing Res Rev 1:705–720

Shen Q, Zhang B, Xu R, Wang Y, Ding X, Li P (2010) Antioxidant activity in vitro of the selenium-contained protein from the Se-enriched *Bifidobacterium animalis* 01. Anaerobe 16:380–386

Shida K, Kiyoshima-Shibata J, Nagaoka M, Watanabe K, Nanno M (2006) Induction of interleukin-12 by *Lactobacillus* strains having a rigid cell wall resistant to intracellular digestion. J Dairy Sci 89:3306–3317

Simmering R, Breves R (2009) Pre- and probiotic cosmetics. Hautarzt 60:809–814

Spigelman M, Ross M (2008) Method of using topical probiotics for the inhibition of surface contamination by a pathogenic microorganisms and composition therefor. US Patent 0107699 A1, 8 May 2008

Sullivan M, Schnittger SF, Mammone T, Goyarts EC (2009). Skin treatment method with *Lactobacillus* extract. US Patent 7,510,734 B2, 31 Mar 2009

Sun J, Shi YH, Le GW, Ma XY (2005) Distinct immune response induced by peptidoglycan derived from *Lactobacillus* sp. World J Gastroenterol 11:6330–6337

Teodorescu R (1999) A natural eubiotic product for maintenance and treatment of teguments. WO Patent 007332, 18 Feb 1999

Terahara M, Kurama S, Takemoto N (2001) Prevention by lactic acid bacteria of the oxidation of human LDL. Biosci Biotechnol Biochem 65:1864–1868

Teraki Y, Shiohara T (2003) IFN-gamma-producing effector CD8+ T cells and IL-10-producing regulatory CD4+ T cells in fixed drug eruption. J Allergy Clin Immunol 112:609–615

Trautmann A, Akdis M, Schmid-Grendelmeier P, Disch R, Bröcker EB, Blaser K, Akdis CA (2001) Targeting keratinocyte apoptosis in the treatment of atopic dermatitis and allergic contact dermatitis. J Allergy Clin Immunol 108:839–846

Valdéz JC, Peral MC, Rachid M, Santana M, Perdigón G (2005) Interference of Lactobacillus plantarum with *Pseudomonas aeruginosa* in vitro and in infected burns: the potential use of probiotics in wound treatment. Clin Microbiol Infect 11:472–479

Volz T, Biedermann T (2009) Outside-in. Probiotic topical agents Hautarzt 60:795–801

Vukmanovic-Stejic M, Agius E, Booth N, Dunne PJ, Lacy KE, Reed JR, Sobande TO, Kissane S, Salmon M, Rustin MH, Akbar AN (2008) The kinetics of CD4 + Foxp3+ T cell accumulation during a human cutaneous antigen-specific memory response in vivo. J Clin Invest 118:3639–3650

Williams NT (2010) Probiotics. Am J Health Syst Pharm 67:449–458
Xu Y, Fisher GJ (2005) Ultraviolet (UV) light irradiation induced signal transduction in skin photoaging. J Dermatol Sci Suppl 1:S1–S8
Yadav H, Jain S, Sinha PR (2007) Production of free fatty acids and conjugated linoleic acid in probiotic dahi containing *Lactobacillus acidophilus* and *Lactobacillus casei* during fermentation and storage. Int Dairy J 17:1006–1010
Zeuthen LH, Fink LN, Frøkiaer H (2008) Toll-like receptor 2 and nucleotide-binding oligomerization domain-2 play divergent roles in the recognition of gut-derived lactobacilli and bifidobacteria in dendritic cells. Immunology 124:489–502

Challenges of Beneficial Health Claims

István Siró

Contents

1 Importance of Beneficial Health Claims 244
2 Health Claim Legislation Worldwide 245
  2.1 United States 245
  2.2 Canada 246
  2.3 European Union 246
  2.4 Japan 247
  2.5 Australia and New Zealand 248
3 Substantiation of Health Claims for Probiotics 248
  3.1 Identification of Probiotic Microorganisms 249
  3.2 Enumeration of Probiotics in Food 250
  3.3 Enumeration of Probiotics in the GIT 251
  3.4 Survival of Probiotics in the GIT 251
  3.5 In Vitro Simulation of the GIT 252
  3.6 In Vitro Assessment of Adhesion and Colonization 253
  3.7 Antimicrobial Activity 254
4 Some Aspects of In Vivo Animal Studies 254
5 Clinical Human Studies 255
  5.1 Identification of Target and Study Population 256
  5.2 Defining Active Components 256
  5.3 Form of Delivery 256
  5.4 Effect of Background Diet 257
  5.5 Effective Dosage 257
  5.6 Long-Term Effects 258
  5.7 Meta-Analyses 258
6 Some Beneficial Claims and Supporting Evidences 258
7 Consumer Attitudes 259
8 Summary and Concluding Remarks 261
References 262

I. Siró (✉)
Faculty of Food Science, Corvinus University of Budapest, Ménesi út 43-45, 1118 Budapest, Hungary
e-mail: istvan.siro@gmail.com

M.-T. Liong (ed.), *Probiotics*, Microbiology Monographs 21,
DOI 10.1007/978-3-642-20838-6_10, © Springer-Verlag Berlin Heidelberg 2011

Abstract Probiotic bacteria are live microorganisms that, when administered in adequate amounts, confer a health benefit on the host. The different beneficial effects of specific probiotic strains may be translated into health claims. The use of health claims is strictly regulated by legislation worldwide. A common element of those is that substantiation of claims should be based on scientific evidences. Providing such evidences requires a long and expensive procedure, while a number of challenges have to be addressed. Different in vitro and in vivo methods are applied for screening and characterizing the putative probiotic strains. Although these tests may be useful as assessment tools, probiotics must be validated by properly designed human clinical studies. Poor prior selection, as well as the limited capacity of in vitro tests and animal models to mimic the human organization may contribute to contradictions between in vitro findings and in vivo feasibility. This chapter reviews the crucial steps of substantiation of health claims associated with probiotics with special emphasis on the related challenges.

1 Importance of Beneficial Health Claims

A number of health claims have been associated with probiotics, such as the control of gastrointestinal infections, improvement in lactose metabolism, antimicrobial and antimutagenic properties, reduction in serum cholesterol, immune system stimulation, and improvement in the infections of bowel disease. In addition, new evidences suggest that probiotics can be used in the prevention and treatment of atopic diseases, obesity, diabetes, and some cancers. The mechanisms by which probiotics exert their effects may involve modifying gut pH, antagonizing pathogens through antimicrobial production, competing for pathogen binding, receptor sites, nutrients and growth factors, stimulating immunomodulatory cells, and producing lactase (Collado 2009; Kaur et al. 2009)

Several selection criteria have been used for novel probiotic strains, which can be summarized into three groups, such as safety, functional, and technological aspects (Dunne et al. 2001). In vitro studies are generally useful for screening purpose and for gaining information on specific functional properties of the tested strains, as well as for assessing their safety. It has been realized, however, that existing tests are not always accurate to predict the functionality and efficacy of probiotic strains in vivo. Selection and validation should therefore be based on both in vitro and in vivo models with a reliable predicted value or function.

Although numerous in vitro tests can be correlated with animal models, standardized, verifiable human clinical studies are necessary to demonstrate the safety (phase I trials) and efficacy (phase II trials) of a putative probiotic, as well as whether it is superior to existing therapies. Phase II human studies should be designed as double-blind, randomized, and placebo-controlled. The experimental evidence to identify probiotic microorganisms and to demonstrate their efficacy in clinical trials is more challenging than for other potential functional foods, because effects are mediated by living microorganisms and may therefore be influenced by the status of these microorganisms. Mechanisms are seldom known, markers are generally not available, and matrix effects may be pronounced.

The credibility of health claims is of crucial importance. Reliable probiotic products with scientifically supported claims should be clearly distinguished from those of doubtful quality. It is essential that the health claims stated on the labels of products supply the consumer with reliable information. In the case of nonsubstantiated claims, the consumer is manipulated into buying a product under false pretences, in the one hand, and there is a potential risk of inappropriate treatment for a condition, on the other hand (Brink et al. 2005). Regulations related to health claims differ among countries, but underlying all is an emphasis on scientific credibility of any statements of health benefits (Sanders et al. 2005).

2 Health Claim Legislation Worldwide

Before the major challenges and gaps associated with establishing health claims are discussed, some of the relevant regulatory issues are highlighted briefly in the followings. However, for more detailed discussion related to the legislation of probiotics and health claims the reader is referred to relevant chapter of this volume written by Abe (Chapter 11) or other relevant publications (Asp and Bryngelsson 2008; Hasler 2008; Heimbach 2008; Jew et al. 2008; Tapsell 2008). Since at the time of the publication of this chapter, changes may have been introduced in the legislations, it is advisable to consult websites of authorities for updates.

2.1 United States

In the USA, health claims are authorized under the Nutrition Labeling and Education Act (NLEA) of 1990. A health claim is statutorily defined as a statement that expressly or by implication characterizes the relationship of any substance in a food or dietary supplement to a disease or health-related condition (Heimbach 2008). Health claims are based on a very high standard of scientific evidence. Before a health claim may be lawfully used on food, Food and Drug Administration (FDA) must authorize that upon review of a petition for the claim. The claim can be approved provided that it is supported by significant scientific agreement among qualified experts. Randomized, controlled clinical intervention trials are considered as appropriate tools for health claim approval (Saavedra and Degnan 2009; Crittenden 2009).

Judicial rulings during the 1990s have led to an additional category of health claims referred as "qualified health claims." FDA has implemented a policy of reviewing and permitting health claims to appear on foods even if data and information in the supporting health claim submission do not meet the "significant scientific agreement" (Saavedra and Degnan 2009). With qualified health claims, the FDA established a ranking system from "B" to "D" level, which reflects good to very low weight of the scientific evidence, respectively, supporting the proposed claim. Unqualified "A" levels claims are those that meet the standard of significant scientific agreement (Hasler 2008).

An additional category of claims that may appear on probiotics is the so-called structure or function claim. With respect to probiotics, examples of such type of claims include "helps maintain intestinal flora" or "helps support a healthy immune system." One may notice that these claims avoid mentioning any disease and that they are directed at healthy consumers who wish to remain healthy.

2.2 Canada

In Canada, the Natural Health Product Directorate (NHPD) is responsible for the regulations concerning probiotics and label claims. NHPD has its own working definition of a probiotic: "a monoculture or mixed-culture of live microorganisms that benefit the microbiota indigenous to humans." It is also stated that probiotics are limited to nonpathogenic microorganisms and that dead microbes or other microbes cannot considered or labeled as "probiotics." Regulations allow therapeutic claims, as well as risk reduction and structure/function claims. The treatment claim refers to the diagnosis, treatment, mitigation, or prevention of a disease, disorder, or abnormal physical state or symptoms. An example of this type of claim might read "clinical trials have shown that probiotics can be used in the treatment and prevention of pouchitis." Risk reduction describes the relationship between using a medicinal ingredient and reducing the risk of developing a specific disease or abnormal physiological condition. An example of this type of claim might be "probiotics reduce the risk of colon cancer." Efficacy data to support any label health claim in Canada are judged based on a hierarchy of criteria used to measure the scientific rigor of the research that generated the data, and which is described in details elsewhere (Sanders et al. 2005). The amount and type of evidence required is depending on the type of claim and severity of the symptoms or conditions. The evidence is assessed on the basis of its strength, credibility, and quality.

2.3 European Union

European Regulation (EC) No. 1924/2006 on nutrition and health claims made on foods came into force in early 2007. The main aim of the regulation was to protect consumers and to provide free movement of products without hindrance from differences in national legislation. Unlike US and Canadian legislation, the EC regulations prohibit any claims referring to the prevention, treatment, or cure of a human disease for a food. However, risk reduction and structure/function claims are allowed. The Regulation defines a health claim in general as "any claim that states, suggests or implies that a relationship exists between a food category, a food or one of its constituents and health." A reduction of disease risk claim is defined as "any health claim that states, suggests or implies that the consumption of a food category, a food or one of its constituents significantly reduces a risk factor in the development of a human disease." Reduction of disease risk claims and claims referring to children's development and health are addressed specifically in Article 14, whereas other health claims belong to Article 13. All of these claims shall be based on generally accepted scientific evidence and well understood by the average consumer (Asp and Bryngelsson 2008). The Regulation states that scientific substantiation should be the main aspect taken into account for the use of nutrition and health claims, and it is the task of the European Food Safety Authority (EFSA) to judge whether the scientific substantiation is sufficient. Consumer relevance is not explicitly mentioned in the Regulation. However, the fact that claims must be "truthful, clear, reliable and useful for the consumer in choosing a healthy diet" could be interpreted as also including the relevance aspect (Asp and Bryngelsson 2007). It is also emphasized that the evidence to support the health claim must be based on human studies, and in vitro and animal studies on their own are not sufficient to substantiate a claim (Ouwehand and Lahtinen 2009).

2.4 Japan

In 1991, the FOSHU (Foods for Specific Health Uses) system was introduced in Japan as the world's first approval system of health-claim labeling for food products. Such "foods with health claims" represent a category between drugs and foods (Fukushima and Iino 2006). FOSHU products have several health-claim categories; however, among all the health claim categories, the claim "improve GI conditions" is the most common health-claim category (Chang 2009). For probiotics, in particular, this is the only allowed claim at present; however, advanced health claims of probiotics are expected to be approved in the future once the efficacy is properly documented based on the scientific evidence. Probiotics represent one of the most common food categories with health benefits. FOSHU is a voluntary system, and thus it is not necessary to obtain FOSHU approval for every probiotic with health benefits. However, all producers may apply for a certification from the Ministry of Health, Labor and Welfare (MHLW) according to the FOSHU system. Conditions for FOSHU approval stipulate that the petitioner show that (1) the food and the incorporated functional component(s) are expected to improve nutrition and contribute to maintain or promote health, (2) indicate the medical and nutritional facts for the health use, and (3) if possible establish the appropriate daily dose. The final product and the "related functional components" in the food should be confirmed as safe upon normal dietary consumption (e.g., not consumed rarely but rather on a daily basis). A human study conducted according to good clinical practice (GCP), in collaboration with a third party, should be carried out to evaluate health benefits of final products. The results of the study should be published in a peer-reviewed scientific journal to ensure the validity of the results from an objective and scientific points of view (Fukushima and Iino 2006).

2.5 *Australia and New Zealand*

Australia and New Zealand share common food standards managed through the authority of Food Standard Australia New Zealand (FSANZ). Health claims are defined as "statements about the nutritional and health benefits of foods used to promote products to consumers." The current legislation on nutrition, health, and related claims is about to be changed and FSANZ is expected to complete the First Review Report for Proposal P293 by April 2011 (http://www.foodstandards.gov.au). The new proposal contains standards related to nutritional content claims, general-level health claims, and high-level health claims. In the case of general-level claims, users need to hold substantiation dossiers and submit them for review on request. These claims cover nutrient content claims and some low-risk functional claims related to nonserious diseases. High-level claims, on the other hand, require formal preapproval. These claims relate to more serious conditions and cover biomarker and risk reduction claims. Besides providing an acceptable level of scientific evidence in support of the claim substantiation, the claims should meet qualifying criteria (i.e., they are relevant to population health and refer to foods with high nutritional value), and the wording conditions should be satisfactory (i.e., addressing the target population and the whole diet). Substantiation requires a systematic review of all the available evidence in humans and a quality assessment of the research. The randomized controlled trial is the study design accepted as providing the ultimate proof of efficacy in trials of both therapeutic agents and behavioral strategies (Tapsell 2008).

3 Substantiation of Health Claims for Probiotics

In order to achieve the desired health benefits and to successfully substantiate health claims, several issues should be considered. As illustrated in Fig. 1, this procedure is rather complex and each of the steps represents a crucial point. As the probiotic effects are strain specific, it is necessary to properly identify the microorganism to species/strain level. During strain selection, different functional, technological, and safety aspects are viewed. To produce the claimed health benefits, a sufficient number of viable microorganisms must be present throughout the entire shelf life of the product. In addition, candidate probiotic strains may possess well-characterized positive traits, such as resistance to gastric acidity and bile toxicity, ability to transiently colonize the gastrointestinal tract (GIT), capacity to inhibit intestinal pathogens, or the ability to stimulate the immune system. All of these criteria have to be demonstrated using in vitro tests and in vivo animal models. Finally, probiotics

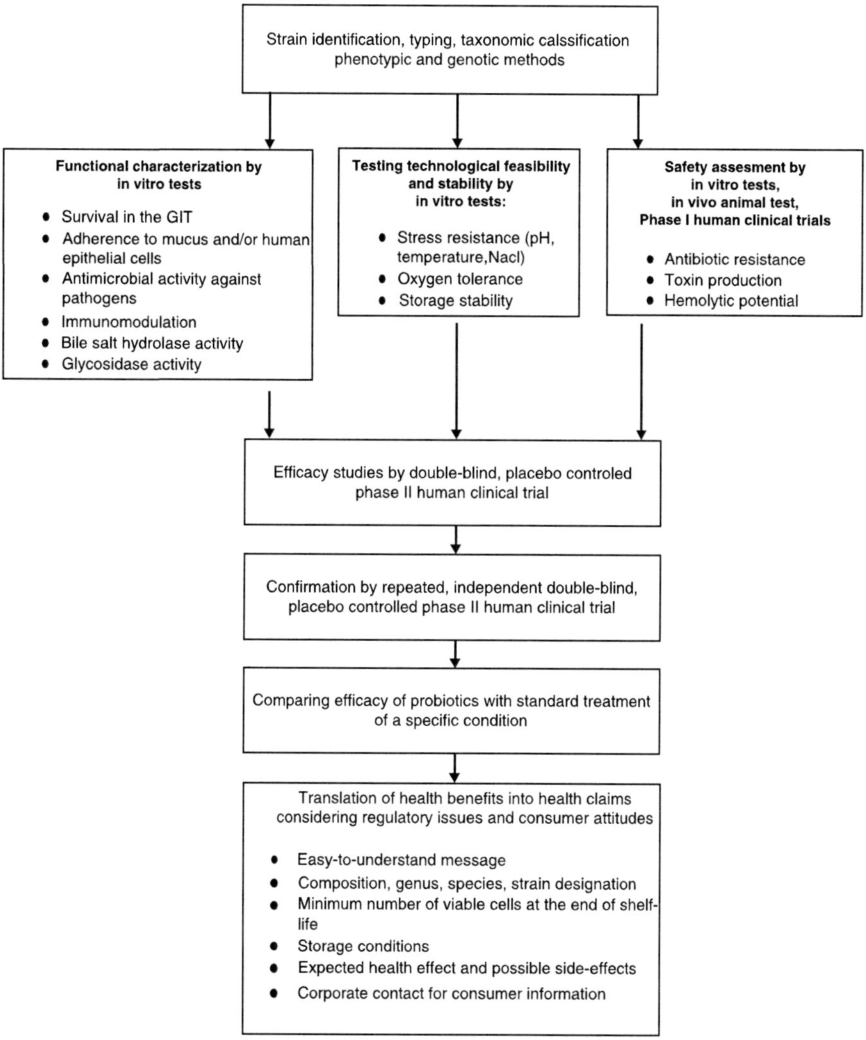

Fig. 1 Schematic process of the substantiation of health claims for probiotics

must be validated with human clinical studies. All of these steps are crucial with respect to establish reliable health claims. The following sections describe the methods applied and highlight their limitations and the related challenges.

3.1 Identification of Probiotic Microorganisms

The strain-specific benefits of probiotics emphasize the need for proper strain identification, which represent a crucial step in successful health claim substantiation. A vast array of specific and reproducible molecular techniques is now available for identification and typing of, for instance, lactobacilli and bifidobacteria. It is recommended to employ a combination of phenotypic and genetic techniques to accomplish proper identification, classification, and typing. Polyphasic characterization combining phenotypic, biochemical, genotypic, and sequencing results is now being used to reliably identify these bacteria to the strain level (Farnworth 2008). Methods based on the polymerase chain reaction (PCR) are extensively used and allow the differentiation between strains of the same species and to some extent, also between species. Random amplification of polymorphic DNA (RAPD), amplified ribosomal DNA restriction analysis (ARDRA), enterobacterial repetitive intergenic consensus sequence (ERIC)-PCR, repetitive extrogenic palindromic (REP)-PCR, triplicate arbitrarily primer (TAP)-PCR, restriction fragment length polymorphism (RFLP)-PCR, multiplex PCR, amplified fragment-length polymorphism (AFLP), pulsed-field gel electrophoresis (PFGE), temperature gradient gel electrophoresis (TGGE), and denaturing gradient gel electrophoresis (DGGE) are some examples of the techniques applied. More detailed presentation of identification and typing methods can be found elsewhere (Gueimonde and Reyes-Gavilan 2009; Margolles et al. 2009).

3.2 *Enumeration of Probiotics in Food*

The definition of probiotic emphasizes that the organism must be taken in adequate amounts to be effective. Health benefits have normally been attributed in clinical studies to doses of probiotics in excess of 10^8–10^9 viable cells per day (Prado et al. 2008). Therefore, it is widely accepted that probiotic food should contain more than 10^6–10^7 cfu/g at the time of consumption (Doleyres and Lacroix 2005; Jayamanne and Adams 2006). To support a health claim, it is thus imperative that the product manufacturer provides data about the total number of live microorganisms in the product when consumed, as well as methodology that can be used to verify these values (Farnworth 2008). At present, however, there is little incentive for manufacturers to state the type, viability, and number of bacteria on their product label due to the high costs for such quality assurance. In spite of the availability of culture-independent molecular tools for quantifications of probiotics in commercial products, most manufacturers apply conventional culture techniques for enumeration purposes (Sánchez et al. 2009). Moreover, culture-dependent techniques are crucial to determine possible physiological or biochemical changes in the population of probiotic bacteria during refrigerated storage of the product (Vinderola et al. 2000). Differential enumeration of probiotic bacteria in food products, in particular in dairy products, is rather problematic due to the presence of several closely related species of lactobacilli. For definite identification and enumeration, some selective media for each targeted species have been developed (Van de Casteele et al. 2006; Kim et al. 2010). The main disadvantage of these selective media is that they can underestimate counts of the microbial group selected (Sánchez et al. 2009).

3.3 Enumeration of Probiotics in the GIT

There is no doubt that it is essential to follow the survival and the colonization ability of probiotic strains in the gut. The detection and enumeration of a certain probiotic strain is challenging due to the presence of high number of other microorganisms in the intestinal environment. The current knowledge of intestinal microbiota composition and interaction with the host is still limited. Further complications may derive from the fact that microorganisms may exist in several physiological states of viability depending on the ecological environment. Bacteria under stressful condition (e.g., acidic conditions in the GIT) may modify their viability maintaining equilibrium between multiplication and survival activities (Nyström 2002). This status often referred as starved or dormant cells, which means that although the cells fail to replicate, they retain some metabolic activities typical of viable cells and may return to be cultivable under favorable conditions. While this phenomenon has extensively been investigated in the case of pathogenic strains, there has been limited interest to study the different states of viability of probiotics and intestinal bacteria (Gueimonde and Reyes-Gavilan 2009).

Unlike methods measuring multiplication as the sole criterion of viability, employment of some fluorescence-based techniques allows discrimination between viable, injured, or dead cells. Often applied criteria of viability are, for instance, membrane integrity, intracellular esterase activity, intracellular pH (Lahtinen et al. 2006), or the quantification of the 16S rRNA (Lahtinen et al. 2008).

3.4 Survival of Probiotics in the GIT

The principle of probiotic activity is that the oral administration of the probiotic organisms allows the specific probiotic bacteria present in the food to enter the GIT. Thus, a vital requirement toward making any health claim is the establishment of safe passage of the probiotics into the lower part of the intestinal tract (Grill et al. 2000). During the transit through the different sections of the GIT, probiotics are exposed to different stress conditions. Losses occur all along the GIT, but it is apparent that the acidic environment of the stomach and the presence of bile in the duodenum are major factors affecting the viability of probiotic bacteria (Mainville et al. 2005). There have been several attempts to mimic these conditions in in vitro models in order to assess the survival of probiotic strains. Several pH and bile concentrations have been tested for variable times. These conventional methodologies involve growing the cells to be studied, centrifugating and resuspending the cells in an acidic milieu, with or without washing, at pH values varying between 1.0 and 3.0 or incubating bacteria in a medium containing ±0.3% Oxgall bile, porcine bile, or bile salts (Vizoso Pinto et al. 2006; Huang et al. 2007; Vinderola et al. 2008; Stack et al. 2010). In general, bifidobacteria are less tolerant to acidic conditions than lactobacilli, while the former seems to be more tolerant to bile (Margolles et al. 2009). It should be emphasized again that acid tolerance is strain dependent. Some studies have been carried out using gastric juice and bile of human origin (Dunne et al. 2001). Besides these basic in vitro tests some more sophisticated model systems have been developed in order to better mimicking the fate of probiotics in the GIT.

3.5 *In Vitro Simulation of the GIT*

The simulator of the human intestinal microbial ecosystem (SHIME) was developed to simulate the entire human gastrointestinal system (Molly et al. 1993). It is inoculated with human fecal material to establish a microbial population that resembles that found in the human GIT. The original model consisted of five reactors representing the duodenum/jejunum, ileum, cecum and ascending colon, transverse colon, and the descending colon. A sixth reactor was later added to simulate the stomach.

TNO's model of the upper GI tract (TIM) contains four chambers to simulate the stomach, duodenum, jejunum, and ileum (Minekus et al. 1999). The simulated parameters included the body temperature, pH, salivary, gastric and intestinal mixing with peristaltic movements, secretions and absorption of water and small molecules. Adhesion of microorganisms to the reactor vessels and tubing, colonization in the microbiota, feed-back mechanisms, as well as the proper ingestion of the feed with the use of saliva-like fluids and ability to select for probiotic strains are neglected in these models (Ji and Xiao 2006). The SHIME is inoculated with human fecal bacteria that are not representative of the microbial composition in different parts of the colon (De Wiele et al. 2004), thus the interaction of probiotics with the host cannot be simulated. The digestion process in TIM is a difficult mathematic model and the prediction of the fate of ingested materials is not possible (Spratt et al. 2005).

The survival of probiotic strains in the GIT might not only depend on their number and physiological state but also on the food matrix and food consumption habits that affect bile excretion. This has taken into consideration in the designation of a dynamic model consisting of two reactors maintained at 37°C, one simulating stomach conditions and the other simulating duodenum conditions (Mainville et al. 2005). This dynamic model was shown to better represent the events during upper GI tract transit than the conventional methods, by incorporation of a food matrix to buffer the gastric acidity and therefore expose bacteria to pH levels found in vivo before, during, and after a meal.

More recently, a single reactor GIT model simulator (GITS) which simulates the environment during the passage of food through the stomach and upper intestinal tract has been developed (Sumeri et al. 2008, 2010). Ritter et al. (2009) compared the survival of *Lactobacillus gasseri* K7 in a similar simulator with in vivo measurements and obtained results that confirmed the reliability of this model.

3.6 *In Vitro Assessment of Adhesion and Colonization*

Adhesion of probiotics to intestinal mucus and epithelial cells has long been considered as one of the most important selection criteria for probiotic microorganisms. Adhesion to the intestinal mucosa enables temporary colonization, immune modulation, and competitive exclusion of pathogens (Lahtinen and Ouwehand 2009). There is controversy as to whether adherence is a prerequisite of a probiotic because adherence to the host's intestinal epithelium may not be required for all probiotic actions (Lin 2003).

The difficulties implicated with measuring probiotics adhesion in vivo may justify the use of in vitro models for the preliminary selection of potentially adherent strains. A major challenge, however, has been the lack of a homogenous cell model that could form a polarized barrier mimicking the physiological and functional morphology of the intestinal epithelia. Although a number of different methods have been used to assess the adhesion of probiotics, the use of epithelial cell lines (e.g., Caco-2, HT-29, or HT-29-MTX cells) obtained from human colorectal cancer is the most common technique applied in the assessment of probiotic adhesion (Lappara 2009) with proven good approximation to the in vivo situation (Walter et al. 1996).

The adhesion properties are strain specific and usually related to cell surface characteristics and composition (Xu et al. 2009). The bacterial adhesion to solvent (BATS) [also referred as bacterial adhesion to hydrocarbon (BATH)] assay can be used to determine cell surface properties (Kos et al. 2003). The different factors affecting the adhesion have been extensively studied and among others include low pH, presence of bile, endogenous microbiota, presence of other probiotic strains, digestive enzymes, presence of lignans, calcium, or Mg^{++} and Zn^{++} ions (Lahtinen and Ouwehand 2009). While some studies suggest that probiotic combinations may have synergistic adhesion effects (Ouwehand et al. 2000; Collado et al. 2007), others presume that other probiotic strains may reduce the adhesion ability of a given strain via competition for the same adhesion sites (Larsen et al. 2007). The age and health status of the host have also been suggested as determinants of the adhesion properties (Ouwehand et al. 2003; Klarin et al. 2005).

Enumeration of the adhered probiotics, in theory, can be performed by conventional methods based on plating or microscopic counting. These methods, however, cannot be used when adhesion of one bacterial strain is studied in an environment where other bacteria are present. In order to distinguish bacteria in a mixed population, radiolabels, fluorochromes, or bacteria-specific antibodies may be applied. Vesterlund et al. (2005a) compared the applicability of such methods for the assessment of adhesion of different probiotic (*L. casei* and *L. rhamnosus* GG) and pathogenic (*E. coli* and *S. enterica* serovar Typhimurium) strains.

Adhesion ability of probiotics has been extensively tested against mucus obtained from human (Vesterlund et al. 2005a; Collado et al. 2007) or animal origin (Lauková et al. 2004). A good correlation between the human mucus model and the adhesion to Caco-2 has been found by Gueimonde et al. (2006) when testing three Lactobacillus strains. Laukova et al. (2004) reported that the adhesion ability of enterococci to human, canine, and porcine intestinal mucus is strain dependent and do not exhibit host specificity. Ex vivo adhesion assay using resected colonic tissue has also been demonstrated (Ouwehand et al. 2002; Vesterlund et al. 2005b).

The correlation of in vitro adhesion tests to in vivo colonization has been assessed in some studies. An early study by Crociani et al. (1995), for example, reported good correlation between the results obtained by using Caco-2 cells and those derived from human trials, when comparing in vitro and in vivo adhesion ability of bifidobacteria. Wang et al. (2009) demonstrated that *Lactobacillus plantarum* L2 possessed good in vitro adhesion ability and showed significant in vivo colonization of rodents GIT.

Although the conclusion of most studies is that the adhesion to intestinal cells found to be associated with temporary colonization, in vitro and in vivo results did not always correlate (Ibnou-Zekri et al. 2003).

3.7 Antimicrobial Activity

It has been demonstrated that probiotics have developed different mechanisms to survive in competition with other microorganisms, including pathogens. In principle, this is exercised via competition for nutrients and physical binding location, as well as through the production of antimicrobial substances (Margolles et al. 2009). This latter ability of LAB and bifidobacteria to produce antimicrobial substances (e.g., organic acids, hydrogen peroxide, carbon dioxide, and bacteriocins) implies an important biological role for bacteria and has been one of the key elements in the selection of these bacteria for probiotic use (Dunne et al. 2001). Testing the interference on adhesion between probiotics and pathogens had been the subject of numerous in vitro studies (Collado et al. 2006; Gueimonde et al. 2006; Strahinić et al. 2007; Xu et al. 2009).

In order to test the antimicrobial activity, cocultures probiotic/enteropathogen are carried out and viability of both types of bacteria is determined. The general conclusion of these in vitro studies is that the inhibition is clearly strain- and culture condition dependent and that several molecules and mechanisms are involved in the process (Margolles et al. 2009).

4 Some Aspects of In Vivo Animal Studies

Preclinical testing of probiotics is often performed in in vivo animal studies. These experiments are strictly regulated by legislation and must be conducted in accordance to scientific, humane, and ethical principles (Frias 2009). Animal trials are relatively cost-effective ways to screen probiotics with the most potential beneficial effect to human host and understand their growth and survival patterns as well as their mode of action. In contrast to human studies, animal studies give researchers the flexibility to use a variety of efficient experimental designs and lead to generalizable inferences about research hypotheses pertaining to potential health effects of probiotics and their mechanisms of action (Ahmedna 2006).

Animal model can be categorized as (1) induced models, (2) spontaneous models, (3) genetically modified (GM) models, (4) negative models, and (5) orphan models. Mice and rats are the most popular laboratory animal models used in the research related to probiotics; however, studies with other species (e.g., pork, sheep, goat, cattle, fish, rabbits, dogs, and cats) have been reported. The ultimate selection of the animal model should be primary based on how well the model explains the specific aims rather than how well it represents the target. Phylogenetic proximity, for instance, is not always a guarantee for valid extrapolation (Frias 2009). Body weight/size and metabolic rate should also be considered upon selection. The suitability of an animal model is, among others, influenced by the species, breed, strain, genotype, and other exogenous factors. Attention should be paid on that the laboratory model is free from any unwanted microbial agents as these may invalidate results and conclusions. It is also important to emphasize that an infection is not necessarily synonymous with disease, and, for example, most rodent infections are latent and do not develop clinical symptoms.

There is a high number of related publication, thus for further information, the reader is referred to published literature on animal models, and animal studies aim to assess the efficacy and safety of probiotics.

5 Clinical Human Studies

In vitro human and animal studies may provide useful insights on probiotic efficacy, action, and safety for probiotics targeted for human use. Such experiments, however, cannot be considered as adequate substantiation for claims for humans (Sanders et al. 2005). It is fundamental that the probiotics for human consumption should undergo well designed and controlled clinical human studies. Although these experiments are expensive, there is no substitute for these trials. Moreover, each strain and product should be documented and tested independently because extrapolation of data from closely related strains is not acceptable. Protocols for human nutrition studies need to be developed for probiotics. The design of clinical studies used in pharmaceutical development can serve as a reference point for human probiotics studies, but specific protocols and criteria relevant to probiotic foods may be needed (Salminen and Gueimonde 2004). Mention should be made on the observation that the different research teams using similar markers often express the results of their trials in different ways; therefore, harmonization in this area is highly recommended (Rijkers et al. 2010).

There have been numerous published clinical studies which aimed to investigate the effect of oral administration of probiotics on the balance of intestinal microbiota and on a variety of disorders. However, several studies were of improper design, resulting in unreliable data. Indeed, unless they are designed with caution, these human studies will fail to adequately substantiate health claims. Some of the most crucial factors to be considered upon the design and implementation of human studies are briefly discussed in the followings.

5.1 *Identification of Target and Study Population*

Assessment of the efficiency and safety of probiotics should ideally be performed directly to the target population, which may be either the general population or a subgroup of subjects with a given condition. It is important to mention that within the healthy population, variations exist in the immune status which is difficult to clearly define. It is generally assumed that the projected functionality of probiotics in risk reduction of infection would mainly be measurable in the segment of the population with a suboptimal immune function, whereas the effect may be undetectable in the fully immunocompetent (i.e., healthy) subjects. Differences in microbiota composition and/or function of the subjects studied may influence the results of clinical trials (Ahmedna 2006). Therefore, a better characterization of the microbiota of the study population, relying on the high-throughput technologies under development, could become an important addition to future trials. The sample population size is always important in proving efficacy in the general population. It is necessary to be able to extrapolate from a small study population to the overall target group. This extrapolation may be carried on the basis of relevant epidemiological background data as well as interaction and intervention modeling methods (Rijkers et al. 2010).

5.2 *Defining Active Components*

It is often difficult to define the active ingredient in a probiotic food, in particular, when the product has undergone fermentation resulting in the development of other component followed by the living bacteria (e.g., cell wall, metabolites, bioactive peptides, etc.). In some cases, these ingredients may be responsible for the beneficial health effect, rather than the living probiotic bacteria (Farnworth 2008).

5.3 *Form of Delivery*

An important aspect of conducting clinical studies is developing the appropriate format for the test product. Many factors enter into product definition, such as inclusion of other potentially bioactive ingredients, use of a blend of probiotic strains, combined use of probiotics and prebiotics (i.e., symbiotics), method of growth and preservation of probiotic strain(s), levels of probiotics used, and delivery matrix. The mode of delivery to the human body will clearly influence the targeted site of the product (e.g., enteric-coated capsules may be less impacted by stomach acid) (Sanders et al. 2005).

Effects of the probiotics may vary when incorporated in different food matrices, thus products ideally should be tested as they are intended to be sold. Food matrix composition and processing conditions can interfere with the efficacy of the probiotic in terms of viability, stability, and the quantity of active biocompounds produced that are responsible for the studied health effects (Rijkers et al. 2010).

Although placebo-controlled trials are required for reliable health claim substantiation, it can be difficult to develop an appropriate placebo for some studies, especially for food delivery systems. However, even if the placebo is not indistinguishable from the test product to the subjects, it is still possible to blind a study as long as none of the participants know which product is test and which is placebo (Sanders et al. 2005).

5.4 *Effect of Background Diet*

Most studies excluded other potential active microorganisms from the diet of the tested subjects during run-in and/or wash-out periods. Potential dietary components that may overestimate the effect of the probiotics must be reported or standardized in the diets of both the control and probiotic group. For instance, the amount of saturated fat or cholesterol ingested when exploring the efficacy of a potential probiotic on blood lipids has to be fixed, or the amount of indigestible carbohydrates when measuring the breath test response to a lactose load (Rijkers et al. 2010).

5.5 *Effective Dosage*

Probiotics need to be delivered to the desired gastrointestinal site in sufficient quantity to achieve desired health effect. The daily dose, frequency of administration, time of administration, and the duration of administration is of high importance. The duration of administration is dependent on the nature of the diseases.

While a treatment for diarrhea is short term, prevention of cancer is of longer term. Neither the influence of frequency of administration nor the effect of timing is apparent so far. It can be assumed that probiotics administered orally before meal should possess good gastric acid and bile tolerance. Probiotics consumed during meal are diluted by food, thus encounter between probiotic organisms and the mucosal receptors is limited. Regarding the influence of the daily frequency of administration, it is generally accepted that a probiotic strain, which is not capable for good adhesion and colonization would need to be administered more frequently (Lee 2009).

5.6 *Long-Term Effects*

It is generally recognized that the influence of nutrition on health is characterized by small effects over a long period of time. Thus, performing long-term intervention trials is also necessary, especially in order to observe the improvement of wellness and whether the effect will persist with time. Most studies with probiotics are short-term (<12 week) studies. Postmarket surveillance is important to monitor the long-term beneficial (or adverse) effects. However, it is difficult to perform since subjects' diet cannot be accurately monitored (Sanders et al. 2005).

5.7 *Meta-Analyses*

Meta-analysis can be used to investigate the effect of probiotics consumption on a specific condition with the aim of confirming the efficacy of that given probiotic. The variability of the effects among the different subject groups can also be followed. Another object can be the comparison of the effect of different probiotic strains on the same specific condition and the identification of "active" and "nonactive" strains.

Collection of data on different probiotic strains, different conditions, and different subject characteristics by meta-analysis often results in an average nonsignificant effect attributed to the heterogeneity of benefits and probiotics studied. This underlines that there is no "generic" benefit common to all probiotics. Therefore, it is important, when using and designing a meta-analysis, to identify the objective and accordingly to select the right inclusion criteria to achieve useful conclusions (Rijkers et al. 2010).

6 Some Beneficial Claims and Supporting Evidences

In general, the beneficial health effects of probiotics are presumed to be realized through (1) balancing the intestinal microbiota, (2) influencing the metabolic activity, or (3) immunomodulation (Collado 2009). The primary clinical interest in the application of probiotics is the prevention and treatment of gastrointestinal infections and diseases. However, other target areas have appeared recently. It is becoming evident that other conditions including allergies, dental carries, respiratory infections, obesity, and urogenital infections, not initially associated with the gut microbiota, might also be affected (Farnworth 2008; Morrow and Kollef 2008).

A scientific consensus was being achieved concerning the beneficial effects of some probiotic bacteria for two application areas including (1) lactose intolerance (De Vrese et al. 2001; Montalto et al. 2006; Rampengan et al. 2010) and (2) diarrhea, in particular, acute-, and antibiotic-associated diarrhea (AAD) (McFarland 2006;

Szajewska et al. 2007a, b; Kale-Pradhan et al. 2010; de Vrese and Marteau 2007; Farnworth 2008). The strains *L. rhamnosus* GG, *Saccharomyces cerevisiae* Boulardii, *L. casei* Shirota, and *B. animalis* Bb-12 are certainly the most investigated probiotic cultures in relation to the management of the above-mentioned disorders.

The evidence of the efficacy of probiotics in patients suffering from constipation is limited, but the evidence seems promising for some strains to bring relief to patients suffering from such discomfort. Studies investigating the preventive effect of probiotics in the context of common cold and flu infections show that the studied strains failed to lower the incidence of episodes but that they have the potential to decrease the duration of episodes, which suggests that the immune system may be more efficient in fighting off common cold and flu infections after consuming these strains (Weichselbaum 2009).

Further clinical studies are needed to verify the beneficial influence of certain probiotics on other disorders including traveler's diarrhea, irritable bowel syndrome (IBS), inflammatory bowel diseases (IBD), *Helicobacter pylori* infection, oral infections, respiratory tract infections, dental carries, allergic diseases, and bacterial vaginosis. Some recent examples of probiotic strains used in the prevention and treatment of certain diseases are listed in Table 1. It was not possible, however, to discuss the health effects of probiotics in detail within the present format. Further information on the subject can be obtained from a vast array of recent reviews (Hsieh and Versalovic 2008; Vasiljevic and Shah 2008; Zuccotti et al. 2008; Kaur et al. 2009; Lomax and Calder 2009; Minocha 2009; Sánchez et al. 2009; Jankovic et al. 2010), as well as from relevant chapters of this volume. Disorder/disease specific reviews and meta-analyses are also available in the scientific literature.

7 Consumer Attitudes

The evidence for health effects must be scientifically assessed and then translated into health claims statements. Health claims have become an established way of communicating the healthiness of a food product. Such claims are widely used for marketing and product differentiation purposes as they potentially affect consumers' responses to foods in general, and to unfamiliar or novel foods in particular (Sabbe et al. 2009). Although, in principle, it is task of the authority bodies to approve food claims, it is the consumer who responds to these claims using a more personal knowledge of food and health. Thus, it is central to understand consumer perceptions, as these perceptions determine the intended outcome of the message: food purchase (Tapsell 2008). Different aspects should be taken into consideration when communicating beneficial health effects. First, scientific results must be translated into common language, which must then be accepted as truthful. This draws on the roles and responsibilities of scientists, governments, and food manufacturers. A recent survey performed in Brazil, for instance, revealed the need for an elementary easy-to-understand educational program using accessible

Table 1 Some examples of probiotic strains used in the prevention and treatment of certain diseases

Disorder	Probiotic strain(s)	Selected references
Acute diarrhea	*L. rhamnosus* GG	Szymański et al. (2006), Basu et al. (2009)
	S. boulardii	Villarruel et al. (2007), Htwe et al. (2008)
Antibiotic-associated diarrhea	*B. clausii*	Nista et al. (2004)
	S. boulardii	Kotowska et al. (2005), Can et al. (2006)
	L. acidophilus + *L. casei*	Beausoleil et al. (2007)
Clostridium difficile colitis	*L. acidophilus*	Plummer et al. (2004)
	L. rhamnosus GG	Lawrence et al. (2005)
Radiation-induced diarrhea	*L. casei* + *L. plantarum* + *L. acidophilus* + *L. delbruekii* spp. *Bulgaricus* + *B. longum* + *B. breve* + *B. Infantis* + *S. salivarius* spp. thermophilus	Delia et al. (2007)
Irritable bowel syndrome (IBS)	*L. paracasei*, ssp. Paracasei + *L. acidophilus* + *B lactis*	Simrén et al. (2010)
	L. acidophilus + *B. lactis* + *B. bifidum*	Williams et al. (2009)
	L. rhamnosus GG + *L. rhamnosus Lc705* + *Propionibacterium freudenreichii* ssp. Shermanii JS + *B. animalis* ssp. lactis	Kajander et al. (2008)
Inflammatory bowel diseases (IBD)	*L. rhamnosus* GG	Schultz et al. (2004)
	L. casei + *L. plantarum* + *L. acidophilus* + *L. delbruekii* spp. Bulgaricus + *B. longum* + *B. breve* + *B. Infantis* + *S. salivarius* spp. thermophilus	Bibiloni et al. 2005
Helicobacter pylori infection	*B. animalis* + *L. casei*	Goldman et al. (2006)
	S. boulardii	Hurduc et al. (2009)
	L. gasseri	Boonyaritichaikij et al. (2009)
Respiratory tract infections	*L. gasseri* + *B. longum* + *B. bifidum*	de Vrese et al. (2005, 2006)
Allergic diseases	*L. acidophilus*	Taylor et al. 2007
	L. rhamnosus GG + *B. breve* + *Propionibacterium freudenreichii* spp. Shermanii	Kukkonen et al. 2007
Oral infection	*L. rhamnosus* GG	Hatakka et al. 2007
Dental caries	*L. rhamnosus* LB21	Stecksén-Blicks et al. (2009)
High serum cholesterol	*L. fermentum*	Simons et al. (2006)
	L. plantarum	Nguyen et al. (2007)
	L. rhamnosus Lc705 + *Propionibacterium freudenreichii* ssp. Shermanii JS	Hatakka et al. (2008)

language in order to fix the concepts related to probiotic products (Viana et al. 2008). The use of proper terminology is of high importance. As an example, according to a survey by Bruhn et al. (2002), US consumers did not want to think about the "gastrointestinal tract," "foodborne illness," or "diarrhea" while buying food, therefore the use of such terms should be avoided in health claims. For increased credibility, promotion of probiotic culture benefits must not overstate the protective effects of probiotic cultures.

The sources of health claims also seem to influence consumers' credibility in probiotic food. US consumers are more likely to trust information provided by recognized health groups, and believe a message if there is scientific consensus on the issue (Bruhn et al. 2002). Hailu et al. (2009) found that Canadian consumers strongly prefer claims verified by government and place little value on "nonverified" claims made by product manufacturers.

All the above-mentioned issues, together with the regulatory background, should be taken into consideration when wording a given health claim.

8 Summary and Concluding Remarks

The present chapter has demonstrated that substantiation of health claims is a complex process, which requires significant efforts and special caution. Some of the requirements, which need to be fulfilled for successful health claim justification, are the followings:

1. Proper identification and enumeration of the tested strain, distinction between the different physiological forms of the cells (i.e., viable, dead, or dormant)
2. Selection of adequate in vitro models to assess the viability of the probiotic strains in the GIT
3. Testing possible modulations of probiotic functionality/viability as consequence of inclusion into diverse food vehicles
4. Considering the limitations of animal models when interpreting the results obtained
5. Careful selection and exact definition of target population to detect the health benefit
6. Conducting human intervention studies (double-blind, randomized, placebo-controlled) according to GCPs, which include monitoring and reporting on confounding factors, e.g., ingestion of other potentially active microorganisms, dietary components, drugs, or lifestyle that may interfere with the explored benefit
7. Addressing issues related to consumer attitude and legislation

There are, however, further challenges which need to be addressed in the near future. Although numerous conditions have been tested with various probiotic products, very few health/metabolism conditions have been studied thoroughly to obtain a label health claim. Much of the current uncertainty with regards to the efficacy of probiotics is attributed to differences in study designs, discrepant study populations, and inconsistencies specific to the probiotic, including the strain used (dose administered, product formulation, and route of administration). Thus, harmonizing study protocols are highly recommended.

It is also evident that there is limited understanding concerning the mechanism of probiotic activity in vivo, which is a key element in resolving multiple controversies. Indeed, most of the publications provide evidences supporting the concept that probiotics improve human health without exploring the causal relationships between their effects on health and mechanisms of action. To provide a scientific rationale for the use of probiotics, the precise mechanism of their action needs to be illustrated.

Demonstration of the beneficial health effects of probiotics in generally healthy humans is also a challenging task due to the lack of validated biomarkers of health and risk factors of diseases. Another difficulty is to prove the long-term effects and safety of probiotic foods. Development of appropriate biomarkers may reduce the burden on human subjects and provide risk indicators and wellness measures.

An area of particular clinical interest is whether combination of probiotic strains in therapy is superior to single-strain therapy, thus possible synergetic or antagonistic effects should be further investigated.

The emerging interest in probiotics for disease management necessitates a review of the above-mentioned issues. Interlinking the expertise and scientific knowledge on food, GIT functionality, and human health is essential for designing probiotics with well-defined health benefits.

References

Ahmedna M (2006) Statistical considerations for testing the efficacy of probiotics. In: Ahmedna M, Goktepe I, Juneja VK (eds) Probiotics in food safety and human health. CRC Press, Boca Raton, pp 383–404

Asp NG, Bryngelsson S (2007) Health claims in the labelling and marketing of food products: the Swedish food sector's Code of Practice in a European perspective. Scand J Food and Nutr 51:107–126

Asp NG, Bryngelsson S (2008) Health claims in Europe: new legislation and PASSCLAIM for substantiation. J Nutr 138:1210S–1215S

Basu S, Paul DK, Ganguly S, Chatterjee M, Chandra PK (2009) Efficacy of high-dose Lactobacillus rhamnosus GG in controlling acute watery diarrhea in Indian children: a randomized controlled trial. J Clin Gastroenterol 43:208–213

Beausoleil M, Fortier N, Guénette S, L'Ecuyer A, Savoie M, Franco M, Lachaîne J, Weiss K (2007) Effect of a fermented milk combining Lactobacillus acidophilus CL1285 and Lactobacillus casei in the prevention of antibiotic-associated diarrhea: a randomized, double-blind, placebo-controlled trial. Can J Gastroenterol 21:732–736

Bibiloni R, Fedorak RN, Tannock GW, Madsen KL, Gionchetti P, Campieri M, De Simone C, Sartor RB (2005) VSL#3 probiotic-mixture induces remission in patients with active ulcerative colitis. Am J Gastroenterol 100:1539–1546

Boonyaritichaikij S, Kuwabara K, Nagano J, Kobayashi K, Koga Y (2009) Long-term administration of probiotics to asymptomatic pre-school children for either the eradication or the prevention of Helicobacter pylori infection. Helicobacter 14:202–207

Brink M, Senekal M, Dicks LMT (2005) Market and product assessment of probiotic/prebiotic – containing functional foods and supplements manufactured in South Africa. South African Med J 95:114–119

Bruhn CM, Bruhn JC, Cotter A, Garrett C, Klenk M, Powell C, Stanford G, Steinbring Y, West E (2002) Consumer attitudes toward use of probiotic cultures. J Food Sci 67:1969–1972

Can M, Beşirbellioglu BA, Avci IY, Beker CM, Pahsa A (2006) Prophylactic Saccharomyces boulardii in the prevention of antibiotic-associated diarrhea: a prospective study. Med Sci Monitor 12:PI19–PI22

Chang WTH (2009) Legal status and regulatory issues – Human Probiotics – Asia. In: Lee YK, Salminen S (eds) Handbook of probiotics and prebiotics, 2nd edn. Wiley, Hoboken, NJ, pp 95–105

Collado MC (2009) Role of probiotics in health and diseases. In: Lee YK, Salminen S (eds) Handbook of probiotics and prebiotics, 2nd edn. Wiley, Hoboken, NJ, pp 257–259

Collado MC, Gueimonde M, Sanz Y, Salminen S (2006) Adhesion properties and competitive pathogen exclusion ability of bifidobacteria with acquired acid resistance. J Food Prot 69:1675–1679

Collado MC, Meriluoto J, Salminen S (2007) Development of new probiotics by strain combinations: is it possible to improve the adhesion to intestinal mucus? J Dairy Sci 90:2710–2716

Crittenden R (2009) Incorporating probiotics into foods. In: Lee YK, Salminen S (eds) Handbook of probiotics and prebiotics, 2nd edn. Wiley, Hoboken, NJ, pp 58–75

Crociani J, Grill JP, Huppert M, Ballongue J (1995) Adhesion of different bifidobacteria strains to human enterocyte-like Caco-2 cells and comparison with in vivo study. Lett Appl Microbiol 21:146–148

De Vrese M, Marteau PR (2007) Probiotics and prebiotics: effects on diarrhea. J Nutr 137: 803S–811S

De Vrese M, Stegelmann A, Richter B, Fenselau S, Laue C, Schrezenmeir J (2001) Probiotics – compensation for lactase insufficiency. Am J Clin Nutr 73:421S–429S

De Vrese M, Winkler P, Rautenberg P, Harder T, Noah C, Laue C, Ott S, Hampe J, Schreiber S, Heller K, Schrezenmeir J (2005) Effect of Lactobacillus gasseri PA 16/8, Bifidobacterium longum SP 07/3, B. bifidum MF 20/5 on common cold episodes: a double blind, randomized, controlled trial. Clin Nutr 24:481–491

De Vrese M, Winkler P, Rautenberg P, Harder T, Noah C, Laue C, Ott S, Hampe J, Schreiber S, Heller K, Schrezenmeir J (2006) Probiotic bacteria reduced duration and severity but not the incidence of common cold episodes in a double blind, randomized, controlled trial. Vaccine 24:6670–6674

De Wiele TV, Boon N, Possemiers S, Jacobs H, Verstraete W (2004) Prebiotic effects of chicory inulin in the simulator of the human intestinal microbial ecosystem. FEMS Microbiol Ecol 51:143–153

Delia P, Sansotta G, Donato V, Frosina P, Messina G, De Renzis C, Famularo G (2007) Use of probiotics for prevention of radiation-induced diarrhea. World J Gastroenterol 13:912–915

Doleyres Y, Lacroix C (2005) Technologies with free and immobilised cells for probiotic bifidobacteria production and protection. Int Dairy J 15:973–988

Dunne C, O'Mahony L, Murphy L, Thornton G, Morrissey D, O'Halloran S, Feeney M, Flynn S, Fitzgerald G, Daly C, Kiely B, O'Sullivan GC, Shanahan F, Collins JK (2001) In vitro selection criteria for probiotic bacteria of human origin: correlation with in vivo findings. Am J Clin Nutr 73:386S–392S

Farnworth ER (2008) The evidence to support health claims for probiotics. J Nutr 138: 1250S–1254S

Frias R (2009) Laboratory animal models in research. In: Lee YK, Salminen S (eds) Handbook of probiotics and prebiotics, 2nd edn. Wiley, Hoboken, NJ, pp 263–267

Fukushima Y, Iino H (2006) Probiotics in food safety and human health: current status of regulations on the use of probiotics in Japan. In: Ahmedna M, Goktepe I, Juneja VK (eds) Probiotics in food safety and human health. CRC Press, Boca Raton, pp 431–463
Goldman CG, Barrado DA, Balcarce N, Rua EC, Oshiro M, Calcagno ML, Janjetic M, Fuda J, Weill R, Salgueiro MJ, Valencia ME, Zubillaga MB, Boccio JR (2006) Effect of a probiotic food as an adjuvant to triple therapy for eradication of Helicobacter pylori infection in children. Nutrition 22:984–988
Grill JP, Perrin S, Schneider F (2000) Bile salt toxicity to some bifidobacteria strains: role of conjugated bile salt hydrolase and pH. Can J Microbiol 46:878–884
Gueimonde M, De Los Reyes-Gavilán CG (2009) Detection and enumeration of gastrointestinal microorganisms. In: Lee YK, Salminen S (eds) Handbook of probiotics and prebiotics, 2nd edn. Wiley, Hoboken, NJ, pp 25–43
Gueimonde M, Jalonen L, He F, Hiramatsu M, Salminen S (2006) Adhesion and competitive inhibition and displacement of human enteropathogens by selected lactobacilli. Food Res Int 39:467–471
Hailu G, Boecker A, Henson S, Cranfield J (2009) Consumer valuation of functional foods and nutraceuticals in Canada. A conjoint study using probiotics. Appetite 52:257–265
Hasler CM (2008) Health claims in the United States: an aid to the public or a source of confusion? J Nutr 138:1216S–1220S
Hatakka K, Ahola AJ, Yli-Knuuttila H, Richardson M, Poussa T, Meurman JH, Korpela R (2007) Probiotics reduce the prevalence of oral Candida in the elderly a randomized controlled trial. J Dent Res 86:125–130
Hatakka K, Mutanen M, Holma R, Saxelin M, Korpela R (2008) *Lactobacillus rhamnosus* LC705 together with *Propionibacterium freudenreichii* ssp shermanii JS administered in capsules is ineffective in lowering serum lipids. J Am Coll Nutr 27:441–447
Heimbach JT (2008) Health-benefit claims for probiotic products. Clin Infect Dis 46:S122–S124
Hsieh MH, Versalovic J (2008) The human microbiome and probiotics: implications for pediatrics. Curr Problems Paed Adolescent Health Care 38:309–327
Htwe K, Khin SY, Tin M, Vandenplas Y (2008) Effect of *Saccharomyces boulardii* in the treatment of acute watery diarrhea in Myanmar children: a randomized controlled study. Am J Trop Med Hyg 78:214–216
Huang HY, Huang SY, Chen PY, King VA, Lin YP, Tsen JH (2007) Basic characteristics of *Sporolactobacillus inulinus* BCRC 14647 for potential probiotic properties. Curr Microbiol 54:396–404
Hurduc V, Plesca D, Dragomir D, Sajin M, Vandenplas Y (2009) A randomized, open trial evaluating the effect of *Saccharomyces boulardii* on the eradication rate of *Helicobacter pylori* infection in children. Acta Paediatrica Int J Paed 98:127–131
Ibnou-Zekri N, Blum S, Schiffrin EJ, Von der Weid T (2003) Divergent patterns of colonization and immune response elicited from two intestinal *Lactobacillus* strains that display similar properties in vitro. Infect Immun 71:428–436
Jankovic I, Sybesma W, Phothirath P, Ananta E, Mercenier A (2010) Application of probiotics in food products-challenges and new approaches. Curr Opin Biotechnol 21:175–181
Jayamanne VS, Adams MR (2006) Determination of survival, identity and stress resistance of probiotic bifidobacteria in bio-yoghurts. Lett Appl Microbiol 42:189–194
Jew S, Vanstone CA, Antoine JM, Jones PJ (2008) Generic and product-specific health claim processes for functional foods across global jurisdictions. J Nutr 138:1228S–1236S
Ji YY, Xiao DC (2006) GIT physicochemical modeling – a critical review. Int J Food Eng 2(4), Art. 4, 1–10
Kajander K, Myllyluoma E, Rajilić-Stojanović M, Kyrönpalo S, Rasmussen M, Järvenpää S, Zoetendal EG, De Vos WM, Vapaatalo H, Korpela R (2008) Clinical trial: multispecies probiotic supplementation alleviates the symptoms of irritable bowel syndrome and stabilizes intestinal microbiota. Aliment Pharmacol Ther 27:48–57

Kale-Pradhan PB, Jassal HK, Wilhelm SM (2010) Role of *Lactobacillus* in the prevention of antibiotic-associated diarrhea: a meta-analysis. Pharmacotherapy 30:119–126

Kaur IP, Kuhad A, Garg A, Chopra K (2009) Probiotics: delineation of prophylactic and therapeutic benefits. J Med Food 12:219–235

Kim ER, Cho YH, Kim YH, Park SO, Woo GJ, Chun HN (2010) Comparison of bifidobacteria selective media for the detection of bifidobacteria in Korean commercial fermented milk products. Korean J Food Sci Animal Res 30:154–162

Klarin B, Johansson ML, Molin G, Larsson A, Jeppsson B (2005) Adhesion of the probiotic bacterium *Lactobacillus plantarum* 299v onto the gut mucosa in critically ill patients: a randomised open trial. Crit Care 9:R285–R293

Kos B, Šušković J, Vuković S, Šimpraga M, Frece J, Matošić S (2003) Adhesion and aggregation ability of probiotic strain *Lactobacillus acidophilus* M92. J Appl Microbiol 94:981–987

Kotowska M, Albrecht P, Szajewska H (2005) *Saccharomyces boulardii* in the prevention of antibiotic-associated diarrhoea in children: a randomized double-blind placebo-controlled trial. Aliment Pharmacol Ther 21:583–590

Kukkonen K, Savilahti E, Haahtela T, Juntunen-Backman K, Korpela R, Poussa T, Tuure T, Kuitunen M (2007) Probiotics and prebiotic galacto-oligosaccharides in the prevention of allergic diseases: a randomized, double-blind, placebo-controlled trial. J Allergy Clin Immunol 119:192–198

Lahtinen S, Ouwehand A (2009) Adhesion to intestinal mucus and epithelium by probiotics. In: Lee YK, Salminen S (eds) Handbook of probiotics and prebiotics, 2nd edn. Wiley, Hoboken, NJ, pp 377–384

Lahtinen SJ, Ouwehand AC, Reinikainen JP, Korpela JM, Sandholm J, Salminen SJ (2006) Intrinsic properties of so-called dormant probiotic bacteria, determined by flow cytometric viability assays. Appl Environ Microbiol 72:5132–5134

Lahtinen SJ, Ahokoski H, Reinikainen JP, Gueimonde M, Nurmi J, Ouwehand AC, Salminen SJ (2008) Degradation of 16 S rRNA and attributes of viability of viable but nonculturable probiotic bacteria. Lett Appl Microbiol 46:693–698

Lappara JM (2009) Cell line models in research. In: Lee YK (ed) Handbook of probiotics and prebiotics, 2nd edn. Wiley, Hoboken, NJ, pp 259–263

Larsen N, Nissen P, Willats WGT (2007) The effect of calcium ions on adhesion and competitive exclusion of *Lactobacillus* ssp. and *E. coli* O138. Int J Food Microbiol 114:113–119

Lauková A, Strompfová V, Ouwehand A (2004) Adhesion properties of enterococci to intestinal mucus of different hosts. Vet Res Commun 28:647–655

Lawrence SJ, Korzenik JR, Mundy LM (2005) Probiotics for recurrent *Clostridium difficile* disease [1]. J Med Microbiol 54:905–906

Lee YK (2009) Effective dosage for probiotic effects. In: Lee YK, Salminen S (eds) Handbook of probiotics and prebiotics, 2nd edn. Wiley, Hoboken, NJ, pp 52–58

Lin DC (2003) Probiotics as functional foods. Nutr Clin Prac 18:497–506

Lomax AR, Calder PC (2009) Probiotics, immune function, infection and inflammation: a review of the evidence from studies conducted in humans. Curr Pharm Des 15:1428–1518

Mainville I, Arcand Y, Farnworth ER (2005) A dynamic model that simulates the human upper gastrointestinal tract for the study of probiotics. Int J Food Microbiol 99:287–296

Margolles A, Mayo B, Ruas-Madiedo P (2009) Screening, identification, and characterization of *Lactobacillus* and *Bifidobacterium* strains. In: Lee YK, Salminen S (eds) Handbook of probiotics and prebiotics, 2nd edn. Wiley, Hoboken, NJ, pp 4–24

McFarland LV (2006) Meta-analysis of probiotics for the prevention of antibiotic associated diarrhea and the treatment of *Clostridium difficile* disease. Am J Gastroenterol 101:812–822

Minekus M, Smeets-Peeters M, Havenaar R, Bernalier A, Fonty G, Marol-Bonnin S, Alric M, Marteau P, Huis In't Veld JHJ (1999) A computer-controlled system to simulate conditions of

the large intestine with peristaltic mixing, water absorption and absorption of fermentation products. Appl Microbiol Biotechnol 53:108–114
Minocha A (2009) Invited review: probiotics for preventive health. Nutr Clin Prac 24:227–241
Molly K, Vande Woestyne M, Verstraete W (1993) Development of a 5-step multi-chamber reactor as a simulation of the human intestinal microbial ecosystem. Appl Microbiol Biotechnol 39:254–258
Montalto M, Curigliano V, Santoro L, Vastola M, Cammarota G, Manna R, Gasbarrini A, Gasbarrini G (2006) Management and treatment of lactose malabsorption. World J Gastroenterol 12:187–191
Morrow LE, Kollef MH (2008) Probiotics in the intensive care unit: why controversies and confusion abound. Crit Care 12(3):160
Nguyen TDT, Kang JH, Lee MS (2007) Characterization of *Lactobacillus plantarum* PH04, a potential probiotic bacterium with cholesterol-lowering effects. Int J Food Microbiol 113:358–361
Nista EC, Candelli M, Cremonini F, Cazzato IA, Zocco MA, Franceschi F, Cammarota G, Gasbarrini G, Gasbarrini A (2004) *Bacillus clausii* therapy to reduce side-effects of anti-*Helicobacter pylori* treatment: randomized, double-blind, placebo controlled trial. Aliment Pharmacol Ther 20:1181–1188
Nyström T (2002) Aging in bacteria. Curr Opin Microbiol 5:596–601
Ouwehand AC, Lahtinen S (2009) Legal status and regulatory issues – human probiotics – Europe. In: Lee YK, Salminen S (eds) Handbook of probiotics and prebiotics, 2nd edn. Wiley, Hoboken, NJ, pp 106–110
Ouwehand AC, Isolauri E, Kirjavainen PV, Tölkkö S, Salminen SJ (2000) The mucus binding of *Bifidobacterium lactis* Bb12 is enhanced in the presence of *Lactobacillus* GG and *Lact. delbrueckii* subsp. bulgaricus. Lett Appl Microbiol 30:10–13
Ouwehand AC, Salminen S, Tölkkö S, Roberts P, Ovaska J, Salminen E (2002) Resected human colonic tissue: new model for characterizing adhesion of lactic acid bacteria. Clin Diagn Lab Immunol 9:184–186
Ouwehand AC, Salminen S, Roberts PJ, Ovaska J, Salminen E (2003) Disease-dependent adhesion of lactic acid bacteria to the human intestinal mucosa. Clin Diag Lab Immunol 10:643–646
Plummer S, Weaver MA, Harris JC, Dee P, Huter J (2004) Clostridium difficile pilot study: effects of probiotic supplementation on the incidence of *C. difficile* diarrhoea. Int Microbiol 7:59–62
Prado FC, Parada JL, Pandey A, Soccol CR (2008) Trends in non-dairy probiotic beverages. Food Res Int 41:111–123
Rampengan NH, Manoppo J, Warouw SM (2010) Comparison of efficacies between live and killed probiotics in children with lactose malabsorption. Southeast Asian J Trop Med Public Health 41:474–481
Rijkers GT, Bengmark S, Enck P, Haller D, Herz U, Kalliomaki M, Kudo S, Lenoir-Wijnkoop I, Mercenier A, Myllyluoma E, Rabot S, Rafter J, Szajewska H, Watzl B, Wells J, Wolvers D, Antoine JM (2010) Guidance for substantiating the evidence for beneficial effects of probiotics: current status and recommendations for future research. J Nutr 140:671S–676S
Ritter P, Kohler C, Ah UV (2009) Evaluation of the passage of *Lactobacillus gasseri* K7 and bifidobacteria from the stomach to intestines using a single reactor model. BMC Microbiol 9:87
Saavedra JM, Degnan FH (2009) Legal status and regulatory issues – human probiotics – The United States of America. In: Lee YK, Salminen S (eds) Handbook of probiotics and prebiotics, 2nd edn. Wiley, Hoboken, NJ, pp 111–123
Sabbe S, Verbeke W, Deliza R, Matta V, Van Damme P (2009) Effect of a health claim and personal characteristics on consumer acceptance of fruit juices with different concentrations of açaí (Euterpe oleracea Mart.). Appetite 53:84–92

Salminen S, Gueimonde M (2004) Human studies on probiotics: what is scientifically proven. J Food Sci 69:M137–M140
Sánchez B, De Los Reyes-Gavilán CG, Margolles A, Gueimonde M (2009) Probiotic fermented milks: present and future. Int J Dairy Technol 62:472–483
Sanders ME, Tompkins T, Heimbach JT, Kolida S (2005) Weight of evidence needed to substantiate a health effect for probiotics and prebiotics: regulatory considerations in Canada, E.U., and US. Eur J Nutr 44:303–310
Schultz M, Timmer A, Herfarth HH, Sartor RB, Vanderhoof JA, Rath HC (2004) *Lactobacillus* GG in inducing and maintaining remission of Crohn's disease. BMC Gastroenterol 4:5
Simons LA, Amansec SG, Conway P (2006) Effect of *Lactobacillus fermentum* on serum lipids in subjects with elevated serum cholesterol. Nutr Metabol Cardio Dis 16:531–535
Simrén M, Öhman L, Olsson J, Svensson U, Ohlson K, Posserud I, Strid H (2010) Clinical trial: the effects of a fermented milk containing three probiotic bacteria in patients with irritable bowel syndrome – a randomized, double-blind, controlled study. Aliment Pharmacol Ther 31:218–227
Spratt P, Nicolella C, Pyle DL (2005) An engineering model of the human colon. Food Bioprod Process 83:147–157
Stack HM, Kearney N, Stanton C, Fitzgerald GF, Ross RP (2010) Association of beta-glucan endogenous production with increased stress tolerance of intestinal lactobacilli. Appl Environ Microbiol 76:500–507
Stecksén-Blicks C, Sjöström I, Twetman S (2009) Effect of long-term consumption of milk supplemented with probiotic lactobacilli and fluoride on dental caries and general health in preschool children: a cluster-randomized study. Caries Res 43:374–381
Strahinić I, Cvetanović D, Kojić M, Fira DJ, Tolinački M, Topisirović LJ (2007) Characterization and antimicrobial activity of natural isolate *Lactococcus lactis* subsp. lactis BGSM1-19. Acta Vet 57:509–521
Sumeri I, Arike L, Adamberg K, Paalme T (2008) Single bioreactor gastrointestinal tract simulator for study of survival of probiotic bacteria. Appl Microbiol Biotechnol 80:317–324
Sumeri I, Arike L, Stekolštšikova J, Uusna R, Adamberg S, Adamberg K, Paalme T (2010) Effect of stress pretreatment on survival of probiotic bacteria in gastrointestinal tract simulator. Appl Microbiol Biotechnol 86:1925–1931
Szajewska H, Skórka A, Dylag M (2007a) Meta-analysis: *Saccharomyces boulardii* for treating acute diarrhoea in children. Aliment Pharmacol Ther 25:257–264
Szajewska H, Skórka A, Ruszczyński M, Gieruszczak-BiaŁek D (2007b) Meta-analysis: *Lactobacillus* GG for treating acute diarrhoea in children. Aliment Pharmacol Ther 25:871–881
Szymański H, Pejcz J, Jawień M, Chmielarczyk A, Strus M, Heczko PB (2006) Treatment of acute infectious diarrhoea in infants and children with a mixture of three *Lactobacillus rhamnosus* strains – a randomized, double-blind, placebo-controlled trial. Aliment Pharmacol Ther 23:247–253
Tapsell LC (2008) Evidence for health claims: a perspective from the Australia-New Zealand Region. J Nutr 138:1206S–1209S
Taylor AL, Dunstan JA, Prescott SL (2007) Probiotic supplementation for the first 6 months of life fails to reduce the risk of atopic dermatitis and increases the risk of allergen sensitization in high-risk children: a randomized controlled trial. J Allergy Clin Immunol 119:184–191
Van de Casteele S, Vanheuverzwijn T, Ruyssen T, Van Assche P, Swings J, Huys G (2006) Evaluation of culture media for selective enumeration of probiotic strains of lactobacilli and bifidobacteria in combination with yoghurt or cheese starters. Int Dairy J 16:1470–1476
Vasiljevic T, Shah NP (2008) Probiotics-from Metchnikoff to bioactives. Int Dairy J 18:714–728
Vesterlund S, Paltta J, Karp M, Ouwehand AC (2005a) Measurement of bacterial adhesion-in vitro evaluation of different methods. J Microbiol Methods 60:225–233

Vesterlund S, Paltta J, Karp M, Ouwehand AC (2005b) Adhesion of bacteria to resected human colonic tissue: quantitative analysis of bacterial adhesion and viability. Res Microbiol 156:238–244

Viana JV, Da Cruz AG, Zoellner SS, Silva R, Batista ALD (2008) Probiotic foods: consumer perception and attitudes. Int J Food Sci Technol 43:1577–1580

Villarruel G, Rubio DM, Lopez F, Cintioni J, Gurevech R, Romero G, Vandenplas Y (2007) *Saccharomyces boulardii* in acute childhood diarrhoea: a randomized, placebo-controlled study. Acta Paediatrica Int J Paed 96:538–541

Vinderola CG, Bailo N, Reinheimer JA (2000) Survival of probiotic microflora in Argentinian yoghurts during refrigerated storage. Food Res Int 33:97–102

Vinderola G, Capellini B, Villarreal F, Suárez V, Quiberoni A, Reinheimer J (2008) Usefulness of a set of simple in vitro tests for the screening and identification of probiotic candidate strains for dairy use. LWT Food Sci Technol 41:1678–1688

Vizoso Pinto MG, Franz CMAP, Schillinger U, Holzapfel WH (2006) *Lactobacillus* spp. with in vitro probiotic properties from human faeces and traditional fermented products. Int J Food Microbiol 109:205–214

Walter E, Janich S, Roessler BJ, Hilfinger JM, Amidon GL (1996) HT29-MTX/Caco-2 cocultures as an in vitro model for the intestinal epithelium: in vitro-in vivo correlation with permeability data from rats and humans. J Pharm Sci 85:1070–1076

Wang B, Li J, Li Q, Zhang H, Li N (2009) Isolation of adhesive strains and evaluation of the colonization and immune response by *Lactobacillus plantarum* L2 in the rat gastrointestinal tract. Int J Food Microbiol 132:59–66

Weichselbaum E (2009) Probiotics and health: a review of the evidence. Nutr Bull 34:340–373

Williams EA, Stimpson J, Wang D, Plummer S, Garaiova I, Barker ME, Corfe BM (2009) Clinical trial: a multistrain probiotic preparation significantly reduces symptoms of irritable bowel syndrome in a double-blind placebo-controlled study. Aliment Pharmacol Ther 29:97–103

Xu H, Jeong HS, Lee HY, Ahn J (2009) Assessment of cell surface properties and adhesion potential of selected probiotic strains. Lett Appl Microbiol 49:434–442

Zuccotti GV, Meneghin F, Raimondi C, Dilillo D, Agostoni C, Riva E, Giovannini M (2008) Probiotics in clinical practice: an overview. J Int Med Res 36:1A–53A

Safety and Regulation of Probiotic Foods and Supplements

Fumiaki Abe

Contents

1 Introduction 270
2 Translocation, Bacteremia, and Sepsis 271
2.1 Bacterial Translocation Associated with Probiotics 271
2.2 Mucin Degradation Activity of Probiotics 272
2.3 Translocation Ability of Probiotics 273
2.4 Translocation of Probiotics and Virulence 274
2.5 Confirmation of Probiotic Safety by Examining Translocation Ability 275
3 Antibiotic Resistant Genes Transfer 276
3.1 Risk of Probiotics Harboring Acquired Antibiotic Resistance Genes 276
3.2 Bifidobacteria Harboring Antibiotics Resistance Genes in the Environment 277
3.3 Probiotics Harboring Antibiotic Resistance Genes Used in Commercial Products 278
3.4 Possibility of Horizontal Transfer of Antibiotics Resistance Genes from Probiotics 278
3.5 Antibiotic Resistance Genes in Lactic Acid Bacteria 280
3.6 Algorithm for Decision Making of Probiotics Selection Related to Antibiotic Resistance 280
4 From Research to Consumer: Quality Assurance and Control 281
4.1 Hygienic Standard of Probiotics 281
4.2 Stability of Probiotics 283
4.3 Regulatory Control 286
5 Global Regulations for Health Claims of Foods 288
5.1 Regulation of Food with Health Claims in Japan 288
5.2 Food with Nutrient Function Claims 289
5.3 Food for Specified Health Uses 290
6 Conclusions 293
References 294

F. Abe (✉)
Food Science & Technology Institute, Morinaga Milk Industry Co., Ltd., 1-83, 5-chome Higashihara, Zama, Kanagawa 252-8583, Japan
e-mail: f_abe@morinagamilk.co.jp

M.-T. Liong (ed.), *Probiotics*, Microbiology Monographs 21,
DOI 10.1007/978-3-642-20838-6_11,

Abstract Probiotics safety has received a lot of attention recently due to some serious issues including bacterial translocation causing sepsis and horizontal transfer of acquired antibiotic resistance gene. To resolve these concerns, manufacturers have to demonstrate safety of probiotics on a strain-by-strain basis because not all probiotics are the same. Also, to prevent the outbreak of pathogenic bacteria with antibiotic resistance, probiotics harboring acquired antibiotic resistance genes such as *tetW* should not be used since recent research cannot rule out the possibility of gene transfer even if the gene is on chromosomal DNA. Furthermore, high hygienic standard to prevent contamination by pathogenic bacteria or allergen during the production of probiotics is another requirement to assure the safety of probiotics. Meanwhile, health claims for probiotics are regulated in Japan, called the Food for Special Health Uses (FOSHU) system. Both proven effectiveness and safety are required for approval as an FOSHU.

1 Introduction

Many researchers have reported the health benefits of probiotics, such as improvement of intestinal environment, stimulation of immune capacity, reduction of allergy risk, alleviation of diarrhea, improvement of constipation, and reduction of cancer risk (Colombel et al. 1987; Reddy and Rivenson 1993; Yaeshima et al. 1997; Fujii et al. 2006; Xiao et al. 2006; van der Aa et al. 2010). Many consumers now take probiotic products in anticipation of these effects. However, some concerns of probiotic safety including infections such as sepsis and horizontal transfer of antibiotic resistance genes have been reported (Courvalin 2006; Snydman 2008; Whelan and Myers 2010). Since probiotics are generally consumed as food, and not as medicine, any factor that may pose a risk to human health must be excluded in probiotic products. Another unique aspect is that probiotics can be consumed any time by various population groups including vulnerable persons such as infants, children, the elderly, and patients with diseases. Therefore, it is necessary to consider not only the effectiveness of probiotics for healthy persons, but also the safety for persons at risk (Liong 2008). In 2008, an important concern regarding probiotic safety was published in the world literature. In a clinical investigation of severe acute pancreatitis in the Netherlands, Besselink et al. (2008) reported that probiotics increased the risk of mortality. This finding raised a warning that the safety of probiotics use has to be considered seriously. Another concern of pathogenic bacteria with antibiotic resistant genes such as New Delhi metallo-β-lactamase 1 (NDM-1) made headlines recently throughout the world (Center for Disease Control and Prevention 2010). Since probiotic products have been used worldwide as foods such as yogurt, infant formula, and dietary supplement, there is a possibility that some risk factors, for instance antimicrobial resistance factor, harbored in probiotics may be disseminated worldwide. Also, it is a fact that many antibiotic resistance genes are already present in some intestinal bacteria. We have to watch these kinds of concerns.

Meanwhile, the quality of probiotic product is another aspect for probiotic safety. Recently, many probiotics are added to powdered formulas for infant or children (Abe et al. 2009a). However, since contamination by pathogenic bacteria such as *Salmonella* is one of the biggest concerns of powdered formulas, high hygienic specifications are needed for probiotics as an ingredient of powdered formula (Cahill et al. 2008). Also, allergen contamination can cause serious problems. These concerns have to be addressed to allow consumers to take probiotic products without anxiety.

On the other hand, health claims for probiotics are being discussed in the world. Japan has already regulated a health claim system, called Food with Health Claims (FHC). One category is Food for Specified Healthy Uses (FOSHU), and many probiotic products including yogurts, milk beverages, and fermented milks have been approved as FOSHU. Proof of both effectiveness and safety aspects is necessary to be approved as FOSHU. This FOSHU system is convenient for consumers because they can understand the functions of the product by reading the label on the food. Therefore, this system could be a good reference for overseas countries that wish to establish a similar system. However, the FHC system is quite complicated and may be slightly difficult to understand for foreign countries.

In this chapter, probiotics safety and regulation for health claim system in Japan will be discussed to consider another aspect of probiotics apart from effectiveness.

2 Translocation, Bacteremia, and Sepsis

2.1 Bacterial Translocation Associated with Probiotics

In the past decade, some clinicians and microbiologists have reported bacterial translocation issues related to probiotics consumption. In 2004, Kunz et al. (2004) observed two cases of *Lactobacillus* bacteremia during treatment with *Lactobacillus* GG. In 2005, another case of *Lactobacillus* GG bacteremia associated with probiotic use was reported (De Groote et al. 2005). Furthermore, Land et al. (2005) proved by molecular analysis that the organism causing *Lactobacillus* sepsis in two pediatric patients was indistinguishable from the probiotic strain ingested by the patients. Also, Ledoux et al. (2006) reported *Lactobacillus acidophilus* bacteremia after use of a probiotic. These cases raised an alarm over the safety of some probiotics regarding the risk of bacterial translocation, sepsis, or bacteremia, although all the cases reported were patients with diseases such as short gut syndrome and AIDS, or premature infants, but not healthy individuals. In a systematic review of safety of probiotics in patients receiving nutritional support, Whelan and Myers (2010) reported that there were 20 reports of adverse events in 32 patients, all of which were infections due to *Lactobacillus rhamnosus* GG or *Saccharomyces boulardii*. They concluded that some probiotic products (strains or combinations) may increase the risk of complications in specific patient groups.

In addition, Besselink et al. (2008) reported that probiotic prophylaxis increased the risk of mortality in severe acute pancreatitis and recommended not to administer probiotics in this category of patients. Recently, bacteremia caused by *Lactobacillus casei*, a species generally used as probiotics, was reported in a heavy dairy consumer (Russo et al. 2010). Also, severe sepsis after treatment with the probiotic *Escherichia coli* Nissle 1917 was observed (Guenther et al. 2010). These reports suggest that the safety issue of probiotics is not limited to isolated incidents but is constantly present in the clinical setting. These observations warn of the possibility of bacterial translocation and subsequent sepsis or endocarditis caused by probiotic strains and highlight the necessity to investigate translocation ability as one aspect of safety evaluation for probiotics before use.

2.2 Mucin Degradation Activity of Probiotics

Generally, the first step of bacterial translocation is the invasion of bacteria through the intestinal wall. Gork et al. (1999) reported that mucin on the surface of the intestinal wall is very important to prevent bacterial translocation, suggesting that mucin decomposition increases the risk. Some investigators have reported mucin degradation activity of probiotic bacteria. Zhou et al. (2001) demonstrated that three probiotic strains of *L. rhamnosus*, *L. acidophilus*, and *Bifidobacterium lactis* did not decompose mucin in vitro. Abe et al. (2010) also studied mucin degradation activities in various bifidobacterial strains including the type strains of *Bifidobacterium longum* subsp. *longum*, *Bifidobacterium breve*, and *B. longum* subsp. *infantis*. Initially, they observed that the bifidobacteria tested could not grow well in basal medium containing mucin as the sole carbon source in comparison with glucose, as shown in Fig. 1. And they confirmed by sodium dodecyl sulfate-polyacrylamide gel electrophoresis (SDS-PAGE) study that the bifidobacteria did not decompose mucin in the medium, although fecal sample as positive control did. On the other hand, Ruas-Madiedo et al. (2008) recently reported that some *Bifidobacterium* species such as *B. bifidum* and a few strains of *B. breve* and *B. longum* possessed mucin degradation activity and also had genes encoding mucin decomposing enzymes, although most strains of *B. longum* and all strains of *Bifidobacterium pseudocatenulatum* tested had no such ability. These findings indicate that mucin decomposition activity should be investigated at strain level.

Furthermore, some investigators also studied the effects of probiotics on intestinal surface including mucin and epithelial cell structures in mice after probiotic administration. Abe et al. (2010) confirmed by morphometric analysis that administration of a high dose of *B. longum* BB536 induced no changes in intestinal surface structures such as mucosal thickness, villus height, crypt depth, and epithelial cell height in the ileum, cecum, and colon (Table 1). These in vivo observations are more realistic to obtain safety information regarding mucin integrity in intestinal surface than in vitro test. Since mucin plays an important role to prevent bacterial invasion in the intestines, it is important to confirm whether probiotics

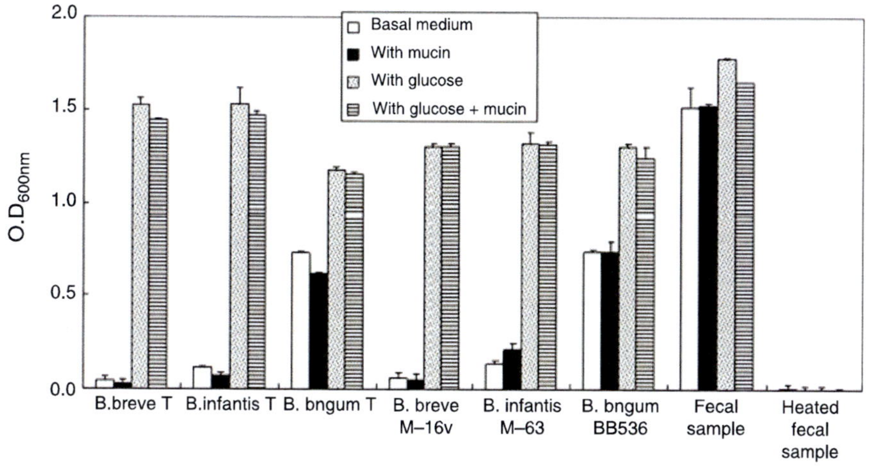

Fig. 1 Growth of *Bifidobacterium* species in medium with/without mucin or/and glucose. Each bifidobacteria strain was cultured for 24 h at 37°C in four kinds of medium: basal medium (no carbon source), basal medium with mucin, basal medium with glucose, and basal medium with mucin and glucose. After cultivation, OD at 660 nm of each culture was measured. All the *Bifidobacterium* species tested could not utilize mucin for growth

Table 1 Morphometry of mucosal structure in ileum[a]

Group	Villus height	Crypt depth	Mucosal thickness	Epithelial cell height
Control	183.7 ± 25.2	97.1 ± 14.6	289.2 ± 34.2	20.9 ± 2.0
Probiotics	188.4 ± 21.1	108.4 ± 15.4	305.0 ± 31.7	19.6 ± 2.0

[a]Values are presented as mean ± SD (μm).

used in food have mucin decomposition activity, in other words, the ability to injure the mucin layer.

2.3 *Translocation Ability of Probiotics*

To examine the translocation ability of probiotics, a simple and straightforward method is to administer probiotic bacteria orally and count the invaded bacteria in various organs and tissues including blood, liver, spleen, kidney, and mesenteric lymph nodes. These investigations have been used to demonstrate probiotic safety. Kabeir et al. (2008) confirmed no translocation ability of *B, pseudocatenulatum* into liver and blood of mice after oral administration. In another translocation analysis after administration of *B. longum* BB536 for 7 days, no bacteria and no bifidobacteria were observed in all organs and tissues including blood, liver, kidney, spleen, and mesenteric lymph nodes (Abe et al. 2010). For lactobacilli, Paturi et al. (2008) observed that *L. acidophilus* and *Lactobacillus paracasei* did not translocate into spleen, liver, and blood in mice. However, the experiments discussed earlier were conducted in conventional mice with no disease, which are models for healthy persons. On the other hand, most cases of sepsis or bacteremia related to the administration of probiotics were observed in patients with diseases, and not in healthy persons. Therefore, use of immunocompromised animals as a model of vulnerable persons including infants or immunocompromised patients such as those with leukemia, premature infants, serious diseases, AIDS, and after undergoing surgery is more appropriate for examining translocation and sepsis. Matsumoto et al. (2008) produced a good immunocompromised mouse model to observe bacterial translocation and subsequent sepsis. They administered cyclophosphamide, which is an immunosuppressive agent sometimes used in cancer patients, to mice to produce an immunocompromised condition. These mice had lower leukocyte counts and became susceptible to translocation of intestinal bacteria. After administration of *Pseudomonas aeruginosa* to this model, these bacteria were observed in blood and liver of the mice, and finally all mice died within a few weeks. When *B. longum* BB536 was administered orally to the immunocompromised mice before and after *P. aeruginosa*, no colonization of *B. longum* BB536 in liver and blood was observed and colonization of *P. aeruginosa* was suppressed resulting in improved mortality. These observations suggest that the probiotic tested is very safe with respect to bacterial translocation even in immunocompromised patients. As shown by this model, using immunocompromised animals is helpful to understand whether the probiotics can be used not only by healthy persons, but also by persons with immunocompromised conditions, such as premature infants, elderly persons, and patients with serious diseases.

2.4 Translocation of Probiotics and Virulence

All probiotic bacteria could become susceptible to bacterial translocation under conditions that seriously compromise the intestinal wall barrier, such as severe gastroenterological injury, serious disease, and heavy use of antibiotics. However, there is a big difference between bacterial translocation and sepsis or endocarditis. Even if bacterial translocation occurs, the bacteria will pose no threat to the host if the translocated bacteria do not cause sepsis, endocarditis, infection, or any other disease, and are cleared from the body immediately (Yazawa et al. 2000). Conversely, if the possibility of causing sepsis or infection is high after probiotic translocation occurs, the probiotics should not be used even if it has low translocation ability. Yamazaki et al. (1985, 1991) reported interesting investigations using germfree mice. When the pathogen *E. coli* O-111 was administered orally to germfree mice, most mice died within a few weeks because *E. coli* O-111 was translocated and produced enterotoxin in organs. On the other hand, when *B. longum* BB536 was administered to germfree mice, no mice died although this strain was observed in various organs such as the liver, kidney, and mesenteric lymph nodes during the first 4 weeks. The administered bifidobacteria disappeared from all organs after *B. longum*-specific IgG and IgA were produced 4–8 weeks

Table 2 Translocation of *Bifidobacterium longum* to internal organs after *B. longum*-monoassociation

Weeks after monoassociation	No. of mice	Isolation of *B. longum*		
		Liver	MLN[a]	Kidney
1	3	2/3[b]	3/3	2/3
2	3	2/3	3/3	3/3
4	4	0/5	1/5	N.D[c]
8	5	1/5	1/5	1/5
12	5	0/5	0/5	0/5
18	5	0/5	0/5	0/5

[a]MLN: Mesenteric lymph nodes[b]Number of positive/Number of tested mice[c]N.D: Not determined

later (Table 2). During the period when translocated bifidobacteria were detected in various organs, no mice exhibited any disease symptom and none died. These results indicate that *B. longum* BB536 strain causes no infection and no adverse effect even if translocation occurs, suggesting that this strain is safe for consumption by patients with increased risk of bacterial translocation. Furthermore, this strain has been administered to immunocompromised patients with leukemia treated with anticancer drugs including cyclophosphamide to confirm the effect of probiotics on *Candida* infection. Kageyama et al. (1984) observed that *Candida* that had proliferated in patients after administration of antileukemic drug was reduced by *Bifidobacterium* administration and no adverse event was induced by the probiotic.

B. longum BB536 was administered to germfree mice and translocation to liver, mesenteric lymph nodes, and kidney was observed by culturing the organs. The administered bifidobacteria were detected in every organ tested during the first 2 weeks. After 4 weeks, bifidobacteria disappeared from all organs. All the mice were not ill and none died.

2.5 *Confirmation of Probiotic Safety by Examining Translocation Ability*

In conclusion, to confirm probiotic safety with respect to bacterial translocation or sepsis, the following investigations are necessary.

1. Mucin decomposition activity
2. Bacterial translocation ability in conventional animal model
3. Observation of intestinal surface structure in animal model
4. Bacterial translocation ability in immunocompromised animal model
5. Infection activity after bacterial translocation
6. Human clinical investigation on healthy persons
7. Human clinical investigation on vulnerable persons

Probiotics are generally used as food and all kinds of people including vulnerable persons have a chance to consume probiotic products. Most people anticipate beneficial health effects after consumption. Therefore, probiotics should have proven effectiveness and must have adequate safety data including acute and chronic toxicity, bacterial translocation, as well as sensitivity in vulnerable persons (Abe et al. 2009b). Reports have implicated some probiotic strains to have the potential to translocate and subsequently causing sepsis, which might constitute a severe health problem. There are many candidates of probiotics, and not all strains are the same. As probiotics have many positive potentials for human health, manufacturers of probiotics have to select the suitable strains and confirm their safety from various aspects before launching, so as not to lose consumers' trust.

3 Antibiotic Resistant Genes Transfer

Around a decade ago, Teuber (1999) raised an alarm on the spread of antibiotic resistance genes in human intestinal microflora and in the food chain. Unfortunately, in spite of his warning, many studies have reported an increase of antibiotics resistance genes in our intestinal microflora and in the environment during the past decade. Recently, *Enterobacteriaceae* harboring the antibiotic resistance gene *NDM-1* have been a subject of great concern (Health Protection Agency 2009; Center for Disease Control and Prevention 2010; Kumarasamy et al. 2010). Since transfer of antibiotic resistance genes to pathogenic bacteria would create serious problems, the environmental trend of resistance gene transfer has to be surveyed carefully and monitored continuously.

3.1 Risk of Probiotics Harboring Acquired Antibiotic Resistance Genes

Probiotic products may contain various bacteria species. Since many probiotics products are sold worldwide in various forms such as yogurt, infant formula, cheese, and dietary supplement (Vinderola et al. 2000; Mattila-Sandholm et al. 2002, Charalampopoulos et al. 2002; Champagne et al. 2005; Philips et al. 2006; Abe et al. 2009a, c), a huge number of probiotic cells are consumed by a vast number of consumers and the probiotics strains are in contact with many species of intestinal microorganisms at high densities. Therefore, if probiotic bacteria harbor antibiotic resistance genes that can transfer to pathogenic bacteria or intestinal microorganisms, there would be a risk of spread of antibiotic resistance genes or antibiotic resistance bacteria in our intestines or environment. Spigaglia et al. (2008) characterized the tetracycline resistance gene *tetW* of a human isolate of tetracycline-resistant *Clostridium difficile*, and found that the gene sequence was

99% homologous with the gene found in the human strain *B. longum* F8. This finding suggests a possibility that *tetW* in both strains might have resulted from exchange between the two strains (the possibility must be quite low) or was acquired from the same origin such as other intestinal bacteria harboring this sequence. Nevertheless, this report may suggest that *tetW* is disseminated by horizontal gene transfer. In fact, Jacobsen et al. (2007) demonstrated in vivo occurrence of horizontal gene transfer of antibiotic resistance plasmid from *Lactobacillus plantarum* to *Enterococcus faecalis* in gnotobiotic rat intestine. From these observations, our intestine may provide a favorable environment for horizontal gene transfer, including the dissemination of antibiotic resistance genes. For these reasons, the presence or absence of antibiotic resistance genes in the cells should be addressed for probiotic strains sold in the market.

3.2 *Bifidobacteria Harboring Antibiotics Resistance Genes in the Environment*

Recently, some reports have indicated that many intestinal bacteria harbor antibiotics resistance genes. Masco et al. (2006) observed that some bifidobacteria isolated from human, animal, and probiotics products had tetracycline-resistant activities and that *tetW* was responsible for the resistance in all the strains. Aires et al. (2007) reported that 26% and 7% of bifidobacteria isolates from human had *tetW* and *tetM* genes, respectively, suggesting that our intestines have already had many bacteria harboring acquired antibiotic resistance genes such as *tetW*, and already carry a risk of transferring antibiotics resistance gene to harmful bacteria. Liu and Pop (2009) created an antibiotic resistant genes database. Bifidobacteria are listed in this database and some antibiotics resistance genes such as *tetW*, *tetC*, *tetQ*, *tetR*, *ermB*, and *cata-1* are documented in some strains of bifidobacteria (Table 3). In particular, *tetW* is present in various *Bifidobacterium* species, for instance, *B. longum*, *B. animalis* subsp. *lactis*, *B. bifidum*, *B. longum* subsp. *infantis*, and *B. thermophilum*. Therefore, evidence from various studies shows that many

Table 3 List of antibiotics resistance genes on bifidobacteria in Antibiotics Resistance Gene Database[a]

Resistance gene	Resistance type	*Bifidobacterium* species listed in on database
tetW	Tetracycline	*B. longum* subsp. *infantis*, *B. bifidum*, *B. animalis* subsp. *lactis*, *B. thermophilum*, *Bifidobacterium spp.*
tetC	Tetracycline	*B. bifidum*
tetL	Tetracycline	*B. thermophilum*
tetQ	Tetracycline	*B. bifidum*
cata-1	Chloramphenicol	*B. adolescentis*
ermX	Erythromycin	*B. thermophilum*

[a]Antibiotics Resistance Gene Database: http://ardb.cbcb.umd.edu/index.html (Liu and Pop 2009)

bifidobacteria in our intestines already harbor antibiotic resistance genes and that we are potentially endangered by pathogenic bacteria with antibiotic resistance as already reported by Spigaglia et al. (2008).

3.3 Probiotics Harboring Antibiotic Resistance Genes Used in Commercial Products

Some reports have shed light on the questions of where the antibiotic resistance genes came from and how these genes are disseminated to bifidobacteria and other bacteria. Xiao et al. (2010) investigated Japanese probiotic products containing bifidobacteria. In their report, among *Bifidobacterium* species including *B. longum* subsp. *longum, B. breve, and B. animalis* subsp. *lactis* isolated from various probiotic products, all isolates of *B. animalis* subsp. *lactis* showed resistance to tetracycline and harbored *tetW*. Kastner et al. (2006) also detected *tetW* from probiotic cultures of *B. lactis*. Furthermore, Gurimonde et al. (2010) reported recently that all *B. animalis* subsp. *lactis* strains isolated from various sources including yogurt, fermented food, or probiotic supplement carried *tetW* responsible for tetracycline resistance. As these reports indicated, there are many probiotic commercial products containing bifidobacteria with *tetW* around us. The presence of an acquired *tetW* gene in several probiotic product isolates stresses the need for a minimal safety evaluation during the selection of *Bifidobacterium* strains for probiotic use. On the other hand, some comparative studies of *tetW* carried by different strains have been conducted. Ammor et al. (2008a) analyzed the *tetW* gene and the flanking sequences in intestinal *Bifidobacterium* species and observed that the *tetW* genes are very well conserved among bifidobacterial strains. Also, Kazimierczak et al. (2006) reported that various intestinal bacteria including *B. longum, Clostridium* spp., *Butyrivibrio fibrisolvens, Mitsuokella multiacida,* and *Roseburia* sp. possessed both *tetW* and the flanking sequences that were very homologous. Considering all the above findings, one cannot rule out the possibility that *tetW* harbored in *B. animalis* subsp. *lactis* may be one of the origins of *tetW* in many bifidobacterial strains of the human intestine.

3.4 Possibility of Horizontal Transfer of Antibiotics Resistance Genes from Probiotics

Next, we question whether *tetW* in *B. animalis* subsp. *lactis* can be transferred to other bacteria including pathogenic bacteria. Kazimierczak et al. (2006) demonstrated that there was horizontal gene transfer between *B. longum* and *B. adolescentis* although at low frequency. However, transfer of *tetW* in *B. animalis* subsp. *lactis* to other bacteria has never been reported. Generally, horizontal gene transfer occurs by transduction, transconjugation, transposon, or natural transformation (Levy and Marshall 2004). Transduction is mediated by bacteriophage, but there is little information on bacteriophage or transduction in the bifidobacteria group. Transconjugation seems to occur often in our intestine but in many cases, plasmid is necessary to transmit the genes to recipient cells in this system. Since most *tetW* genes are located on chromosomal DNA, and not plasmid, the transfer of *tetW* by transconjugation would be unlikely (Ammor et al. 2008a, b). Transposons such as Tn916 and Tn5276 have been reported to mediate horizontal gene transfer in lactic acid bacteria (Mathur and Singh 2005). Although a recent study has demonstrated a transposase gene upstream of *tetW* in *B. animalis* subsp. *lactis,* and cotranscription of *tetW* and the transposase gene in tandem (Gurimonde et al. 2010), there is no report that *tetW* is involved in a transposon. On the other hand, natural transformation allows transfer of all chromosomal sequences to a recipient cell if the recipient cell is competent, because competent cells can accept naked DNA. Although natural transformation is documented only across a restricted number of species, Meibom et al. (2005) reported very surprising results regarding natural transformation of antibiotic resistance gene in *Vibrio cholerae*, which had never been shown to be competent for natural transformation. In their report, antibiotic resistant gene could be transferred to *V. cholerae* by natural transformation only when the growth conditions mimicked the natural habitat, such as the surface of crab shell, whereas no transformation was observed under normal in vitro conditions even with the same strain. It was postulated that natural habitat conditions including the availability of key substances (chitin in this case), high cell density, and nutrient limitation or stress are necessary to produce competent cells and to activate the natural transformation activity. Considering our intestinal environment, the cell density is much higher than the in vitro liquid cultures commonly used in the laboratory, and numerous naked DNA from dead bacterial cells are present around intestinal bacterial cells. Also, nutrient limitation is perhaps a norm in the intestine because most of the easily utilizable nutrients have been taken up by humans and a huge number of intestinal bacteria compete for the residual nutrients. Furthermore, bile salt which is normally present in the intestine has been reported to induce tetracycline resistance phenotype in bifidobacteria (Noriega et al. 2005). Therefore, the intestinal environment may be conducive to natural transformation. In streptococci, it has been known for decades that only a few streptococcal species develop the competent state spontaneously when grown under laboratory conditions (Tomasz and Hotchkiss 1964). However, the recent discovery of a new mechanism regulating natural transformation in *Streptococcus* species suggests that this property might be more widespread among streptococci than previously thought (Johnsoborg and Havarstein 2009; Havarstein 2010). Also, Chen and Dubnau (2004) reported that various bacteria could be competent and had unique DNA uptake systems, suggesting that various bacteria in our intestine could take up naked DNA including *tetW* and the flanking sequence. Although further discussions and investigations are needed to prove the possibility that *B. animalis* subsp. *lactis* is one of the origins of *tetW* that is widespread in the food chain, this possibility cannot be denied at the present moment.

3.5 *Antibiotic Resistance Genes in Lactic Acid Bacteria*

On the other hand, some lactic acid bacteria such as *L. crispatus, L. plantarum, L. johnsonii, and L. reuteri,* as well as *E. faecalis* have been reported to be antibiotic resistant and harbor antibiotic resistance genes including *ermB*, *tetM*, and *tetW* (Mathur and Singh 2005; Klare et al. 2007). Because most of these lactic acid bacteria are generally used as starter culture for meat and dairy products or as probiotics, they may pose a hazard of transferring these genes horizontally to intestinal bacteria and pathogenic bacteria. In particular, some lactobacilli and enteroccci have plasmids, and in vivo study has demonstrated that plasmid harboring antibiotics resistance gene in lactobacilli is transferred horizontally to other bacteria (Jacobsen et al. 2007). As our intestinal flora contains many bacteria that have the ability to acquire foreign DNA by transconjugation, horizontal gene transfer may occur regularly between bacteria living in our intestine. Considering the above concerns and the fact that many isolates from dairy and meat products harbor antibiotics resistance genes, strains of bacteria carrying acquired resistance to antimicrobial(s) should not be used as feed additives, unless it can be demonstrated that it is a result of chromosomal mutation(s), as recommended by the FEEDAP panels in the EFSA technical guidance (EFSA 2008).

3.6 *Algorithm for Decision Making of Probiotics Selection Related to Antibiotic Resistance*

Courvalin (2006) reviewed probiotics safety focusing on antibiotic resistance, and proposed an algorithm for decision making of probiotics selection related to antibiotic resistance, as given in Table 4. According to this proposal, as the first step we should know the phenotype of antibiotic resistance. If there is no unique resistance, the strain can be used. If present, it is necessary at the second step to investigate whether the antibiotic resistance is transferable or not. If it is transferable, the strain is not suitable for use as probiotics. If no transferability is estimated,

Table 4 Steps for decision making of probiotic strain selection with respect to antibiotic resistance

Step	Questions	Answer	Judgment
1	Does the probiotic strain have unique antibiotics resistance on phenotype?	Yes	To step 2
		No	OK for use
2	Are the antibiotic resistances transferable?	Yes	Not OK for use
		No	To step 3
3	Are the antibiotic resistance genes acquired genes?	Yes	Not OK for use
		No	To step 4
4	Are the antibiotic resistance genes stabilized in the chromosome and can the convincing evidence be shown?	Yes	OK for use
		No	Not OK for use

then as the next step, whether the antibiotic resistant gene is acquired or not has to be addressed. If the gene is acquired, the strain cannot be used because of the possibility that the gene may be further transferred to other bacteria. If the gene is not acquired, then evidence is needed to show in a positive way that the gene is stabilized in the chromosome, such as by demonstrating colocalization with a house-keeping gene on the same DNA fragment. If convincing evidence is obtained, the strain becomes OK for probiotic use. This algorithm is not difficult to understand and makes a whole lot of sense.

With the widespread of many antibiotic resistance genes in recent years, there is increasing concern of possible outbreak of pathogenic bacteria with antibiotic resistance genes acquired from probiotics carrying the genes. As FAO/WHO or EFSA recommended, probiotics harboring antibiotic resistance gene should not be used in the food chain and feeding area (Joint meeting of FAO/WHO 2002; EFSA 2008). To play a role in halting the spread of antibiotic resistant genes across the environment including our intestine, all manufacturers of probiotic products should review their probiotics from the antibiotic resistance point of view and choose their probiotic bacteria according to the algorithm proposed by Courvalin (2006).

4 From Research to Consumer: Quality Assurance and Control

Currently, probiotics are used in various commercial products such as yogurt, milk, powdered formula, and dietary supplement. To introduce these probiotic products into food chain, the manufacturers have to guarantee the hygienic quality, probiotic stability, and regulatory compliance. In this section, each of the above items will be discussed.

4.1 Hygienic Standard of Probiotics

Recently, some powdered formulas for infants or children contain probiotics including bifidobacteria and lactobacilli (Abe et al. 2009a; Muto et al. 2010). In case of infant formula, the quality of probiotic ingredient has to be considered carefully, because infants are usually much more susceptible to various infections by pathogenic bacteria than healthy persons, and probiotic ingredients cannot be sterilized before or after mixing with powdered formula since probiotics are living bacteria. In particular, contamination of powdered infant formulas with *Salmonella* and with *Cronobacter sakazakii* has been the cause of infections in infants, sometimes with serious sequelae (EFSA 2004; Reiji et al. 2009). Therefore, infant formulas and the powder ingredients such as probiotic powder in infant formulas, which cannot be sterilized, should not contain any pathogenic contaminants including *C. sakazakii* and *Salmonella*. However, as Kandhai et al. (2004) observed, *C. sakazakii* is present in various food production environments and households, suggesting that *C. sakazakii* has many chances to contaminate probiotics powder in the probiotic manufacturing process.

Normally, probiotics powder is produced through a series of steps including the culturing step, powdering step, and packaging step, as shown in Fig. 2. In order to produce probiotics powder hygienically, hygienic concept should be maintained in every step. In the culturing step, probiotics are cultured in an appropriate medium that has been sterilized. The sterilization temperature is crucial to avoid contaminants. However, excess heat treatment induces denaturation of the medium, which may repress growth of the probiotics. Therefore, appropriate sterilization

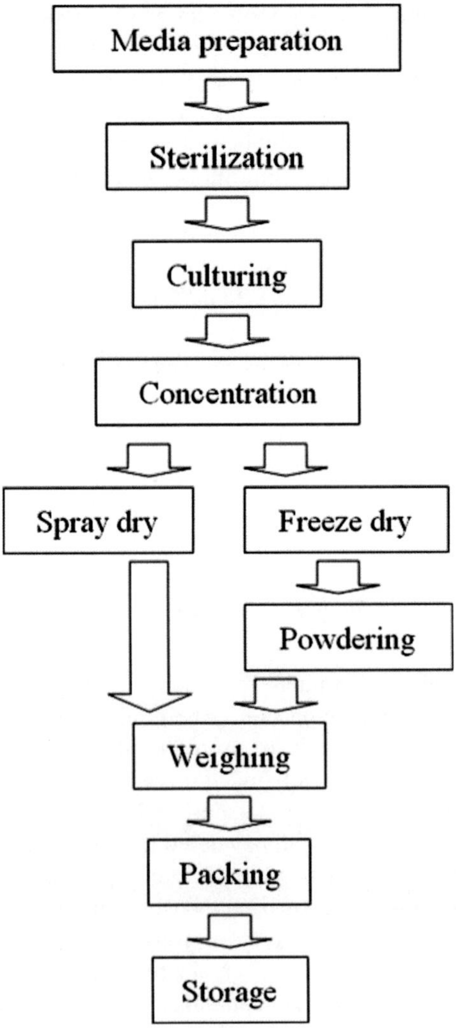

Fig. 2 Schematic presentation of the process of probiotic powder production

condition (such as 115°C for 10 min or 130°C for 2 s) should be decided on a strain-by-strain basis. Also, the culture process can grow not only the target probiotic bacteria, but also every contaminant, if present. Therefore, even if an insignificant number of contaminants are present in the culturing medium before culturing, the small number may become a big problem after this process. In the powdering process for probiotics production, both spray-dry and freeze-dry processes are available. Comparatively, the heat shock exerted on probiotic bacteria is stronger with spray-drying than freeze-drying, even when spray conditions such as inlet and outlet air temperatures are well controlled. However, spray-drying is sometimes preferred because of cheaper cost and easier control. On the other hand, the freeze-dry process is more complicated. Because the heat shock exerted during freeze-drying is less than spray-drying, contaminants are not reduced during the process. After freeze-drying, powdering is necessary. Equipment for freeze-drying and powdering has to be controlled under hygienic condition. Since it is difficult to sterilize the equipment completely and there are many steps operated by workers, hygienic control to avoid bacterial contamination may be more difficult than the spray-dry process. As final steps, weighing and packaging are necessary after either the spray-dry or the freeze-dry process. The packaging step handles powder and sometimes the powder is open in the production room. Therefore, strict hygienic environments are necessary to avoid the chance of contamination by pathogenic bacteria. In particular, because many steps involve manual handling, hygienic education for workers is important.

Although the Hazard Analysis and Critical Control Point (HACCP) concept is needed throughout the whole production process, specifying target contaminants is important in probiotic production. For example, to produce probiotic powder for infant formula, contamination by *Salmonella* and *C. sakazakii* has to be considered in every step. If the product is general food and not infant formula, strict hygienic condition may be not necessary. Appropriate HACCP plan should be designed depending on the usage of the probiotic. These hygienic concepts will ensure that high quality probiotic products reach the consumers.

Pathogenic bacteria are not the only contaminant that manufacturers of probiotics should control, other contaminants such as pesticides, heavy metals including lead and arsenic, foreign bodies, and allergens also need to be excluded from the products. These contaminants sometimes cause serious problems to consumer's health. To prevent these accidents, hygienic systems such as HACCP or ISO2200 should be maintained during the manufacture of probiotics. Assessment by a third party using the good quality certification system is useful to ensure product quality.

4.2 Stability of Probiotics

The stability of probiotics in food is very important because according to the definition, probiotics are live microbials. However, it is not easy to add the living cells to yogurts, beverages, or powder foods and to maintain adequately high viable counts during the shelf-life. Although many probiotics suppliers are investigating intensively on how to produce stable probiotics by various technologies, there are large differences in stability between *probiotic* strains.

4.2.1 Stability of Probiotics in Powder Products

Stability is particularly important for probiotic powder used in infant formulas or dietary supplements. These products have long shelf-lives such as 18 and 24 months, and are stored under ambient temperature. These conditions are harsh storage conditions for probiotics. The stability of probiotics is different between strains (Abe et al. 2009a). Also, product characteristics such as water activity, storage temperature, and packaging material affect the stability (Abe et al. 2009d). Figure 3 compares the stability of two kinds of probiotic powders (*Bifidobacterium* powder A and *Bifidobacterium* powder B) which are being used in commercial products. At lower water activity, both probiotic powders were stable and almost the same. However, at higher water activity of greater than 0.13, viable counts of *Bifidobacterium* powder B decreased sharply from the early storage period. In particular, when these probiotic powders were used to supplement infant formula having water activity of 0.25, *Bifidobacterium* powder A maintained survival rate similar to that of lower water activity, although *Bifidobacterium* powder B survival decreased to less than 1/100 after 1 year storage. These results suggest that choosing the appropriate probiotic material upon considering characteristics including stability is very important for the production of probiotic products because there are great differences in stability between probiotic materials.

4.2.2 Stability of Probiotics in Fermented Milk Products

Probiotic yogurt has been marketed worldwide (Holzapferl et al. 2001; Heller 2001; Gueimonde et al. 2004). Many yogurts supplemented with *Bifidobacterium* are being sold in the market, but many of the products use *B. animalis* group, which does not reside in human intestine (Jayamanne and Adams 2006). The main reason is that *B. animalis*, which lives in animal intestine, can survive stably in yogurt because of its robust characteristics (Jayamanne and Adams 2006). Generally, human bifidobacteria including *B. longum*, *B. breve*, or *B. infantis* are weaker than *B. animalis*, so that human bifidobacteria cannot survive in yogurts during long shelf-life. However, new starter culture technologies have been developed recently and *B. longum* BB536 has been reported to survive well in yogurt (Abe et al. 2009c). The stability data of *B. longum* BB536 in drinking fermented milk produced by the new technologies confirms that *B. longum* BB536 can survive stably during long storage period even when sucrose concentration in the fermented milk is changed from 2 to 10% (Fig. 4). Using these starter technologies, any yogurt

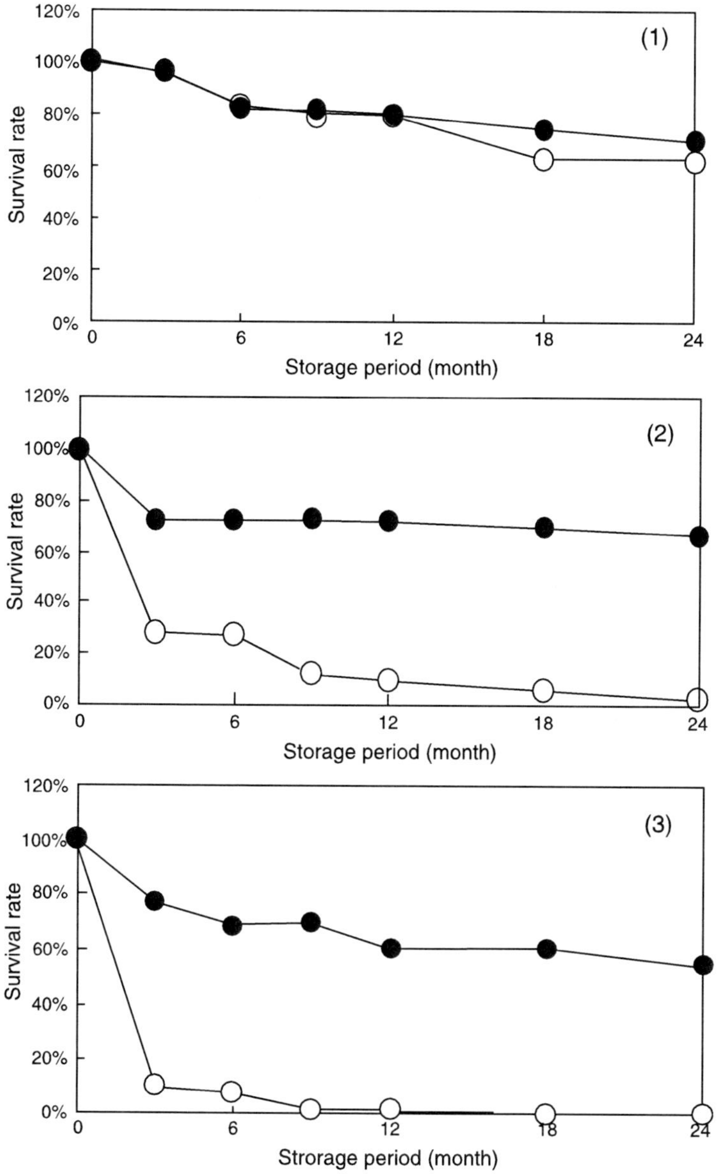

Fig. 3 Effect of water activity on stability of *Bifidobacterium* powder A and B. *Closed symbols* are *Bifidobacterium* powder, A. *Open symbols* are *Bifidobacterium* powder, B. (1) and (2) are bifidobacteria stability in starch powder with water activity of 0.1 and 0.13, respectively. (3) is the stability in infant formula with water activity of 0.25

manufacturer can produce probiotic yogurt containing human bifidobacteria with no concern about stability, and consumers can anticipate the probiotic effects by eating these yogurts.

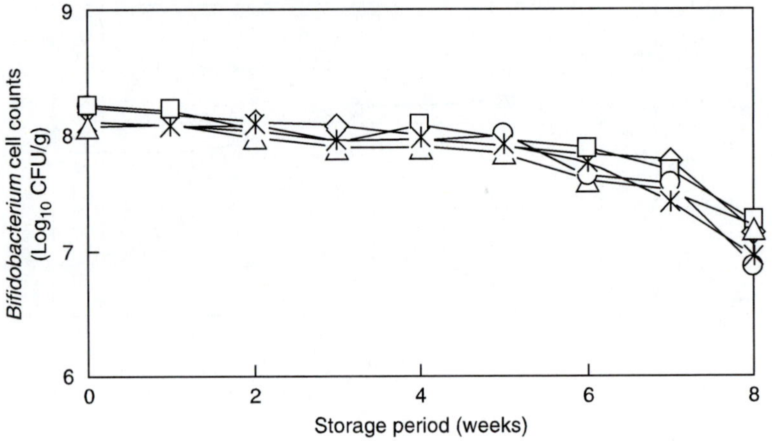

Fig. 4 Survival rates of bifidobacteria in fermented milk drink manufactured by new technology. Sucrose concentrations: *open diamond*: 2%, *open square*: 4%, *open triangle*: 6%, *open circle*: 8%, *asterisk*: 10%

4.3 Regulatory Control

Good probiotics promote the health of consumers. Probiotics are consumed in various countries of the world. However, there are some differences in the regulation of foods between countries, such as the allergen list, specification of genetically modified food (GMO), residual pesticides, the food additive list, and the labeling system. To market probiotics in various countries or regions, the national or regional food regulations should be understood and respected. In particular, allergen contamination may have serious health consequence because persons with allergy are very sensitive and sometimes respond to very small amount of allergen. Probiotics are usually cultured in culture media as shown in Fig. 2 and the ingredients of the culture medium may remain in the final product. Since the culture medium may contain some allergens such as milk and soy, declaration of allergen may be needed for some probiotics, if any allergen is used during production and its complete elimination from the final product cannot be confirmed (Rasic and Kurmann 1983; Tang et al. 2007). However, because of the differences in the allergen list between countries, manufacturers of probiotics need to know the list for each country or region where probiotic products will be marketed.

Table 5 shows the lists of allergen regulated in USA, EU, Japan, and by Codex. The lists of USA, EU, and Codex are almost same except for some items such as lupin and molluscs. Milk, egg, soy, wheat, and crustaceans are common allergens in the world. In particular, milk ingredients including casein peptide, whey peptide, and lactose are often used as medium ingredient for culturing probiotics, although the number of persons with milk allergy patients is increasing worldwide. However, due to recent reports that administration of probiotics is effective to alleviate allergic symptoms in infants with cow's milk allergy, the opportunity of using

Table 5 Allergen lists for regulatory purpose used in EU, USA, Japan, and by Codex (2010)

Regulatory body	Allergens in the list
EU	Cereals, crustaceans, eggs, fish, peanuts, soybeans, milk (lactose includes), nuts, celery, mustard, sesame seeds, lupin, molluscs, sulfur dioxide and sulfite at concentrations of more than 10 mg/kg or 10 mg/l expressed as SO_2
USA	Milk, egg, fish, crustacean shellfish, tree nuts, wheat, peanuts, soybeans
Japan	1. Ministerial ordinance Milk, egg, wheat, buckwheat (soba), peanuts, shrimp/prawns, crab 2. Notice Abalone, squid, salmon roe, oranges, kiwifruit, beef, walnuts, salmon, mackerel, soybeans, chicken, banana, pork, "matsutake" mushroom, peaches, yams, apples, gelatin
Codex	Cereals containing gluten, crustacean, eggs, fish, peanuts, soybeans, milk (lactose includes), tree nuts, sulfite in concentration of 10 mg/kg or more

probiotics in people with allergy is expected to increase (Taniuchi et al. 2005, van der Aa et al. 2010). Therefore, in the production of probiotics for allergic patients, manufacturers have to exclude allergen-containing ingredients such as milk in the whole probiotic production process or remove them completely and confirm their nonexistence if allergen ingredients are used.

Also, cross-contamination in the probiotic production line has to be considered. Sometimes a production line is not dedicated to only one probiotic product, but produce different kinds of probiotic products. Since different probiotic products may be produced by different methods including various culture media, there is a possibility of allergen contamination thorough inadvertent incorporation of residual allergens in the production line. Therefore, it is necessary to check for allergen residues in the lines and to validate the washing method to remove all residues. Enzyme-linked immunosorbent assay (ELISA) or polymerase chain reaction (PCR) is generally used to check for residual allergen. Because persons with milk allergy may be sensitive to very low milk contents, ELISA analysis should be completely negative in probiotic products that may be consumed by persons with milk allergy.

The regulatory allergen lists for food in Japan are different from those of other countries. There are two kinds of allergen lists in Japan: the Ministerial Ordinance and the Notice. The Ministerial Ordinance lists seven kinds of allergens that manufacturers are mandated to indicate on the label of final food products. These allergens include milk, egg, wheat, buckwheat (soba), peanuts, shrimp/prawns, and crab. The Notice lists 18 kinds of allergens that manufacturers are not mandated but are recommended to indicate on the label. These allergens include abalone, squid, salmon roe, oranges, kiwi fruit, beef, walnuts, salmon, mackerel, soybeans, chicken, banana, pork, "matsutake" mushroom, peaches, yams, apples, and gelatin. Also, the definitions or the scopes of some allergens are different from EU or USA. For example, wheat in the Japanese list does not contain other cereal ingredients such as rye, oats, and spelt, although wheat in the EU or USA list contains all cereals related to gluten. Also, although seafood or fish includes all kinds of seafood or fish in the EU or USA list, the scope of each is separately defined in the Japanese list, i.e., salmon and mackerel as fish, and abalone, squid, and salmon roe as seafood. Actually, the Japanese regulation targets the ingredient names more individually. Therefore, due to the differences in regulation of allergen indication between countries or regions, manufacturers of probiotics have to select appropriate ingredients to produce probiotics and to declare allergens to avoid serious allergic events in consumers in case of worldwide use. Although only the food allergen lists are discussed in this section, other food regulations such as those on GMO, pesticides, and labeling system may also be different between regions or countries. To protect consumer's safety, correct information such as declaration of allergen should be given on the label of probiotic products because most consumers including vulnerable persons anticipate healthy effects from probiotics and do not expect health hazards.

5 Global Regulations for Health Claims of Foods

Health or function claim of food has attracted interest recently. In particular, the EU health claim system has just started and many food manufacturers and consumers are watching its development. Also, USA has a function claim system for food. However, to the best of the author's knowledge, no probiotic products have ever been approved for health or function claim in EU and USA as of October 2010. Japan is the first country that has a health claim system, called Food with Health Claims (FHC) including Food for Specified Health Use (FOSHU). Some probiotic products such as yogurt and milk containing probiotics have already been approved as FOSHU and sold with labels indicating health claims in the Japanese market. In this section, the Japanese food health claim system and FOSHU products focusing on probiotics will be described.

5.1 Regulation of Food with Health Claims in Japan

In Japan, there are traditionally two broad categories: medicine and food. Before 1991, there was no way to advertise or indicate any function or effect on the label of products belonging to the food category, although it was possible to do so for products in the medicine category. Therefore, food manufactures were not able to send any message to consumers even if the functions and benefits of the products are supported by abundant scientific and clinical evidence. Also consumers could not obtain correct information about the functions of food products in the market. The FOSHU regulatory system was first introduced in 1991 and was modified in 2001. The current health claim system in Japan is shown in Fig. 5. The category of "Food with Health Claims (FHC)" is positioned between the "Food" and "Medicine" categories. Similar to medicinal products, FHC products may indicate health claims on the label provided the food products have proven effectiveness as

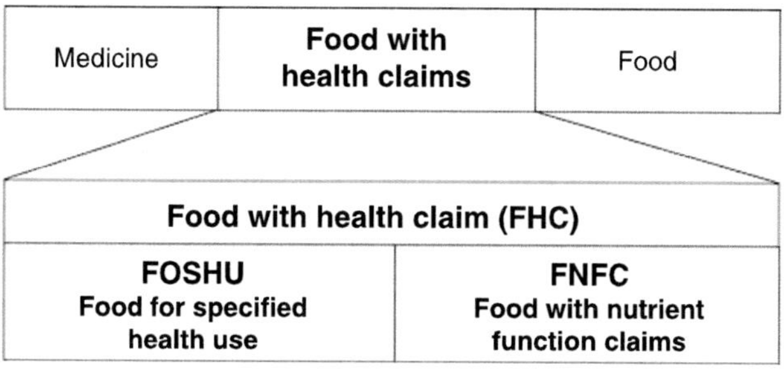

Fig. 5 Health claim system in Japan

demonstrated by clinical investigations or contain specified nutrients provided by the regulation.

The policies of labeling of FHC are as follows.

1. A FHC product should conform to national nutritional goal and health policy.
2. A FHC product should clearly specify that the product is useful in supplying nutritional substances or for specific health purpose.
3. An indication on the label should be based on scientific evidence and be true.
4. An indication on the label should be easily comprehensible and clear, and correct words and sentences should be used to provide appropriate information to consumers.
5. A FHC product has to show warning indications including appropriate intake to prevent health hazard from overdose.
6. A FHC product should comply with other relevant laws including the Japan Food Sanitation law and Pharmaceutical law.
7. A FHC product should be labeled clearly that the product belongs to FHC.
8. A FCH product should not refer to the diagnosis, cure, or prevention of disease.

FHC is divided into two categories, "Food with Specified Healthy Uses" (FOSHU) and "Food with Nutrient Function Claims" (FNFC). A food product belonging to FNFC may indicate claims on the label without permission from the Ministry of Health, Labour and Welfare (MHLW) if the food contains specified vitamins or minerals provided by the FNFC regulation. As of September 2010, 13 vitamins and 5 minerals are approved as nutrients for FNFC. On the other hand, FOSHU products are foods that contain functional ingredient and can show some claims on the label. However, labeling of an FOSHU product has to be approved by the MHLW after evaluation on the stated effectiveness, physiology, and safety of the product. A total of 960 FOSHU products have been approved by September 2010.

5.2 *Food with Nutrient Function Claims*

Food with Nutrient Function Claims is a food labeled with the function of supplying nutrients (vitamins or minerals) necessary for healthy growth, development, and maintenance of health. At present 13 vitamins (niacin, pantothenic acid, biotin, vitamin A, beta-carotene, vitamin B1, vitamin B2, vitamin B6, vitamin B12, vitamin C, vitamin D, vitamin E, and folic acid) and five minerals (zinc, calcium, iron, copper, and magnesium) are approved as ingredients of FNFC. The products labeled as FNFC are distributed by various sales channels such as supermarkets, drug stores, convenience stores, and mail order in Japan. FNFC products are allowed to show nutrition function claims on the label, and official clearance is not required if the products adhere to the standards and specifications of FNFC. Therefore, it is relatively easy for manufactures to produce these products for the Japanese health food market. However, in order to sell a product as FNFC, various points should be considered, such as the content, warning indication, indication of FNFC characteristics, and recommended dosage. Under FNFC, 13 vitamins and 5 minerals have been standardized in terms of claims on the label including the functions and warning indication, as well as the highest or lowest advisable intake.

5.3 Food for Specified Health Uses

FOSHU refers to foods containing ingredient with functions for health and officially approved to claim its physiological effects on the human body. FOSHU is intended to be consumed for the maintenance/promotion of health or special health uses by people who wish to control health conditions. In order to sell a food as FOSHU, assessments of the safety of the food and the effectiveness of the functions on health are required, and the claim should be approved by the MHLW. Since the enactment of this regulation in 1991, various kinds of products have been approved and the number reached 960 in September 2010. To obtain approval as FOSHU, some requirements are listed as follows:

1. Effectiveness on the human body is clearly proven.
2. Absence of any safety issues (animal toxicity tests, confirmation of effects in the cases of excess intake, etc.).
3. Use of nutritionally appropriate ingredients (e.g., no excessive use of salt, etc.).
4. Guarantee of compatibility with product specifications by the time of consumption.
5. Established quality control methods, such as specifications of products and ingredients, processes, and methods of analysis.

The approved FOSHU products can be divided into the following nine categories:

1. Foods to modify gastrointestinal conditions
2. Foods related to blood cholesterol level
3. Foods related to blood sugar levels
4. Food related to blood pressure
5. Foods related to dental hygiene

6. Cholesterol plus gastrointestinal conditions, triacylglycerol plus cholesterol
7. Foods related to mineral absorption
8. Foods related to osteogenesis
9. Foods related to triacylglycerol

5.3.1 Classification of FOSHU Approval

Currently FOSHU are divided into four classes as follows:

1. Regular FOSHU
   Regular FOSHU is the initial "FOSHU" established in 1991. Most of the FOSHU products belong to this class. To be approved as regular FOSHU, adequate documented evidence or certificates related to various parameters including effectiveness, safety, production, and quality are required.
2. Qualified FOSHU
   Food with health function not substantiated on scientific evidence that meets the level of FOSHU, or food with certain effectiveness but without established mechanism of the effective element for the function will be approved as qualified FOSHU.
3. Standardized FOSHU
   Standards and specifications have been established from foods that meet FOSHU approvals and accumulation of scientific evidence. A food is approved as standardized FOSHU when it meets these standards and specifications. Now some oligosaccharides (fructo-oligosaccharide, galacto-oligosaccharide, etc.) and dietary fibers (polydextrose, indigestible dextrin, etc.) are approved in this category.
4. Reduction of disease risk FOSHU
   Reduction of disease risk claim is permitted when reduction of disease risk is clinically and nutritionally established in an ingredient. Currently, only calcium and folic acid are approved in this category.

At the end of 2009, a total of more than 880 food items were approved as FOSHU. Among them, there were only 1 item of qualified FOSHU, 28 items of standardized FOSHU, and 10 items of reduction of disease risk FOSHU. One of the reasons is that these FOSHU classes are still in the early days of establishment. Like FNFC, approval for standardized FOSHU is relatively easy to obtain. Therefore, an increase in foods of this class is anticipated in the near future.

5.3.2 Probiotics and FOSHU Products

FOSHU products contain functional ingredients. Basically, the products are approved as FOSHU because they contain ingredients with beneficial functions for health. Since the regulation started, many FOSHU products have been approved. This means that various functional ingredients have been developed and their functions verified through many investigations in accordance with regulations. Therefore, the FOSHU system has triggered the production of many functional ingredients such as probiotics.

Foods associated with gastrointestinal condition occupy more than 50% of the FOSHU market and there are many products of this category in the Japanese market. Currently, three kinds of ingredients related to gastrointestinal condition are used in FOSHU products; namely, probiotics, oligosaccharides, and dietary fiber. As shown in Fig. 6, probiotics are the dominant ingredient in this category. Once the products are approved as FOSHU, claims can be made on the label, such as "Since this product contains live *Bifidobacterium longum* BB536, this product increases intestinal bifidobacteria and improves the intestinal condition."

Probiotics used in FOSHU products can be divided broadly into lactic acid bacteria and bifidobacteria. Both bacteria are known to be effective in improving gastrointestinal condition or constipation through increasing the intestinal bifidobacterial counts, repressing harmful bacteria, and reducing harmful substances in the intestine (Yaeshima et al. 1997). Under the FOSHU system, each strain of probiotic is approved independently after the effectiveness and safety of each strain are reviewed.

The principal probiotic strains approved for FOSHU products are as follows:

1. Bifidobacteria
   B. longum BB536, *B. breve* Yakult, *B. lactis* Bb-12, *B. lactis* FK120, *B. lactis* LKM512
2. Lactic acid bacteria
   Lactobacillus bulgaricus 2038, *Streptococcus thermophilus* 1131, *Lactobacillus* GG, *L. casei* YIT 9029 (Shirota strain), Lactic acid bacteria LC1, *L. acidophilus*

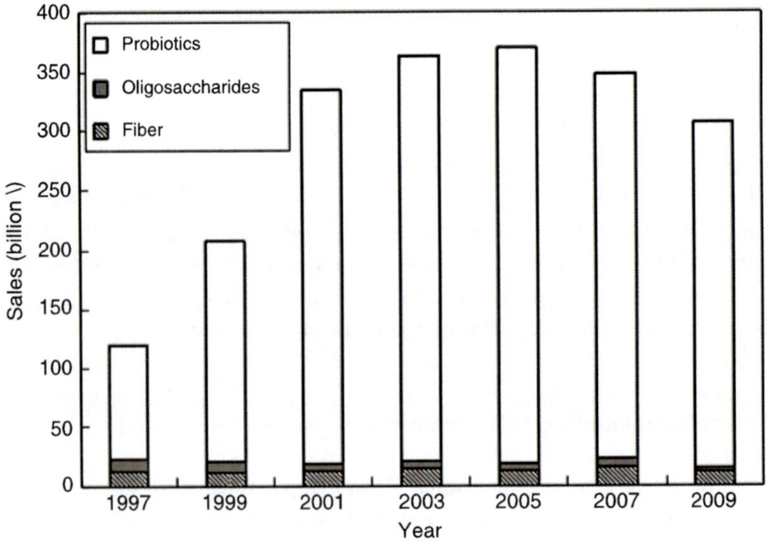

Fig. 6 Japanese market of FOSHU related to gastrointestinal condition

CK92, *Lactobacillus helveticus* CK60, *Lactobacillus gasseri* SP, *Bifidobacterium* SP, *L. casei* NY1301
3. Others
Bacillus subtilis K-2

Most of the FOSHU products using the above probiotics are yogurt, fermented milk, or milk beverages in the Japanese market. Various studies on probiotics have been conducted by many researchers in Japan during the last few decades, and many probiotic products have been developed since the 1970s. The share of probiotics in the FOSHU market is large and the functions and effects of probiotics are well recognized by Japanese consumers. Furthermore, many manufacturers continue to investigate the original probiotics to explore new functions. Therefore, Japanese probiotics will continue to develop in the FOSHU market and expand globally in the future.

As of September 2010, no probiotic food has been approved for function or health claim in USA or EU. Many manufacturers and consumers desire a good labeling system in the food category. Probiotics are one of the most promising ingredients to confer health benefits and these effects have to be communicated clearly to consumers. By officially indicating the healthy benefits on the label, this guarantees that the effects are proven scientifically and statistically. However, the confirmation of safety may be more important. In Japan, proof of both effectiveness and safety is necessary for FOSHU approval. As safety problem has been reported for some probiotics and not all probiotics are the same, manufacturers of probiotics have to demonstrate not only the beneficial effect but also the safety at strain level to apply for health claim.

6 Conclusions

In this chapter, probiotic safety, probiotic quality, and the related regulation typified by FOSHU were described. In the discussion of probiotics, although many investigators and consumers have focused on the beneficial effects, the safety aspect is an important scientific aspect that has to be emphasized. Since numerous people in the world including infants, elderly, and patients with diseases are taking probiotics and anticipating the beneficial health effects, safety should be addressed even more than the beneficial effects. In particular, bacterial translocation inducing sepsis and horizontal gene transfer of antibiotic resistance genes harbored in probiotics are serious problems. Reports have shown that some probiotic bacteria have a potential to cause infection in patients with serious diseases. Continuous surveillance even after launching is necessary to prevent health hazards through bacterial translocation. Also, horizontal gene transfer of antibiotic resistance is becoming a serious issue. Since the emergence of pathogenic bacteria with antibiotic resistance genes such as *NDM-1* may be a fatal problem, probiotics harboring acquired antibiotic resistance gene should not be used as food. To prevent further spread of antibiotic resistance genes, review of current probiotics as well as antibiotics use is required. On the other hand, the quality of probiotics products for consumption by vulnerable persons such as infants should be discussed. It is very important to control contamination by pathogenic bacteria such as *Salmonella* and *C. sakazakii* during the production process by introducing hygienic strategies such as HACCP. Furthermore, there are some differences in food regulations including labeling system between countries or regions. Especially, allergen can cause severe health problem. Given the different allergen lists regulated in various countries, manufacturers of probiotics targeting worldwide markets should consider the regulations on a region or country basis. Another aspect that has attracted increasing attention worldwide is the health claim system for food. While no probiotic products with health claim have been approved in EU and USA, Japan has established the FHC system including FOSHU. Some probiotic products have already been approved as FOSHU. In the near future, similar systems or regulations will be developed worldwide, because many probiotics are being investigated energetically for various health effects. In future development of health claim systems also, it is necessary not to focus on efficacy only, but also to consider safety aspects including bacterial translocation and acquired antibiotic resistance gene.

References

Abe F, Miyauchi H, Uchijima A, Yaeshima T, Iwatsuki K (2009a) Stability of bifidobacteria in powdered formula. Int J Food Sci Technol 44:718–724

Abe F, Yaeshima T, Iwatsuki K (2009b) Safety evaluation of two probiotic bifidobacterial strains, Bifidobacterium breve M-16V and Bifidobacterium infantis M-63, by oral toxicity test using rats. Biosci Microflora 28(1):7–15

Abe F, Tomita S, Yaeshima T, Iwatsuki K (2009c) Effect of production conditions on the stability of a human bifidobacterial species *Bifidobacterium longum* in yogurt. Lett Appl Microbiol 49:715–720

Abe F, Miyauchi H, Uchijima A, Yaeshima T, Iwatsuki K (2009d) Effects of storage temperature and water activity on the survival of bifidobacteria in powder form. Int Dairy Technol 62(2):234–239

Abe F, Muto M, Yaeshima T, Iwatsuki K, Aihara H, Ohashi Y, Fujisawa T (2010) Safety evaluation of probiotic bifidobacteria by analysis of mucin degradation activity and translocation ability. Anaerobe 16(2):131–136

Aires J, Doucet-Populaire F, Butel MK (2007) Tetracycline resistance mediated by tet(W), tet(M), and tet(O) genes of *Bifidobacterium* isolates from humans. Appl Environ Microbiol 73(8):2751–2754

Ammor MS, Florez AB, Alvarez-Martin P, Margolles A, Mayo B (2008a) Analysis of tetracycline resistance tet(W) genes and their flanking sequences in intestinal *Bifidobacterium* species. J Antimicrob Chemother 62:688–693

Ammor MS, Florez AB, Alvarez-Martin P, Margolles A, Mayo B (2008b) Molecular characterization of intrinsic and acquired antibiotic resistance in lactic acid bacteria and bifidobacteria. J Mol Microbiol Biotechnol 14:6–15

Besselink MG, van Santvoort HC, Buskens E, Boermeester MA, van Goor H, Timmerman HM, Nieuwenhuijs VB, Bollen TL, van Ramshorst B, Witteman BJ, Rosman C, Ploeg RJ,

Brink MA, Schaapherder AF, Dejong CH, Wahab PJ, van Laarhoven CJ, van der Harst E, van Eijck CH, Cuesta MA, Akkermans LM, Gooszen HG, Dutch Acute Pancreatitis Study Group (2008) Probiotic prophylaxis in predicted severe acute pancreatitis: a randomised, double-blind, placebo-controlled trial. Lancet 371(9613):651–659

Cahill SM, Wachsmuth IK, Costarrica Mde L, Ben Embarek PK (2008) Powdered infant formula as a source of Salmonella infection in infants. Clin Infect Dis 46(2):268–273

Center for Disease Control and Prevention (2010) Detection of Enterobacteriaceae isolates carrying metallo-beta-lactamase-United States, 2010. MMWR Morb Mortal Wkly Rep 59(24):750

Champagne CP, Gardner NJ, Roy D (2005) Challenges in the addition of probiotic cultures to foods. Crit Rev Food Sci Nutr 45:61–84

Charalampopoulos D, Wang R, Pandiella SS, Webb C (2002) Application of cereals and cereal components in functional foods: a review. Int J Food Microbiol 15:131–141

Chen I, Dubnau D (2004) DNA uptake during bacterial transformation. Nat Rev Microbiol 2(3):241–249

Colombel JF, Cortot A, Neut C, Tomond C (1987) Yoghurt with Bifidobacterium longum reduces erythromycin-induced gastrointestinal effects. Lancet 330(8549):43

Courvalin P (2006) Antibiotic resistance: the pros and cons of probiotics. Dig Liver Dis 38:S261–S265

De Groote MA, Frank ND, Dowell E, Glode MP, Pace NR (2005) *Lactobacillus rhamnosus* GG bacteremia associated with probiotic use in a child with short gut syndrome. Pediatr Infect Dis J 24:278–280

EFSA (2004) Opinion of the Scientific Panel on Biological Hazards on the request from the Commission related to the microbiological risks in infant formulae and follow-on formulae. EFSA J 113:1–35

EFSA (2008) Technical guidance – update of the criteria used in the assessment of bacterial resistance to antibiotics of human or veterinary importance. EFSA J 732:1–15

Fujii T, Ohtsuka Y, Lee T, Kudo T, Shoji H, Sato H, Nagara S, Shimizu T, Yamashiro Y (2006) *Bifidobacterium breve* enhances transforming growth factor beta1 signaling by regulating Smad7 expression in preterm infants. J Pediatr Gastroenterol Nutr 43(1):83–88

Gork AS, Usui N, Ceriati E, Drongowski RA, Epstein MD, Coran AG, Harmon CM (1999) The effect of mucin on bacterial translocation in I-407 fetal and Caco-2 adult enterocyte cultured cell lines. Pediatr Surg Int 15:155–159

Gueimonde M, Delgado S, Mayo B, Ruas-Madiedo P, Margolles A, De los Reyes-Gavilan CG (2004) Viability and diversity of probiotic *Lactobacillus* and *Bifidobacterium* populations included in commercial fermented milks. Food Res Inter 37:839–850

Guenther K, Straube E, Pfister W, Guenther A, Huebler A (2010) Sever sepsis after probiotic treatment with Escherichia coli NISSLE 1917. Pediatr Infect Dis J 29(2):188–189

Gurimonde M, Florez AB, van Hoek AHAM, Stuer-Lauridsen B, Stroman P, de Los Reyes-Gacilan CG, Margolles A (2010) Genetic basis of tetracycline resistance in *Bifidobacterium animalis* subsp. *lactis*. Appl Environ Microbiol 76(10):3364–3369

Havarstein LS (2010) Increasing competence in the genus *Streptococcus*. Mol Microbiol 78(3):541–544

Health Protection Agency (2009) Multi-resistant hospital bacteria linked to India and Pakistan. Health Prot Rep 3(26):3–4

Heller KJ (2001) Probiotic bacteria in fermented foods: product characteristics and starter organisms. Am J Clin Nutr 73:374S–379S

Holzapferl WH, Haberer P, Geisen R, Bjorkroth J, Schillin U (2001) Taxonomy and important features of probiotic microorganisms in food and nutrition. Am J Clin Nutr 73:365S–373S

Jacobsen L, Wilcks A, Hammer K, Hys G, Gevers D, Andersen SR (2007) Horizontal transfer of tet(M) and erm(B) resistance plasmids from food strains of *Lactobacillus plantarum to Enterococcus faecalis* JH2-2 in the gastrointestinal tract of gnotobiotic rats. FEMS Microbiol Ecol 59:158–166

Jayamanne VS, Adams MR (2006) Determination of survival, identity and stress resistance of probiotic bifidobacteria in bio-yoghurts. Lett Appl Microbiol 42:189–194

Johnsoborg O, Havarstein LS (2009) Regulation of natural genetic transformation and acquisition of transforming DNA in *Streptococcus pneumoniae*. FEMS Microbiol Rev 33(3):627–642

Joint FAO/WHO Working Group on Drafting Guidelines for the Evaluation of Probiotics in Food (2002) Guidelines for the evaluation of probiotics in food, London, Ontario, Canada, April 30 and May 1, 2002. Available at: http://www.who.int/foodsafety/publications/fs_management/probiotics2/en/index.html

Kabeir BM, Yazid AM, Stephenie W, Hakim MN, Anas OM, Shuhaimi M (2008) Safety evaluation of *Bifidobacterium pseudocatenulatum* G4 as assessed in BALB/c mice. Lett Appl Microbiol 46:32–37

Kageyama T, Tomoda T, Nakano Y (1984) The effect of *Bifidobacterium* administration in patients with leukemia. Bifidobacteria Microflora 3(1):29–33

Kandhai MC, Reij MW, Gorris LG, Guillanume-Gentill O, van Schothors M (2004) Occurrence of *Enterobacter sakazakii* in food production environments and households. Lancet 363 (9402):39–40

Kastner S, Perreten V, Bleuler H, Hugenshmidt G, Lacroix C, Meile L (2006) Antibiotic susceptibility patterns and resistance genes of starter cultures and probiotic bacteria used in food. Syst Appl Microbiol 29(2):145–155

Kazimierczak KA, Flint HJ, Scott KP (2006) Comparative analysis of sequences flanking tet(W) resistance genes in multiple species of gut bacteria. Antimicrob Agents Chemother 50(8):2632–2639

Klare I, Konstabel C, Werner G, Hus G, Vankerckhoven V, Kahlmeter G, Hildebrandt B, Muller-Bertling S, Witte W, Goossens H (2007) Antimicrobial susceptibilities of *Lactobacillus*, *Pediococcus* and *Lactococcus* human isolates and cultures intended for probiotic or nutritional use. J Antimicrob Chemother 59:900–912

Kumarasamy KK, Toleman MA, Waish TR, Bagaria J, Butt F, Balakrishnan R, Chaudhary U, Doumith M, Giske CG, Irfan S, Krishnan P, Kumar AV, Maharjan S, Mushtag S, Noorie T, Paterson DL, Pearson A, Perry C, Pike R, Rao B, Ray U, Sarma JB, Sharma M, Sheridan R, Thirunarayan MA, Turton J, Upadhyay S, Warner M, Welfare W, Livemore DM, Woodford N (2010) Emergence of a new antibiotic resistance mechanism in India, Pakistan and the UK: a molecular, biological, and epidemiological study. Lancet Infect Dis 10(9):597–602

Kunz AN, Noel JM, Firchok MP (2004) Two case of *Lactobacillus* bacteremia during probiotic treatment of short gut syndrome. J Pediatr Gastroenterol Nutr 38:457–458

Land MH, Rouster-Stevens K, Woods CHR, Cannon ML, Cnota J, Shetty AK (2005) *Lactobacillus* sepsis associated with probiotic therapy. Pediatrics 115:178–181

Ledoux D, Labombardi VJ, Karter D (2006) Lactobacillus acidophilus bacteraemia after use of a probiotic in a patient with AIDS and Hodgkin's disease. Int J STD AIDS 17(4):280–282

Levy S, Marshall B (2004) Antibacterial resistance worldwide: causes, challenges and responses. Nat Med 10(12):S122–S129

Liong MT (2008) Safety of probiotics: translocation and infection. Nutr Rev 66:192–202

Liu B, Pop M (2009) ADRB-Antibiotic resistance genes database. Nucleic Acids Res 37:D443–D447

Masco L, van Hoorde K, De Brandt E, Swings J, Hus G (2006) Antimicrobial susceptibility of Bifidobacterium strains from humans, animals and probiotic products. J Antimicrob Chemother 58:85–94

Mathur S, Singh R (2005) Antibiotic resistance in food lactic acid bacteria – a review. Int J Food Microbiol 105:281–295

Matsumoto T, Ishikawa H, Tateda K, Yaeshima T, Ishibashi N, Yamaguchi K (2008) Oral administration of *Bifidobacterium longum* prevents gut-derived *Pseudomonas aeruginosa* sepsis in mice. J Appl Microbiol 104:672–680

Mattila-Sandholm T, Myllärinen P, Crittenden R, Mogensen G, Fondén R, Saarela M (2002) Technological challenges for future probiotic foods. Int Dairy J 12:173–182

Meibom KL, Blokesch M, Dolganov NA, Wu CY, Schoolnik GK (2005) Chitin induces natural competence in *Vibrio cholerae*. Science 310:1824–1827

Muto M, Abe F, Yaeshima T, Iwatsuki K (2010) Effect of enumeration method on *Bifidobacterium* cell counts in commercial powder products. Biosci Microflora 29(3):143–148

Noriega L, de los Reyes-Gavilan CG, Margolles A (2005) Acquisition of bile salt resistance promotes antibiotic susceptibility changes in *Bifidobacterium*. J Food Prot 68(9): 1916–1919

Paturi G, Phillips M, Kailasapathy K (2008) Effect of probiotic strains *Lactobacillus acidophilus* LAFTI L10 and *Lactobacillus paracasei* LAFTI L26 on systemic immune functions and bacterial translocation in mice. J Food Prot 71(4):796–801

Philips M, Kaliasapathy K, Tran L (2006) Viability of commercial probiotic cultures (*L. acidophilus, Bifidobacterium* sp., *L. casei, L. paracasei* and *L. rhamnosus*) in cheddar cheese. Int J Food Microbiol 108:276–280

Rasic JL, Kurmann JA (1983) Bifidobacteria and their role. Birkhauser AG Graphisches Untermehmen, Basel, Switzerland

Reddy BS, Rivenson A (1993) Inhibitory effect of *Bifidobacterium longum* on colon, mammary, and liver carcinogenesis induced by 2-amino-3-methylimidazo [4,5-f]quinoline, a food mutagen. Cancer Res 53:3914–3918

Reiji MW, Jongenburger I, Grogka E, Gorris LG, Zwietering MH (2009) Perspective on the risk to infants in the Netherlands associated with *Cronobacter* spp. occurring in powdered infant formula. Int J Food Microbiol 136(2):232–237

Ruas-Madiedo P, Gueimonde M, Fernadez-Garcia M, de los Reyes-Gavilan CG, Margolles A (2008) Mucin degradation by *Bifidobacterium* strains isolated from the human intestinal microbiota. Appl Environ Microbiol 74:1936–1940

Russo A, Angeletti S, Lorino G, Venditti C, Falcone M, Dicuonzo G, Venditti M (2010) A case of *Lactobacillus casei* bacteraemia associated with aortic dissection: is there a link? New Microbiol 33(2):175–178

Snydman DR (2008) The safety of probiotics. Clint infec Dis 46(Supple 2):S104–S111

Spigaglia P, Barbanti F, Mastrantonio P (2008) Tetracycline resistance gene tet(W) in the pathogenic bacterium *Clostridium difficile*. Antimicrob Agents Chemother 52(2):770–773

Tang AL, Shah NP, Wilcox G, Walker KZ, Stojanovska L (2007) Fermentation of calcium-fortified soymilk with lactobacillus: effects on calcium solubility, isoflavone conversion, and production of organic acids. J Food Sci 72(9):M431–M436

Taniuchi S, Hattori K, Yamamoto A, Sasai M, Hatano Y, Kojima T, Kobayashi Y, Iwamoto H, Yaeshima T (2005) Administration of *Bifidobacterium* to infants with atopic dermatitis: changes in fecal microflora and clinical symptoms. J Appl Res 5:387–396

Teuber M (1999) Spread of antibiotic resistance with food-borne pathogens. Cell Mol Life Sci 56:755–763

Tomasz A, Hotchkiss RD (1964) Regulation of the transformability of pneumococcal cultures by macromolecular cell products. Proc Natl Acad Sci USA 51(3):480–487

van der Aa LB, Heymans HS, van Aalderen WM, Sillevis Smitt JH, Knol J, Ben Amor K, Goossens DA, Spikkelman AB, the Synbad study group (2010) Effect of a new synbiotic mixture on atopic dermatitis in infants: a randomized-controlled trial. Clin Exp Allergy 40:795–804

Vinderola CG, Prosello W, Ghiberto TD, Reinheimer JA (2000) Viability of probiotic (*Bifidobacterium, Lactobacillus acidophilus* and *Lactobacillus casei*) and nonprobiotic microflora in Argentinian Fresco cheese. J Dairy Sci 83:1905–1911

Whelan K, Myers CE (2010) Safety of probiotics in patients receiving nutritional support: a systematic review of case reports, randomized controlled trials, and nonrandomized trials. Am J Clin Nutr 91:687–703

Xiao JZ, Kondo S, Yanagisawa N, Takahashi N, Odamaki T, Iwabuchi N, Miyaji K, Iwatsuki K, Togashi H, Enomoto K, Enomoto T (2006) Probiotics in the treatment of Japanese cedar pollinosis: a double-blind placebo-controlled trial. Clin Exp Allergy 36:1425–1435

Xiao JZ, Takahashi S, Odamaki T, Yaeshima T, Iwatsuki K (2010) Antibiotic susceptibility of bifidobacterial strains distributed in the Japanese market. Biosci Biotechnol Biochem 74(2):336–342

Yaeshima T, Takahashi S, Matsumoto N, Ishibashi N, Hayasawa H, Iino H (1997) Effect of yogurt containing *Bifidobacterium longum* BB536 on the intestinal environment, fecal characteristics and defecation frequency: a comparison with standard yogurt. Biosci Microflora 16(2):73–77

Yamazaki S, Machii K, Tsuyuki S, Momose H, Kawashima T, Ueda K (1985) Immunological responses to monoassociated *Bifidobacterium longum* and their relation to prevention of bacterial invasion. Immunology 56:43–50

Yamazaki S, Tsuyuki S, Tsuyuki S, Akashiba H, Kamimura H, Kimura M, Kawashima T, Ueda K (1991) Immune response of *Bifidobacterium*-monoassociated mice. Bifidobacteria Microflora 10:19–31

Yazawa K, Fujimori M, Amano J, Kano Y, Taniguchi S (2000) *Bifidobacterium longum* as a delivery system for cancer gene therapy: selective localization and growth in hypoxic tumors. Cancer Gene Ther 7(2):269–274

Zhou JS, Gopal PK, Gill HS (2001) Potential probiotic lactic acid bacteria *Lactobacillus rhamnosus* (HN001), *Lactobacillus acidophilus* (HN017) and *Bifidobacterium lactis* (HN019) do not degrade gastric mucin in vitro. Int J Food Micro 63:81–90

Current Market Trends and Future Directions

Caroline T. Yamaguishi, Michele R. Spier, Juliano De Dea Lindner, Vanete T. Soccol, and Carlos R. Soccol

Contents

1 Introduction 300
2 Economic Relevance of the Probiotic Products 300
3 Current-Commercially Available Probiotic Products 301
  3.1 Liquid Preparations 302
  3.2 Solid Probiotic Preparations 302
  3.3 Products for Feed 306
4 Ongoing Innovations: Improvement of Probiotic Viability and Bioactivity 307
  4.1 Selection of Suitable Product Carrier 308
  4.2 Improvement in Culture Stability 308
  4.3 Microencapsulation 310
5 Future Larger Markets: Infants and the Elderly 313
  5.1 Probiotic for Infants and Children 313
  5.2 Probiotic for the Elderly 315
  5.3 Safety Aspects 316
6 Concluding Remarks 317
References 317

Abstract Probiotic products have been used worldwide and they are increasingly gaining popularity. Current trends in the consumption of probiotics are associated with increased levels of health-consciousness, and the availability of probiotics in the form of dietary supplements. Several companies have profited by marketing these products in different forms, with different purposes, and with recommendation

C.T. Yamaguishi • M.R. Spier • V.T. Soccol • C.R. Soccol (✉)
Bioprocess Engineering and Biotechnology Department, Federal University of Paraná (UFPR), 81531-990 Curitiba, Paraná, Brazil
e-mail: soccol@ufpr.br

J. De Dea Lindner
Food Engineering Department, Santa Catarina State University (UDESC), 89870-000 Pinhalzinho, Santa Catarina, Brazil

M.-T. Liong (ed.), *Probiotics*, Microbiology Monographs 21,
DOI 10.1007/978-3-642-20838-6_12,

for all ages. Important aspects in maintaining the viability and bioactivity of probiotic strains during processing and storage are also discussed in this chapter. The probiotic consumption by infants and the elderly has been supported by scientific evidences and represents a potential niche markets that are in developing and has led to the growth segment of probiotic products.

1 Introduction

Probiotics have achieved a prominent position in the global food market. The consumption of functional foods, such as probiotics, has heightened healthcare awareness and has led to better economic development of this market segment. Functional foods provide nutrients and promote health benefits; probiotics can be consumed in the daily diet and used as preventive medicine against diseases affecting all age groups.

Among the countries that have shown growth in the probiotic market, Europe represents the largest and fastest growing market, followed by Japan (Global Industry Analystic 2010). Currently, there is a wide range of probiotic products offered by companies such as BioGaia Biologics AB, Christian Hansen A/S, ConAgra Functional Foods, Danisco, Groupe Danone, Lifeway Foods, Inc., Nestle S.A., Seven Seas Ltd., Valio and Yakult Honsha Co. Ltd.

This market segment is established mainly in dairy products, but other products have shown relative importance (e.g., dietary supplements, probiotic baked products, probiotic ice creams, probiotic chocolates) leading to a diversification and popularisation of these products. The main claims of the probiotic products are based on several studies that support their potential use in the prevention of inflammatory diseases, allergies, cholesterol reduction, gastrointestinal disorders and some types of cancer.

2 Economic Relevance of the Probiotic Products

The main factors that promote growth in the global market of probiotics are increasing levels of health-consciousness and the availability of probiotics in the form of dietary supplements. The market of probiotic products generated US$ 15.9 billion in 2008 and is forecast to reach US$ 28.8 billion in 2015. The major factors that have facilitated market growth of probiotics are the appropriate components for formulation and the scientific knowledge of the provided benefits (Markets and Markets 2009; Global Industry Analystic 2010).

Probiotics are already used extensively in Europe and in some parts of Asia, and they are gaining in popularity worldwide, particularly in the United States. Europe represents the largest and fastest growing market for them. In addition,

Europe, Germany and the United Kingdom represent approximately 45% of the total market. Japan is the second largest market and is expected to grow at more moderate rates (Global Industry Analystic 2010).

More than 500 probiotic food and beverage products have been introduced in the past decade. These products have been commercialised on different levels, depending on their relevance for health. For instance, probiotic chocolates have garnered more market share than probiotic cheese and butter because probiotic chocolates offer more health benefits without fortification. The leading developers and suppliers of probiotic strains include Danisco, Morinaga and BioGaia. These products are used by companies such as Nestle and Attune. The United States market is growing rapidly due to the general affinity of the population towards probiotic dietary supplements. Probiotic products may include functional foods and beverages, dietary supplements, specialty nutrients and animal feed products. The most common applications are in regular consumption, probiotic therapy and disease prevention.

The main participants who leads the market for probiotics include BioGaia Biologics AB, Chr. Hansen A/S, ConAgra Functional Foods, Danisco, Groupe Danone, Institut Rosell, Lifeway Foods, Inc., Natren, Inc., Nestle S.A., Seven Seas Ltd., Stonyfield Farm and Yakult Honsha Co. Ltd. There is a lot companies which commercialises probiotic products such as Alpharma, Alltech Biotechnology, Amerifit Brands, Arla Foods, Attune Food Inc., Biogaia AB, Bomac Vets Plus, Calpis Company, Cargill, CD Pharma, China-Biotics Inc, CHR Hansen, Danisco, Danone, DSM Food Specialities, Ganeden, Garden of Life, General Mills, Jamieson Laboratories, Arrow Formulas, Kashi Company, Kirkman, Kraft Foods Inc., Lallemand Inc., Life Way Food Inc., Morinaga Milk Industry Co. Ltd, Mother Dairy, Muller Dairy Ltd, Natren, Nebraska Cultures, Nestle, Now Foods, Nutraceutix, Inc, Probi AB, Skanemejerier, Stonyfield Farm, Valio, Well´s Dairy Inc., Wild Wood, Yakult Honsha, Yeo Valey, Yo Cream International Inc. The more recent idea has been based on fortifying fermented desserts with proteins, minerals and probiotic cultures (Markets and Markets 2009; Global Industry Analystic 2010).

3 Current-Commercially Available Probiotic Products

There are new probiotic products available in the market which can present several applications. Industries may use a mixture of strains in several products such as cereal bars, breads, dairy products, Asian fermented foods and some juices and soy beverages. They are customisable and stable, and the benefits have been proven.

Probiotics are available in foods, dietary supplements and in some other forms (e.g., capsules, tablets, pills and powders). The bacteria may have been present originally or added during preparation. Commercial probiotic products are available in liquid preparations and solid formulations (freeze-dried, spray-dried and pills). The market also has offered consumer probiotic products for animal feed.

3.1 Liquid Preparations

Probiotic preparation in liquid form is the most common product that is commercially available to consumers because these products originated and developed from yoghurt formulations. Several liquid formulations, commercial names, companies, probiotic cultures, compositions and applications are presented in Table 1.

3.2 Solid Probiotic Preparations

Different products are commercially available in solid form. The manufacturing of probiotics varies with purpose and product. The most commonly found probiotics are freeze-dried and spray-dried powders and pills, which may or may not be encapsulated. Still, probiotic strains can be incorporated into food matrices (Table 2) promoting the enrichment of the product in nutritional and market value.

3.2.1 Freeze-Dried and Spray-Dried Preparations

Probiotic preparations that are used for food applications are commonly supplied in dried form or are frozen as either freeze-dried (Table 3) or spray-dried powders. Materials used for coating are water-soluble polymers for spray-drying. Besides freeze-dried or spray-dried the material for coating may be waxes, fatty acids, water-soluble, water-insoluble polymers and monomers.

These processes have a high yield for application in several areas, but the exposure of the microorganism to high temperatures during spray-drying can be detrimental to the integrity of the live bacterial cells. The effect of spray-drying on the cell membrane can lead to increased cell permeability, which may result in leakage of intracellular components from the cell into the surrounding environment. The cytoplasmic membrane is one of the most susceptible sites to the stresses associated with spray-drying in bacterial cells; the cell wall, DNA and RNA are also known to be affected, leading to loss of metabolic activity (Teixeira et al. 1997).

To solve this problem a variety of protectants have been added to the drying media before freeze-drying or spray-drying to ensure the viability of probiotics during dehydration, including skim milk powder, whey protein, trehalose, glycerol, betaine, adonitol, sucrose, glucose, lactose and polymers such as dextran and polyethylene glycol (Hubalek 2003; Morgan et al. 2006). Growth-promoting factors including various probiotic/prebiotic combinations and granular starch have been shown to improve culture viability during drying and storage. Recently, the incorporation of soluble gum in a milk-based medium prior to spray-drying has been studied with probiotic cultures.

Table 1 Probiotic liquid products that are commercially available, probiotic cultures, composition and its applications

Product (company)	Probiotic culture	Composition	Application
Yakult® (Yakult Honsha Co. Ltd.)	*Lactobacillus casei* Shirota 10^9 CFU/bottle,	Water, sugar, skim milk powder, glucose, natural and artificial flavours	Probiotic beverage, intestinal flora reposition, improve digestion
Activia® Yogurt (Danon)	*Bifidus regularis* (*Bifidobacterium animalis*), *L. bulgaricus*, *S. thermophilus*	Different variation (strawberry, natural, peaches, vanilla)	Help regulate digestive system
Gefilus® Valio Gefilus® and Gefilus® (FinFood)	*Lactobacillus* GG® (*L. rhamnosus* GG, ATTCC 53103 or LGG)	Presents in the form of milk, buttermilk, yoghurts, fermented milks, milks, daily dose drinks, juices, berry soups, cheese and capsules	Improve digestion
SCD Essential Probiotics™ (Sustainable Community Development)	*Bacillus subtilis var natto*, *Bifidobacterium animalis*, *B. bifidum; B. longum*, *L. acidophilus* *L. bulgaricus*, *L. casei*, *L. delbrueckii*, *L. fermentum*, *L. plantarum*, *Lactococcus lactis*, *L. lactis subsp. Diacetylactis*, *S. cerevisiae*, *Streptococcus thermophilus*	Purified water, sugar cane molasses, rock salt, sea salt, blueberry, cherry and pomegranate juice concentrates, Xtra Immunity™ brown rice extract, SCD Probiotics cultures	Regulate digestion and supporting personal wellness
SVELTY® Gastro Protect (Nestle)	*Lactobacillus johnsonii* La1	A fermented drink milk, flavour, sugars	Controls *H. pylori* infection and stomach discomfort
LC1 Yoghurt® (Nestle)	*L. johnsonii* La1 and acidophilus bacteria	A probiotic yoghurt, fermented milk, flavours, sugars	Regulates digestion, protection against pathogens
Actimel® (Danon)	*L. casei* (*Defensis*)	Milk, sugar, flavours	Protection against pathogen

Available at online database of each manufacturer

Table 2 Probiotic solid products that are commercially available, probiotic cultures, composition and its applications

Product	Probiotic culture	Composition
Probiotic Chewy Cereal Bars	Ganeden BC30	5 g fibre, 2 g protein, prebiotics, omega-3 fatty acids, total 110 cal
Chocolate Probiotic Bars Chocolate Crisp® ATTUNE	*Lactobacillus acidophilus*, *L. casei*, *Bifidobacterium lactis*	Milk Chocolate Coating (evaporated cane juice, chocolate, cocoa butter, inulin, non-fat milk, calcium carbonate, anhydrous milk fat, soy lecithin, vanilla), organic brown rice crisps (organic brown rice flour, organic molasses, calcium carbonate)
XoBiotic™ squares (MXI Corp.™)	*L. helveticus* R0052 and *B. longum* R0175	Dark Chocolate (unsweetened chocolate, sugar, cocoa powder, lecithin, vanilla extract), antioxidant blend (natural cocoa, açaí, blueberry powders) and probiotics
Heini's Yogurt Cultured Cheese (Bunker Hill Cheese Company)	*L. acidophilus*, *L. casei*, *B. lactis*	Milk, yoghurt cultures, coagulants, probiotic cultures and salt

Database from companies' websites

Table 3 Freeze-dried products

Product	Probiotic culture	Application
FloraFIT® (Danisco A/S)	*Bifidobacterium breve* Bb-03, *B. lactis* Bi-07, *B. lactis* Bl-04, *B. longum* Bl-05, *Lactobacillus acidophilus* La-14, *Lactobacillus bulgaricus* Lb-64, *L. Brevis* Lbr-35, *L. casei* Lc-11, *Lactococcus lactis* Ll-23, *L. plantarum* Lp-115, *L. paracasei* Lpc-37, *L. rhamnosus* Lr-32, *L. salivarius* Ls-33, *Streptococcus thermophilus* St-21	Food and beverages
HOWARU® Premium Probiotics (Danisco A/S)	*L. acidophilus* NCFM™	A probiotic product that can be applied in beverages, confectionery, dairy, dietary supplements and frozen desserts
Yógourmet Products (Lyo-San, Inc.)	*L. casei*, *B. Bifidus*, *L. acidophilus*	Starters for yoghurt manufacture
Biorich® (Chr. Hansen A/S)	*L. acidophilus* LA-5 and *Bifidobacterium* BB-12	Starters for yoghurt manufacture

Database from companies' websites

3.2.2 Pills

Some probiotic pills that are commercially available (Table 4) contain probiotic strains and prebiotics, which help with the growth and colonisation of good bacteria within the intestinal flora.

Table 4 Probiotic pills commercially available

Product	Probiotic culture	Composition	Benefits
Nutraelle® DigestiveCare	*Lactobacillus acidophilus*: 5×10^9 CFU/pill. *Bifidobacterium bifidum*: 5×10^9 CFU/pill	Prebiotic (inulin Powder (Chicory Root) – Prebiotic	Body benefits
Align®	10^9 live bacteria per capsule of *Bifidobacterium infantis* 35624 found exclusively in Align	Only *B. Infantis*	Help in the restoration of the digestive system
Florastor®	*Saccharomyces boulardii* lyo (250 mg or 5 billion live cells)	32.5 mg of lactose monohydrate (not derived from milk), 2.85 mg of Mg distearate and titanium dioxide E171 used to distribute the yeast evenly	Promote intestinal health and maintain normal bowel function
Culturelle®	*Lactobacillus rhamnosus* GG (10^9 billion cells)	Inulin	Promote good digestion, support immune system and restore the natural intestinal balance
ReNew Life Ultimate Flora Adult® (Wellness Nutrition)	*B. bifidum* 7.5×10^9; *L. acidophilus* 3×10^9; *L. rhamnosus* 1.35×10^9; *B. breve* 75×10^7; *B. longum* 75×10^7; *L. casei* 75×10^7; *L. plantarum* 45×10^7; *L. lactis* 3×10^8; *L. bulgaricus* 75×10^6; *L. salivarus* 15×10^6		Provides probiotic support for the small intestine, which produces enzymes to help digest foods
Ultimate Flora Women's Care® 25 Billion (Renew Life Formula's, Inc.)	*L. rhamnosus* 8.5×10^9, *L. casei* 5.87×10^9, *L. plantarum* 3.75×10^9, *L. acidophilus* 2.5×10^9, *L. brevis* 1.25×10^9, *B. lactis* 1.25×10^9, *B. longum* 1.25×10^9, *L. paracasei* 25×10^7, *L. salivarius* 25×10^7, *L. bulgaricus* 125×10^6, FOS (fructooligosaccharide), Total Probiotic Cultures 125×10^9	FOS, vegetable capsule (vegetable fibre, water)	Application: restores a healthy yeast balance; promotes vaginal & urinary tract health; helps replenish healthy vaginal flora levels; strengthens natural defences

(continued)

Table 4 (continued)

Product	Probiotic culture	Composition	Benefits
Rapid Yeasts Relief® (part I and II) (Renew Life)	Part I (*L. acidophilus*, *L. rhamnosus*, *L. reuteri*, *L. paracasei*, *L. salivarius*, FOS) Total probiotic cultures 20 × 10^9 CFU/capsule	Part II (Enzyme blend containing protease, papain, cellulase, hemicellulase, lysozyme, lactoferrin, amylase, invertase, beta glucanase, malt diastase, glucoamylase and FOS	Against yeast infection
Flora Bear® for Kids Renew Life	Probiotic blend (*L. acidophilus*, *B. bifidum*, *B. infantis*). Total 1 × 10^9 CFU/tablet	FOS 100 mg, L-glutamina 100 mg, sugars 1 g/tablet. Natural orange flavour, citric acid, vegetable stearate and vegetable stearine	Helps digestion
Buddy Bear® Probiotic	*L. acidophilus* 9 × 10^8 CFU, *B. bifidus* 50 × 10^6 CFU, *B. infantis* 50 × 10^6 CFU	FOS 100 mg, organic cane extract, Orange fruit crystals, natural orange flavour, citric acid, vegetable stearate	Improves intestinal digestion
INNEOV Solaire skin probiotic (Nestle and L'Oreal,)	Strain not reported	Licopene, b-carotene	Capsules, helps to improve skin protection against UV-rays
AZO Yeast	*Lactobacillus sporogenes*	Boneset (Eupatorium perfoliatum, Listletoe Leaf (Viscum album), cellulose, lactose, maltodextrina	Prevents yeast infection and helps to relieve the symptoms caused by infection in women
BIOGAIA® AB probiotics drops (Biogaia)	*Lactobacillus reuteri* 10^8 FCU		Drops used to reduce colic; improves digestive health and function and boosts immunity

Database from companies' websites

3.3 Products for Feed

Animal feed companies and researchers have been looking for alternative products and strategies that can help maintain animal gut health by preventing or reducing

Table 5 Probiotic product for feed

Product	Probiotic culture	Application	Company
BioPlus 2B	*Bacillus subtilis, B. licheniformis*	Growth promoter for pigs and broilers	Chr. Hansen
Primalac	*Lactobacillus acidophilus, L. casei, Bifidobacterium bifidum, Enterococcus faecium, contains rice mill by-product, calcium carbonate*	Poultry, beef and aquaculture (fish)	Star Labs
Yea-Sacc®1026	*Saccharomyces cerevisiae* 1026	Dairy, beef, equine, pigs, sheep	Alltech
All-Pro Biotic™	*Lactobacillus acidophilus, L. plantarum, L. casei, Enterococcus faecium*	Cattle, swine, sheep, goats, poultry, horses	Bomac Vets Plus sp.
Power Punch™	Lactic acid bacteria DDS-1	Cattle and other ruminants. Animals when birthing, weaning, vaccinating, handling, undergoing changes in weather, shipping or in antibiotic treatment	Bomac Vets Plus sp.
Probiotic Paste™	Lactic acid bacteria DDS-1, contains inulin	Animals during shipping, adverse weather, overcrowding or ratio changes, all of which can trigger a stress reaction in the animal.	Bomac Vets Plus sp.

Database from companies' websites

damages caused by attacks from pathogens. An alternative and effective approach to antibiotic administration in livestock is the use of probiotics, which can help improve gut microbial balance and, therefore, the natural defence of the animal against pathogenic bacteria (Modesto et al. 2009; Patterson and Burkholder 2003). In recent years, there has been considerable interest in using probiotic microorganisms and organic acids as alternatives to the use of antibiotics in feed. Table 5 shows some examples of products that contain probiotic cultures for application in feed.

4 Ongoing Innovations: Improvement of Probiotic Viability and Bioactivity

Technological characteristics such as probiotic viability, bioactivity and safety must be taken into consideration during the selection process of probiotic microorganisms. Functional aspects include the following: viability and persistence in the gastrointestinal tract, immunomodulation and antagonistic and antimutagenic properties. Probiotic strains must first be able to be manufactured under industrial conditions. Furthermore, they have to survive and retain their functionality during storage and in the foods into which they are incorporated, all without producing off-flavours. Factors related to the technological and sensory aspects of probiotic food production are important; the food industry can only succeed in promoting the consumption of functional probiotic products in the future by satisfying the demands of the consumer.

Good viability and activity of probiotics are considered prerequisites for optimal functionality. It may be sufficient for certain probiotic strains to grow well during the initial production steps (to obtain high enough numbers in the product), but they do not necessarily retain good viability during storage.

The product carrier, the conditions of processing and storage and the use of techniques that protect the probiotic microorganisms from harsh environments are several variables that tend to influence the viability and stability of the strains, and some aspects of these factors are described below.

4.1 Selection of Suitable Product Carrier

In products such as yoghurt, acidity of the product (pH), storage temperature and storage time influence the viability of probiotic bacteria. The storage temperature at 4°C, thus reflecting refrigerator conditions, was the most important factor in maintaining the viability of the probiotic bifidobacteria during the 4-week storage period. Room temperature is the most damaging factor in storing probiotic bacteria. Suitable product carriers can be based on formulations that require refrigeration temperature (Technical Research Centre of Finland 2003).

Food is a good carrier to deliver probiotic bacteria. Milk is a favourable medium for the growth of lactic acid bacteria (e.g., *Lactobacillus* and *Bifidobacterium*) because it contains lactose, which is easily metabolised by these bacteria. However, concern with the choice of an appropriate product carrier has been less relevant due to the availability of highly concentrated commercial formulations (e.g., freeze-dried preparations). The suitable carrier should not be dairy-based product necessarily, but it should have the ability to deliver high viable populations.

4.2 Improvement in Culture Stability

During the steps of processing, food manufacturing, transport and storage, probiotic strains are susceptible to conditions of heat, humidity or exposure to oxygen. Most research to date has been on the physiological effects and stability of probiotics in the gut. However, researchers have developed a protection system combining processing technology and a mix of nutrients that protect probiotics during processing, transport and storage. This protection system is applied in a variety of products, including powders.

Unlike most probiotic strains currently found in popular brands of yoghurt, some probiotics that have been developed are able to survive the heat and pressure of manufacturing processes and the acidic stomach environment. These new probiotic strains also remain viable without refrigeration, making it ideal for inclusion in shelf-stable products such as cereal bars, biscuits, breads, cereals formulations. One example is the Sunny Crunch bar, a new product containing the probiotic strain GanedenBC30 (Ganeden Biotech Co. Mayfield Heights, Ohio, USA), which can remain viable without refrigeration; this makes it ideal for inclusion in products stored at room temperature.

There are several strains of probiotic bacteria, and their ability to survive through different stress factors is highly variable. One alternative for probiotic bacteria might be protection through combining them with growth-promoting substances. These so-called "prebiotics" can support the survival of certain strains, but finding the suitable probiotic and prebiotic pairs requires further study (Technical Research Centre of Finland 2003).

When spray-drying is used for the preservation of potential probiotic cultures, much of their activity is typically lost after a few weeks of storage at room temperature. This is associated with stress that is induced by temperature changes, phase changes and drying, a combination of which tends to damage cell membranes and associated proteins. The use of protectant substances can improve culture viability during the drying and storage steps.

One of the greatest drawbacks associated with the use of dietary cultures in fermented milk products is the lack of acid tolerance of some species and strains. "Overacidification" or "post-acidification" is mainly due to the uncontrolled growth of strains such as *L. bulgaricus* at low pH values and refrigerated temperatures. The effect of pH on the viability of some microorganisms has been investigated by some researchers. Studies show that the pH must be maintained above 4.6 to prevent the decline of bifidobacteria populations. Also, *L. casei* cultures seem to be more adaptable to the acidic environment.

The use of probiotic cultures as an additive to cheese is another point to be considered. The effect of different concentrations of NaCl on the viability of some microbial strains has been evaluated, and the results show that there is an opposite relationship between the viable counts of all microorganisms and the salt concentration along the storage period. Therefore, salt-tolerant strains can be used as a criterion in selecting probiotic bacteria for cheese manufacturing.

During the manufacture of semi-hard and hard cheese, whey is entrapped in the curd particles and is expulsed by the scalding step, which is performed at temperature ranging from 41 to 55°C. It was noted that the temperatures under 50°C had no significant changes in viability of various probiotics tested. By the way, at 55°C the viability of the non-microencapsulated species decreased significantly (Tamime 1993). Therefore, microencapsulation may have potential for increasing the survival of probiotics during the scalding of cheeses.

In addition, high concentrations of sugar added to milk before fermentation may result in conditions that are inhibitory to the growth of probiotics in yoghurt and other products; this will lead to long fermentation times and poor acidity development. This fact may occur due to the adverse osmotic effects of the solutes in the product. Shah and Ravula (2000) demonstrated that the addition of sugar may be deleterious to the growth of probiotic bacteria especially in products such as fermented frozen dairy desserts, which contain approximately 16% sugar. Walker (2001) also showed that yeast strains are more osmotolerant than bacterial strains due to the elasticity of yeast cell walls; yeast cell walls also buffer against water loss.

4.3 *Microencapsulation*

The survival of probiotic bacteria in different matrices is a constant challenge. Many studies have demonstrated low viability in products such as yoghurts, fermented milks and dairy-fermented desserts. Microencapsulation techniques can provide protection for probiotics through the entrapment of cells in hydrocolloid beads, increasing the viability in food matrices and stability during storage life and protecting against the adverse conditions of the gastrointestinal tract.

Microencapsulation can be employed by several techniques, including spray-drying, freeze-drying and fluidised bed-drying, which converts the cultures to a concentrated powered form (Krasaekoopt et al. 2003). These techniques allow for the release of bacteria in the product.

Considering the method used to form the beads, microencapsulation techniques can be divided into two groups: extrusion (droplet method) and emulsion or two-phase system (Fig. 1). Using these methods, the survival of probiotic bacteria can be increased by up to 70–95% (Heidebach et al. 2009; Pimentel-González et al. 2009).

4.3.1 Extrusion

The supporting material used in extrusion is alginate. It is a natural polymer, also known as alginic acid, that is extracted from seaweed and composed of various proportions of 1–4 linked β-D-mannuronic (M) and α-L-guluronic (G) acids, which may vary in composition and sequence depending on the source. The binding of divalent cations (e.g., Ca^{+2}) and subsequent gel formation depend on the composition and arrangement of the blocks of alginic acid residues (Gemeiner et al. 1994). The droplets may be formed when sodium alginate solution is added to a calcium solution with an instantaneous polymerisation (precipitation) of calcium alginate followed by a gradual gelation of the interior as cross-linked calcium ions permeate the cells. Among the advantages of this support material are its cheapness, simplicity and biocompatibility (Martinsen et al. 1989). In general, authors have used a concentration of sodium alginate ranging from 1 to 2% and 0.05 to 1.5 M $CaCl_2$ (Krasaekoopt et al. 2003). The increasing concentration or viscosity of sodium alginate leads to a decrease in bead size. In addition, the composition of the alginate

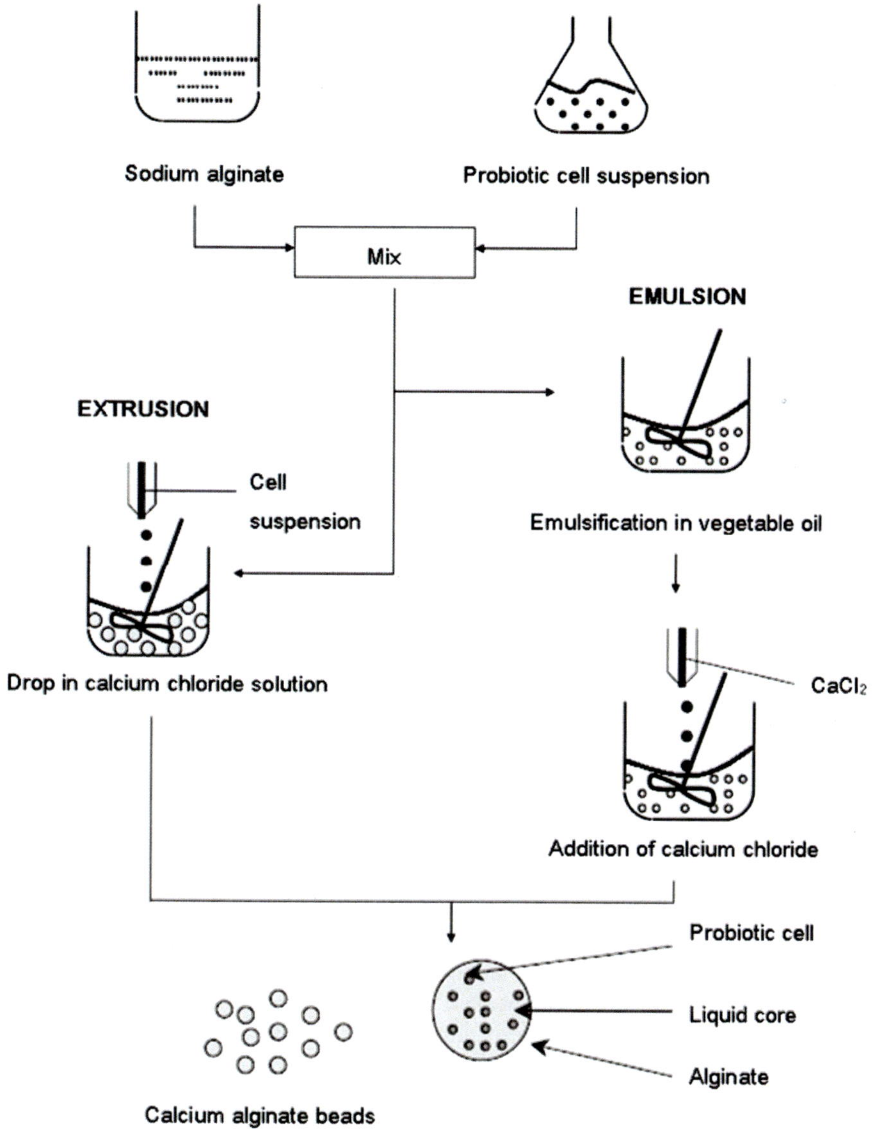

Fig. 1 Microencapsulation of probiotic bacteria in sodium alginate by extrusion and emulsion techniques. Based on Krasaekoopt et al. (2003)

may influence the bead size (e.g., "low guluronic" alginates result in small beads) (Martinsen et al. 1989).

In addition, this method presents some disadvantages because it can be difficult to scale up, and the speed of gel formation is slow compared to the speed of the emulsion technique.

4.3.2 Emulsion

In this method, a small volume of cell polymer suspension (discontinuous phase) is added to a large volume of vegetable oil (continuous phase), which could be soybean, sunflower, canola or corn oil. The mixture is homogenised, and there is formation of a water-in-oil-emulsion. Afterwards, the water-soluble polymer is insolubilised (cross-linked) to form gel beads within the oil phase (Fig. 1). The beads can be harvested later by filtration. The bead size depends on the speed of agitation and varies from 25 to 2 mm (Krasaekoopt et al. 2003).

Other emulsifiers can also be used as the continuous phase (e.g., Tween 80, white light paraffin oil and mineral oil). A mixture of κ-carrageenan and locust bean gum, cellulose acetate phthalate (CAP), alginate, gelatin and chitosan (Anal and Singh 2007) can be used as supporting material in microencapsulation by the emulsion technique.

This method is relatively new for the food industry, and it is easy to scale up aside from providing encapsulation and entrapment of core materials. The bead size is smaller (25–2 mm) than beads formed in the extrusion technique (2–5 mm). However, the cost of using the emulsion technique can be higher due to the use of vegetable oil (Krasaekoopt et al. 2003).

4.3.3 Enhancement of Survival by Microencapsulation Techniques

Recent advances have been achieved with the encapsulation of probiotic cultures. There are several systems employed, and they have been successful because encapsulated bacteria have better survival rates than non-encapsulated bacterial cells, even in simulated gastric conditions. Nevertheless, in some situations it is necessary to cover the beads and proceed with encapsulation with polymer. In general, measures such as cross-linking with cationic polymers (polyethyleneimine, polypropyleneimine and glutaraldehyde), coating with other polymers (chitosan), mixing with starch and using additives (glycerol) help to increase the stability of beads.

The use of entrapped probiotic microorganisms has been employed in the production of dairy products (e.g., yoghurt, cheese and frozen milk products) and biomass production. Encapsulated microorganisms present advantages such as constant characteristics, decrease of incubation time, survival in harsh conditions (e.g., freezing temperatures and acidic environments), protection against attack by bacteriophages, improved stability during storage and productivity.

There are many polymers currently used in the encapsulation of cells that are of non-dairy origin, which limits their use in dairy products. Also, they result in low-density gels with relatively large bead production (1–3 mm) (Heidebach et al. 2009). Furthermore, they can change the sensorial quality of the enriched products. As an alternative to all of these drawbacks, milk proteins have been proposed for use as encapsulant agents because they offer appropriate physico-chemical properties for food applications.

A novel microencapsulation process of probiotic cells that is based on emulsification and rennet-induced gelation of skim milk concentrates has been developed by Heidebach et al. (2009). The researchers noted a high encapsulation yield by this method for *L. paracasei* F19 (105%) and *B. lactis* Bb12 (115%). Moreover, when tested in simulated gastric conditions (pH 2.5 at 37°C for 90 min), both encapsulated strains presented higher viable cell numbers than the free cells; this result was despite the influence of encapsulation and was more pronounced for *B. lactis* Bb12 than for *L. paracasei* F19 (2.8 log units $CFU.g^{-1}$ and 0.8 log unit $CFU.g^{-1}$ higher compared to free cells, respectively).

The efficiency of encapsulation can be enhanced by employing prebiotics as coating materials. Chen et al. (2005) optimised proportions of sodium alginate (1% w/v), peptide (1% w/w) and fructooligosaccharides (3% w/w) as coating materials for encapsulating cells of *L. acidophilus*, *L. casei*, *B. bifidum* and *B. longum*. They observed a high survival rate of microencapsulated cells compared to free cells after storage in refrigerated milk for 2 weeks. The use of prebiotics was found to be beneficial because probiotic microcapsules presented counts of 10^5–10^6 CFU/g in contrast to the 10^2–10^3 CFU/g count of free cells after storage for 12 weeks. However, there are reports that prebiotics can also be detrimental to the encapsulation yield (Chávarri et al. 2010).

5 Future Larger Markets: Infants and the Elderly

The trends of the market have led probiotic companies to develop formulations and products directed to specific target populations such as infants, elderly, athletes and immunocompromised patients. Some studies have demonstrated positive effects when probiotics are administered orally in in vivo trials. Investment in this sector has proven interesting, and the main appeal is using these products as preventive measures against various types of diseases by enhancing the defence system, which decreases the susceptibility of these populations during more aggressive treatments. An overview is presented below regarding the relevance and marketing of probiotic products for infants and the elderly.

5.1 *Probiotic for Infants and Children*

The first bacterial source for newborns is through contact with the mother. The intestine is colonised first by facultative Gram-positive cocci (staphylococci, streptococci and enterococci) and enterobacteria. It is possible that anaerobic bacteria start to colonise on the second day (Harmsen et al. 2000). The genera found in the intestinal microbiota are *Bacteroides*, *Lactobacillus*, *Clostridium*, *Fusobacterium*, *Bifidobacterium*, *Eubacterium*, *Peptococcus*, *Peptostreptococcus*, *Streptococcus* and *Veillonella* (Savage 2005).

In addition, the composition of the indigenous intestinal microbiota of the infants may be influenced by factors such as mode of birth and breast feeding. In preterm infants, for example, the flora is characterised by the presence of coliforms, *Enterococcus* and *Bacteroides* species, where bifidobacteria are scarce (Savage 2005). Breast milk contains substances such as peptides and oligosaccharides, which promote the growth of bifidobacteria (Liepke et al. 2002). The intestinal bifidobacteria species that are influenced by breast feeding are *Bifidobacterium breve*, *Bifidobacterium infantis* and *Bifidobacterium longum*, whereas *Bifidobacterium adolescentis* are more common in the coming years of the children's development (Rautava 2007).

Studies have assessed the efficacy and safety of probiotics in neonates, but due to large differences between studies (e.g., probiotic strains used, age of the target population, dosage of the probiotic product, duration of probiotic supplementation), the results appear to be inconclusive. Nevertheless, many studies have been published suggesting the beneficial action of probiotics. Generally, probiotics provide resistance against colonisation by pathogenic bacteria through competition for nutrients or adhesion sites, production of antimicrobial substances or modulation of immune system. However, the action of probiotics must be strain-specific.

Some mechanisms may be related to the reduction of morbidity during infancy. Probiotics can promote intestinal barrier function, inhibit pathogen adherence and colonisation, degrade allergens, enhance immune maturation through the induction of IgA production and induction of regulatory T cells (Rautava 2007).

The use of probiotics in infancy has been most suggested in diarrhoeal cases. In this sense, *Lactobacillus* GG was shown to be effective in the prevention of acute infantile diarrhoea, including nosocomial spread of infection (Szajewska et al. 2001). Also, nosocomial spread of gastroenteritis and diarrhoea was prevented by bifidobacteria strains (Saavedra et al. 1994; Chouraqui et al. 2004).

Necrotising enterocolitis (NEC) is a gastrointestinal disease that affects mainly premature infants. Trials in animal models have showed that *B. infantis* reduces the risk of NEC in newborn rats (Caplan et al. 1999). In further studies, very low birth weight infants received supplementation based on combinations of probiotics (*L. acidophilus*, *B. infantis*, *S. thermophilus*, *B. bifidus*) that resulted in the reduction of the incidence of NEC (Lin et al. 2005; Bin-Nun et al. 2005).

In the first year of life, children can manifest atopic diseases (food allergy, atopic eczema, allergic rhinoconjunctivitis and atopic asthma), which cause discomfort and may even pose a risk to the infant's life, depending on the degree of inflammation. According to Romagnani (2004), the immune pathology of atopic diseases is characterised by T helper (Th) 2-driven inflammatory responsiveness against environmental conditions or dietary allergens. Probiotics can help reduce the risk of developing this type of disease. Rautava et al. (2002) tested the use of probiotics in breastfeeding mothers and neonatal infants with a high risk of developing atopic disease. Mothers received the supplementation with *Lactobacillus* GG 2–4 weeks before parturition. After birth, breastfeeding mothers and infants consumed probiotics for 6 months. In the group that was administered probiotics, a significant reduction of atopic eczema was found during the first 2 years of life. Interestingly, it

Table 6 Current probiotic products for infants and children

Mark product	Company	Probiotic strain	Health claims
BioGaia® Probiotic Drops	BioGaia®	*L. reuteri*	Protects the gastrointestinal tract, enhances immunity and improves colic symptoms
BioGaia® Probiotic Chewable Tablets	BioGaia®	*L. reuteri*	Promotes gut health and enhances immunity
Good Start® Probiotic	Nestlé®	*B. lactis*	Promotes healthy tract flora
Baby Cereals	Nestlé®	*B. lactis*	Assists the development of the digestive system, promotes a healthy digestive tract
A.B. Pre & Pro	Bio-LiFE	*L. acidophilus* LA-5 and *B. lactis* BB-12	Increases intestinal bifidobacteria and promotes a healthy digestive tract
Dicoflor 30	Vitis Pharma	*L. rhamnosus* GG	Reduction of allergic reactions, supports immunity, acts as an antibiotic treatment and prevents diarrhoea
OptiBac Probiotics For your Child's health	OptiBac® Probiotics	*L. acidophilus* Rosell-52, *B. infantis* Rosell-33 and *B. bifidum* Rosell-51	Promotes good digestive health and supports natural immune defences
EvoraKids™	Oragenics, Inc.	Probiora 3® (*S. oralis*, *S. uberis* and *S. rattus*)	Supports gum and tooth health by maintaining the oral health
Babio (biomilk with probiotics)	Goldim	BifiComplex® (*B. breve*, *B. longum*, *B. bifidum* and *B. infantis*)	Enhances the immune system
Happybaby	Nurture, Inc.	*L. acidophilus*, *B. lactis*, *L. casei*, *L. rhamnosus*	Protects against the development of allergies

Available at the online database of each manufacturer

was observed that the protective effect was most prominent in infants whose mothers received the probiotic supplementation.

According to Scientific Committee on Food of the European Commission, only probiotic bacteria that has been identified molecularly and has been found to have genetic stability can be added to formulae and infant foods. The stated level of viable bacteria should be 10^6–10^8 colony forming units (CFU) per gram of the formulation upon consumption.

Table 6 shows the commercialised probiotic products for infants and children.

5.2 *Probiotic for the Elderly*

As humans age, the immune system tends to fail (known as immunosenescence). This leads to a higher frequency of illnesses among the elderly. Dietary supplementation with probiotics can enhance the immunity of the elderly, and they can be administered during vaccination and following surgical intervention.

This method of boosting immunity was demonstrated by Gill and Rutherfurd (2001), where 13 healthy elderly volunteers (age range 62–77 years) received a supplement with *Lactobacillus rhamnosus* HN001 (DR20^{TM}) for 3 weeks. Afterwards, the elderly subjects had enhanced phagocytic activity of peripheral blood polymorphonuclear (PMN) cells and monocytes (average of 35%).

The seasonal epidemic caused by the influenza virus represents one of the greatest threats to the elderly. Therefore, the use of probiotics can serve as a preventive measure against infection caused by the influenza virus because the clinical efficacy of the vaccine is lower in the elderly. For young healthy adults, the influenza vaccine provides clinical protection in 70–90% of cases; in the elderly, this protection is found in only 17–53% of cases. This may be related to the influenza vaccination seroconversion and seroprotection rates, which are two- to fourfold lower in the elderly when compared to the rates in younger people (Goodwin et al. 2006). Studies performed by Boge et al. (2009) showed that the consumption of two bottles containing 100 g per day of the commercial probiotic drink Actimel® (containing *L. casei* DN-114, combined with *Streptococcus thermophilus* and *Lactobacillus bulgaricus*) helps to increase specific antibody titres after vaccination against influenza in elderly over 70 years of age. In the pilot and confirmatory study, the seroconversion rates were higher in the probiotic group when compared to the rates in the control group for three strains tested (H1N1, H3N2 and B). Moreover, antibody levels in the probiotic group remained higher up to 9 weeks after vaccination.

Nowadays, some probiotics are targeted for the elderly and contain *Lactobacillus* and/or *Bifidobacterium*, mineral and other components. Just as aging and health factors change bone density, they also can alter the body's microbial content. Using probiotic dairy products that contain these strains can help restore and improve the intestinal flora balance, leading to a variety of associated health benefits, including mineral supplementation (Ca, Zn, Mg). These products include beverages such as milk and soymilk, drinkable yoghurts and other solid probiotic products (e.g., breakfast cereals, baked goods, snack foods, breakfast bars and crackers). An example is the product PuroFIT® calcium (FloraFIT company).

5.3 Safety Aspects

There are more safety concerns regarding side effects and bacteraemia (presence of bacteria in the blood) when considering the commercialisation of probiotic products, especially for children and elderly. In recent years, bacteraemia cases have not been reported. In addition, clinical trials have demonstrated the safety of bifidobacteria and lactobacilli, such as *L. casei* in infants and children (Borriello et al. 2003; Srinivasan et al. 2006). Although most of the published studies have not been designed to assess the safety of probiotics. However, competent organisations (e.g., European Food Safety Authority – EFSA) have performed studies, which legally regulate the use of probiotic microorganisms.

6 Concluding Remarks

Probiotics can have a high potential for combating common diseases that afflict specific target populations. This field requires more studies to demonstrate potential adverse effects in neonates and infants. Also, it is important to consider that some properties (e.g., immunomodulation) appear to be strain-specific. More studies on the mechanisms of action of probiotics on the molecular level may indicate target strains (or combinations of strains) for specific functions and optimal dosages for each treatment. In people with genetic predispositions for developing certain diseases, prevention can be accomplished by administering probiotics daily. The discovery of new probiotic strains or engineered strains and the applications in other areas are the trends in probiotic research.

References

Anal AK, Singh H (2007) Recent advances in microencapsulation of probiotics for industrial applications and targeted delivery. Trends Food Sci Tech 18:240–251

Bin-Nun A, Bromiker R, Wilschanski M, Kaplan M, Rudensky B, Caplan M, Hammerman C (2005) Oral probiotics prevent necrotizing enterocolitis in very low birth weight neonates. J Pediatr 147:192–196

Boge T, Rémigy M, Vaudaine S, Tanguy J, Boudet-Sicard R, Van der Werf S (2009) A probiotic fermented dairy drink improves antibody response to influenza vaccination in the elderly in two randomised controlled trials. Vaccine 27:5677–5684

Borriello SP, Hammes WP, Holzapfel W, Marteau P, Schrezenmeir J, Vaara M, Valtonen V (2003) Safety of probiotics that contain lactobacilli or bifidobacteria. Clin Infect Dis 36:775–780

Caplan MS, Miller-Catchpole R, Kaup S, Russell T, Lickerman M, Amer M, Yu X, Thomson R Jr (1999) Bifidobacterial supplementation reduces the incidence of necrotizing enterocolitis in a neonatal rat model. Gastroenterology 117:577–583

Chávarri M, Marañón I, Ares R, Ibáñez FC, Marzo F, Villarán MDC (2010) Microencapsulation of a probiotic and prebiotic in alginate-chitosan capsules improves survival in simulated gastro-intestinal conditions. Int J Food Microbiol 142:185–189

Chen KN, Chen MJ, Liu JR, Lin CW, Chiu HY (2005) Optimization of incorporated prebiotics as coating materials for probiotic microencapsulation. J Food Sci 70:260–266

Chouraqui JP, Van Egroo LD, Fichot MC (2004) Acidified milk formula supplemented with *Bifidobacterium lactis*: impact on infant diarrhea in residential care settings. J Pediatr Gastroenterol Nutr 38:288–292

Gemeiner P, Rexová-Benková L, Švec F, Norrlöw O (1994) Natural and synthetic carriers suitable for immobilization of viable cells, active organelles and molecules. In: Veliky IA, McLean RJ (eds) Immobilized biosystems: theory and practical applications. Chapman and Hall, London, UK, pp 67–84

Gill HS, Rutherfurd KJ (2001) Probiotic supplementation to enhance natural immunity in the elderly: effects of a newly characterized immunostimulatory strain *Lactobacillus rhamnosus* HN001 (DR20™) on leucocyte phagocytosis. Nutr Res 21:183–189

Global Industry Analytic (2010) Global probiotics market to exceed US$28.8 billion by 2015, New Report by Global Industry Analysis, Inc. San Jose, CA. Sep. 14, 2010. http://www.strategyr.com/Probiotics_Market_Report.asp. Cited 21 Oct 2010

Goodwin K, Viboud C, Simonsen L (2006) Antibody response to influenza vaccination in the elderly: a quantitative review. Vaccine 24:1159–1169

Harmsen HJM, Wildeboer-Veloo ACM, Raangs GC, Wagendorp AA, Klijn N, Bindels JG, Welling GW (2000) Analysis of intestinal flora development in breast-fed and formula-fed infants by using molecular identification and detection methods. J Pediatr Gastroenterol Nutr 30:61–67

Heidebach T, Först P, Kulozik U (2009) Microencapsulation of probiotic cells by means of rennet-gelation of milk proteins. Food Hydrocolloid 23:1670–1677

Hubalek Z (2003) Protectants used in the cryopreservation of microorganisms. Cryobiology 46:205–229

Krasaekoopt W, Bhandari B, Deeth H (2003) Evaluation of encapsulation techniques of probiotics for yoghurt. Int Dairy J 13:3–13

Liepke C, Adermann K, Raida M, Magert HJ, Forssmann WG, Zucht HD (2002) Human milk provides peptides highly stimulating the growth of bifidobacteria. Eur J Biochem 269:712–718

Lin HC, Su BH, Chen AC, Lin TW, Tsai CH, Yeh TF, Oh W (2005) Oral probiotics reduce the incidence and severity of necrotizing enterocolitis in very low birth weight infants. Pediatrics 115:1–4

Markets and Markets (2009) Probiotic market- advanced technologies and global market (2009–2014). http://www.marketsandmarkets.com/Market-Reports/probiotic-market-advanced-technologies-and-global-market-69.html. Cited 21 Oct 2010

Martinsen A, Skjak-Braek C, Smidsrod O (1989) Alginate as immobilization material: I. Correlation between chemical and physical properties of alginate gel beads. Biotechnol Bioeng 33:79–89

Modesto M, D'Aimmo MR, Stefanini I, Trevisi P, De Filippi S, Casini L, Mazzoni M, Bosi P, Biavati B (2009) A novel strategy to select *Bifidobacterium* strains and prebiotics as natural growth promoters in newly weaned pigs. Livest Sci 122:248–258

Morgan CA, Herman N, White PA, Vesey G (2006) Preservation of micro-organisms by drying: a review. J Microbiol Meth 66:183–193

Patterson JA, Burkholder KM (2003) Application of prebiotics and probiotics in poultry production. Poultry Sci 82:627–631

Pimentel-González DJ, Campos-Montiel RG, Lobato-Calleros C, Pedroza-Islas R, Vernon-Carter EJ (2009) Encapsulation of *Lactobacillus rhamnosus* in double emulsions formulated with sweet whey as emulsifier and survival in simulated gastrointestinal conditions. Food Res Int 42:292–297

Rautava S (2007) Potential uses of probiotics in the neonate. Semin Fetal Neonatal Med 12:45–53

Rautava S, Kalliomäki M, Isolauri E (2002) Probiotics during pregnancy and breast-feeding might confer immunomodulatory protection against atopic disease in the infant. J Allergy Clin Immunol 109:119–121

Romagnani S (2004) Immunologic influences on allergy and the TH1/TH2 balance. J Allergy Clin Immunol 113:395–400

Saavedra JM, Bauman NA, Oung I, Perman JA, Yolken RH (1994) Feeding of *Bifidobacterium bifidum* and *Streptococcus thermophilus* to infants in hospital for prevention of diarrhoea and shedding of rotavirus. Lancet 344:1046–1049

Savage DC (2005) Mucosal microbiota. In: Mestecky J, Lamm ME, Strober W, Bienenstock J, McGhee JR, Mayer L (eds) Mucosal immunology. Elsevier, Boston, pp 19–33

Shah NP, Ravula RR (2000) Influence of water activity on fermentation, organic acid production and viability of yogurt and probiotic bacteria. Austr J Dairy Technol 55:127–131

Srinivasan R, Meyer R, Padmanabhan R, Britto J (2006) Clinical safety of *Lactobacillus casei* shirota as a probiotic in critically ill children. J Pediatr Gastroenterol Nutr 42:171–173

Szajewska H, Kotowska M, Mrukowicz JZ, Armanska M, Mikolajczyk W (2001) Efficacy of *Lactobacillus* GG in prevention of nosocomial diarrhea in infants. J Pediatr 138:361–365

Tamime AY (1993) Modern cheesemaking: hard cheeses. In: Robinson RK (ed) Modern dairy technology: advances in milk products, vol 2. Elsevier, New York, pp 49–220

Technical Research Centre of Finland (VTT) (2003) Biology-Online.org. Jan, 2003. http://www.biology-online.org/articles/make_sure_beneficial_probiotic.html Cited 21 Oct 2010

Teixeira P, Castro H, Mohácsi-Farkas C, Kirby R (1997) Identification of sites of injury in *Lactobacillus bulgaricus* during heat stress. J Appl Microbiol 83:219–226

Walker GM (2001) Yeast growth. In: Yeast physiology and biotechnology. Wiley, New York, p 255

Index

A
Aberrant crypt foci (ACF), 155, 156
Acidic environment, 309
Acidic sphingomyelinase, 227
Acidification, 194, 196
Acidophiles, 6
Acidosis, 158
Acid tolerance, 252
Active transport, 24
Activity of probiotics, 308
Adenosine triphosphate (ATP), 13, 24
Adhesion, 141, 144, 150, 153, 154, 253–254
  synergistic, 253
Adipocytes, 13
Administration
  duration, 257
  frequency, 257
  time, 257
Aldohols, 23
Alenine recemose, 124
Alginate, 310
Alkaline-shock proteins, 117
Allergenecity, 130
Allergic contact dermatitis, 235
Allergies, 315
Alpha-1-antitrypsin deficiency (A1AD), 184
Amylase A (amyA), 123
Angiotensin-I converting enzyme (ACE)
    inhibitors, 178
Antibacterial, 150–154, 160
Antibiotic-associated diarrhea (AAD),
    141–145
Antibiotic-resistant markers, 132
Antibiotics, 15, 23, 307
Antibiotics resistant gene, 276–281
Anticolorectal cancer, 140, 154–157
Anti-E7 IgG, 123
Anti-inflammatory, 182
Antimicrobial substances, 223–225
Antioxidant enzymes, 233
Anti-PspA-IgA, 126
Apoptosis, 144, 149, 155–157, 159, 234, 235
Archaea, 2–6, 10, 16, 17
Asexual reproduction, 5, 16, 17
Aspergillus niger, 31, 32, 39
Aspergillus oryzae, 31, 32, 39
Assimilation of cholesterol, 176
Atopic dermatitis, 223, 228, 229, 235
Atopic diseases, 314
Atopic eczema, 229
ATP-dependent binding cassette (ABC), 116
Attenuated pathogens, 122
Aujeszky disease virus (ADV), 123
Avicin A and B, 117
Azoreductase, 155

B
Bacilli, 3, 6–8, 14, 18, 19
Bacillus, 7, 8, 14, 15, 20
  B. brevis, 121
Bacteremia, 157
Bacteria, 2
Bacterial translocation, 271–272, 274–276,
    293, 294
Bacteriocin-like substances, 223
Bacteriocins, 117, 147, 152, 223, 224, 227
Bariatric, 157–158
Barrier function, 227
Bead size, 312
Beta defensins, 225–226
β-galactosidase, 146, 147
β-glucocerebrosidase, 227, 228
β-glucuronidase, 155

M.-T. Liong (ed.), *Probiotics*, Microbiology Monographs 21,
DOI 10.1007/978-3-642-20838-6, © Springer-Verlag Berlin Heidelberg 2011

BetL, 121
Bifidobacteria, 140, 143, 146, 151–155, 158
Bifidobacterium, 10, 110, 314
Bifidobacterium longum, 110
Bifidobacterium sp., 89
Bifurcated Y-shaped rods, 20
Bile, 251
Bile acids, 119, 148
Bile exclusion system (BilE), 121
Bile salt hydrolases (BSH), 116, 120, 176
BilE system *See* Bile exclusion system (BilE)
Bile tolerant, 120
Binary fission, 5, 16, 25
Binder, 95
Biolog YT microplate and chromosomal PFGE karyotyping, 35
Biomarkers, 262
Biotherapeutic agents, 40
Bronchoalveolar lavage (BAL), 126
BSH *See* Bile salt hydrolases (BSH)
BusA *(opuA)*, 120

C
Caco-2 cells, 149, 150, 153, 156, 157
Calcium alginate, 93, 94
Candida albicans, 127
Candida humilis, 31, 34, 38
Capsule, 10, 22, 25
Carbohydrates, 90
Carboxysome, 2, 3, 14
Carcinogen, 155–157
Carrier, 191–215
Catalase, 18, 20
Catalase-negative, 18, 20, 22
cdp-A, 116
Cell adhesion, 12
Cell cortex (exctoplasm), 13
Cell entrapment, 93–94
Cell lines, epithelial, 253
Cell membrane, 2, 5, 10, 12–13, 16, 17, 23, 24
Cell pellet, 95, 97
Cell powder, 92, 96, 97, 100–104
Cells
  dead, 251
  dormant, 251
  injured, 251
  viable, 251
Cell signalling, 12, 13
Cell tablet, 96, 100–102, 104
Cell viability, 88–89, 96, 101
Ceramidase, 228
Ceramides, 227–230
Cereal, 192, 193, 204–208, 215
Cereal-based products, 192, 204–208
Chemokines, 126
Chemo-organotrophic, 19, 20
Chemotherapy, 144
Chimeric proteins, 130
Cholesterol, 172
Chronic obstructive pulmonary disease (COPD), 184
Chronic wound, 225, 227
Claims
  functional, 246
  nonverified, 261
  risk reduction, 246
  structure, 246
Classification, 2–4, 6, 17, 18
Clostridium difficile, 143, 144
Coating materials, 95–100, 103
Codex Alimentarius Commission, 130
Coexpression of antigens, 126
Colitis, 143, 148–151, 159, 160
Collagen, 230–231
Colon, 143, 146, 149, 154–157
Colon cancer, 154–157
Colonic drug delivery, 98, 99
Colonization, 251, 253–254
Colonization resistance, 149
Colony forming units (CFU), 100–105
Competitive exclusion, 153
Compressible filler, 96
Compression coated tablet, 98, 100, 103–105
Compression coating, 94–98, 105
Compression pressure, 96–97
Conjugated linoleic acid (CLA), 157
Conjugation, 10, 16
Constipation, 259
Constitutive promoters, 123
Consumer attitudes, 259–261
Consumer perceptions, 110
Consumer preference, 133
Continuous Multi-Purpose Melt Technology (CMT), 93
Controlled release, 98–105
Coprecipitation, 176
Coronary heart disease (CHD), 172
Counterstain (safarine/fuchsine), 8
Crohn's disease, 148, 151, 159
Cutaneous pH, 227–228
Cyanovirin-N (CV-N) protein, 125
Cyclopropane fatty acid synthase, 116
Cytokines, 126, 144, 148, 149, 156
Cytoplasmic inclusion, 13
Cytoplasmic membrane, 12–13, 24

Cytosol, definition, 13
Cytotoxic T lymphocytes, 122
Cytotoxins, 155

D
Dairy-based products, 193–199, 308
D-alanine-dependent growth, 132
Debaryomyces hansenii, 31, 34, 36, 38
Debaryomyces occidentalis, 34, 38
Denaturing gradient gel electrophoresis (DGGE), 59, 67, 68
Dermal applications, 221–237
Dermatitis, 222, 227, 229, 235, 236
Dermatosis, 229
Diarrhea, 140–146, 148, 149, 151, 158, 160
  acute, 258
  antibiotic-associated, 258
  traveler's, 259
Diet, 257
Dietary supplements, 301
Digestive system, 305
Dilution plating, 89, 101–103
Directed gene alterations, 130
Directed mutagenesis, 131
Disintegrants, 96, 100, 104–105
Dlt operon, 116
DNA double strand, 16
DNA vaccines, 127
DSS (dextran sulfate sodium)-induced colitis, 125

E
Effects
  antagonistic, 262
  synergetic, 262
Elderly, 313
Embden–Meyerhof pathway (EMP), 111
EMP *See* Embden–Meyerhof pathway (EMP)
Emulsion, 312
Encapsulation, 91–95, 105
Endospores, 12, 15
Enteric coating materials, 95
Enterococcus, 140, 143, 152, 314
  E. avium, 117
Enterocytes, 141, 146, 159
Enteropathogenic *E. coli* (EPEC), 155, 156
Entrapment, 312
Enumeration, 250
Environmental stress, 232–234
Epithelial, 141, 144, 149, 150, 154, 155, 157
EPS clusters, 118
Ethical concern, 133
Ethyl methyl sulphonate, 129
Eudragit, 95–97, 99, 100, 103–105
Eukaryotic, 2–5, 16, 24, 25
Eukaryotic probiotics, 29–45
European Food Safety Authority, 44
Exopolysaccharides, 197, 206, 211
Extremophiles, 5
Extrusion, 310
Extrusion method, 93, 94

F
Facilitated diffusion, 24
Facultative anaerobes, 22
FAO/WHO *See* Food and Agriculture Organization/World Health Organization (FAO/WHO)
Feed, 306
Fermentation byproducts, 177
Fermentative type of metabolism, 20
Ferrisiderophores, 154
F_1F_0-ATPase, 117
Film coating, 92–93
Fimbriae, 10
flagellum, 3, 10
Flatulence, 146, 148
Flow cytometer, 100
Fluorescence In Situ Hybridization (FISH), 64–66
Food and Agriculture Organization/World Health Organization (FAO/WHO), 3, 30, 32, 34
Food and Drug Administration (FDA), 32
Food grade markers, 132
Food matrix, 252
Foods for Specific Health Uses (FOSHU), 247, 271, 288–294
Foot and mouth antigen, 129
FOS *See* Fructose-oligosaccharides (FOS)
FOSHU *See* Foods for Specific Health Uses (FOSHU)
Fragmentation/budding, 5, 16, 17
Free-fatty acids (FFAs), 223, 227, 228
Freeze-dried, 302
Freeze-drying, 91
Fructose-oligosaccharides (FOS), 116
Fructose-6-phosphate pathway, 20
Fructose-6 phosphoketolase, 20
Fruit, 193, 195, 196, 208–212, 215
Functional foods, 300
Fungemia, 44
Fusion proteins, 130

G
Galactooligosaccharides (GOS), 5, 6, 116
Gametes, 16
Gastrointestinal route, 123
Gastrointestinal tract (GIT), 248
Gas vacuoles, 14
GC-content *See* Guanine-cytosine (GC)-content
Gelatine capsules, 92
Gelation, 94
Gel-forming polymers, 98–99
Generally recognized as safe (GRAS) status, 32
Genetically modified food (GMF), 130
Genotoxins, 155
Glutathione transferase (GSH), 156
GMO, 78–79
GM probiotics, 110
GOS *See* Galactooligosaccharides (GOS)
gp50, 123
gp120, 125
Gram negative bacteria, 8, 11, 22
Granulation, 92
Group translocation, 24
Growth factors, 126
Growth promoter, 307
Guanine-cytosine (GC)-content, 19, 20
Gut-associated lymphoid tissue (GALT), 171

H
Halophiles, 6
Haloquadra walsbyi, 5
Health benefits, 248
Healthcare, 300
Health claims, 271, 288–294
  advanced, 247
  beneficial, 244
  qualified, 245
Helicobacter pylori, 123
Heterocyclic amines, 156
Hierarchal classification, 18
High-density lipoproteins (HDLs) cholesterol, 172
High resolution melt (HRM) analysis, 63–64
High-throughput sequencing, 70–74
High throughput techniques, 131
HIV-1, 125
HMG-CoA reductase, 172, 177
H_2O_2 *See* Hydrogen peroxide (H_2O_2)
Homeostasis, 13
Homofermentative, 22
Horizontal gene transfer, 131–132
Horizontal transfer, 270, 278–279
Hot-melt coating, 93
HPV-16 E7, 123, 126
Hyaluronic acid (HA), 230–231
Hydrogen peroxide (H_2O_2), 152, 223
Hypersensitivity, 181
Hypertension, 177
Hyperthermophiles, 5, 6
Hypertonic surroundings, 23
Hypolactasia, 146

I
Ichtyosis, 229
Identification, 6, 8–10, 17, 18, 20, 249–250
IFN-γ, 127
IFNs, 126
IgA *See* Immunoglobulin A (IgA)
IL-2, 126
IL-6, 126
IL-10, 127
Illnesses, 315
Immobilisation, 93
Immobilized, 203, 204, 206, 207, 211
Immune, 141, 143, 144, 148, 153, 154, 156, 157
Immunity, 306
Immunoglobulin A (IgA), 122, 141, 149, 156
Immunomodulation, 258, 317
Immunomodulatory, 169
Immunostimulation., 158
Industrial excipients, 97
Infantile diarrhoea, 314
Infants, 313
Infections gastrointestinal, 258
Inflammation, 143, 144, 147–149
Inflammatory bowel diseases (IBD), 124, 140, 148–151, 160, 259
Insertion sequence elements, 129
Interferon-γ (IFN-gamma), 156
Interleukin-10 (IL-10), 148, 149, 156
Intestinal barrier, 144, 147, 149
Intestinal dysmotility, 157, 158
Intestinal flora balance, 316
Intranasal route, 123
Inulin, 196, 198, 203
Ion conductivity, 12
Iron-siderophore, 150, 154
Irritable bowel syndrome (IBS), 140, 147–148, 160, 259
Isotonic, 22, 23
Isotonic surroundings, 22, 23

K

120 kDa protein of *S. cerevisiae* var *boulardii*, 44
Keratinocytes, 228, 230, 231, 234, 235
Kluyveromyces lactis, 31, 34, 38, 42
Kluyveromyces lodderae, 31, 34, 38
Kluyveromyces marxianus, 31, 34, 38
Kluyveromyces species, 31, 32, 34, 37, 38

L

LAB, 120
LABGC *See* Lactic acid bacterium genome consortium (LABGC)
lacF, 132
lacG, 132
L. acidophilus, 89, 90, 93, 95, 99–105, 111
Lactacin B, 117
Lactic acid bacteria (LAB), definition, 18
Lactic acid bacterium genome consortium (LABGC), 110
Lactobacillus, 6, 110, 140–143, 145, 146, 149, 152, 153, 155, 158–160, 314
  L. iners, 110
  L. juhnsonii, 111
  L. plantarum, 110
  L. reuteri, 111
  L. salivarius, 122
Lactococcus, 110
Lactoferrin, 154
Lactose, 140, 146–147, 158, 160
Lactose intolerance, 7, 140, 146–147, 158, 160, 258
Legislation, 245–248
Lipase expression, 125
Lipid inclusions, 14
Liposomes, 128
Liquid preparations, 301
Listeria monocytogenes, 120
Livestock, 307
Low density lipoprotein (LDL), 7
Lysis, 22
Lysozyme, 22, 23

M

Magnesium stearate, 96
Magnetosomes, 2, 3, 14
Maldigestion, 146, 147
Market of probiotics, 300
Meat, 193, 212–215
Meat derivative, 212–215
Meiosis, 16, 25
Meloxicam, 125
Membrane lipid oxidation, 97
Messenger RNA (mRNA), 14, 15
Meta-analysis, 258
Metabolic pathways, 5, 13, 20
Metabolic processes, 5, 11, 13, 24
Metabolomics, 131
Metachromatic granules, 13
Metagenomics, 45, 59, 61, 80
Metalloproteinases (MMP), 234
Methanogens/methanogenic archaea, 5, 6
Methicillin-resistant *Staphylococcus aureus* (MRSA), 152
Microaerophylic, 19
Microarray, 74–78, 118
Microencapsulation, 195, 198, 210, 215, 310
Microvilli, 150, 155
Minimum therapeutic dose, 88, 104, 105
Models
  animal, 244
  genetically modified, 255
  induced, 255
  negative, 255
  orphan, 255
  spontaneous, 255
  in vitro, 244
  in vivo, 244
Morphological features, 6
Msm operon, 116
Mucosa, intestinal, 253
Mucosal delivery vectors, 122–129
Mucosal immunity, 122
Murine systems, 124
Mutualists/commensals, 5
Mycoplasma, 11

N

N-acatyleglucosamine (NAG), 10
N-acetylmuramic acid (NAM), 10
Neutral sphingomyelinase, 228, 229
*nis*A, 121
Nisin RK two-component system (NICE), 121
Nitric oxide (NO), 226, 234
N-methyl-N′-nitro-N-nitrosoguanidine (NNG), 129
Nomenclature, 17
Nosocomial infections, 143, 159
Nucleiod, 12, 14

O

Oral bacteriotherapy, 170
Organelle, definition, 2

Organic acids, 195, 202, 203
Organoleptic properties, 195, 207, 208, 215
Osmosis, 23
Osmotolerant, 120
Oxidative stress, 233, 236
Oxygen scavenging capabilities, 90

P

Pamy, 123
Pathobiotechnology, 120
Pathogenic, 147–149, 152–155
Pathogens, 254
PCR, 121
Pectin, 96, 98–100, 104, 105
Pectinolytic enzymes, 98, 100
PEG, 128
Pepridoglycan (Glycocalyx), 11, 13, 25
Peptidoglycan layer, 8, 10, 11, 22
Percentage identity, 130
Persistence, 121
Phagocytosis, 11, 25
Phosphoketolase pathway (PKP), 111
Phospholipids, 12
Phosphotidyl glycerophosphatase, 116
pH sensitive materials, 99–100
Phytoestrogens, 179
Pichia anomala, 127
Pills, 304
PKP *See* Phosphoketolase pathway (PKP)
Plasmids, 3, 12, 15
Pldh, 123
pLET, 124
Pneumococcal surface protein A (PspA), 125
Polysaccharide granules, 14
Population size, 256
Postmenopause, 179
Postoperative, 140, 145, 157–160
Pouchitis, 148–151, 159, 160
Prebiotics, 90, 177, 256, 309
Preservatives, 152
Probiotic bacteria, 88, 91, 92, 99, 105
Probiotic bacteria, definition, 3
Probiotic cultures, 303
Probiotics, 221–237, 300
Product carrier, 308
Proinflammatory, 170, 171
Prokaryotic fission, 16
Protectant substances, 309
Proteolytic, 193, 201, 208, 213
Proteomics, 131
Protoplast, 22
Protoplast fusion technique, 45
Pseudomurein, 10
Psoriasis, 222, 223, 229
Psychrophiles, 6
Ptgs2, 125
pTREX, 124
Pulse field gel electrophoresis (PFGE), 35, 59, 67–68
Pyrococcus furiosus, 6
Pyrosequencing, 59, 71

Q

QacT, 120
Quaternary ammonium, 23

R

Radiation, 129
Radiation-induced diarrhea (RID), 141, 144–145
Randomly amplified polymorphic DNA (RAPD), 35, 68–69, 79
Reactive nitrogen species (RNS), 233
Reactive oxygen species (ROS), 233
Reactive skin, 231, 232
Refrigeration, 91
Regulatable promoters, 123
Regulation of GM probiotics, 133
Renin–angiotensin system, 178
Repetitive genomic element PCR (RepPCR), 69
Restriction fragment length polymorphism and terminal restriction fragment length polymorphism, 69–70
Ribosomes, 2, 3, 12, 13, 15, 25
Roux-en-Y gastric bypass (RNYGB), 157, 158

S

Saccharomyces cerevisiae, 31, 32, 34–43
Saccharomyces cerevisiae (NCYC SC 47, NCYC 1026, CNCM I-1077, CNCM I-1079 and MUCL 39885), 32
Saccharomyces cerevisiae var *boulardii*, 32, 34, 35, 37, 38, 40–44
Safety, 269–294
Salmonella spp., 122
Secretory scFv, 127
Selection criteria, 244
Shelf life, 88, 91
Short bowel syndrome (SBS), 158–159
Short-chain fatty acids (SCFAs), 146, 147, 149, 152

Simple diffusion, 23
Simulated gastric fluid (SGF), 89, 93, 96, 100–105
Simulation, 252
Single-chain variable fragment antibody (scFv), 127
Skin aging, 229, 230, 233
Skin inflammation, 222, 224
Skin lipids, 228
Skin microflora, 223–227, 236
Skin pH, 225, 227
Skin reactivity, 231–232
S-layer proteins, 124, 154
Small intestine bacterial overgrowth (SIBO), 147
Sodium-proton antipracters, 117
Solid formulations, 301
Sources of energy, 5
Soy, 193, 199–204, 215
Soy-based products, 192, 199–204
Spheroplast, 23
Sphingomyelin (SM), 227–229
Sphingomyelinases (SMase), 229, 230
Spirochaete, 8
Spray-dried powders, 302
16S rRNA, 61–63, 67, 68, 70–72, 74, 75
16S rRNA sequencing, 61–62, 70–72, 74
Stability, 257, 308
Staphilococcus hyicus, 125
Stomach, 251
Storage stability, 91, 92, 94, 97
Strain-specific, 317
Stratum corneum lipids, 228, 229
Streptococcus, 110
  S. gordini, 124
  S. mutants, 127
  S. thermophilus, 228, 229, 231, 233
Studies, clinical, 255
Sub-lethal injury, 89, 102, 103
Sub-lethal stress, 90
Sulfubolus solfataricus, 6
synbiotic products, 90
Synbiotics, 155, 177

T
Target group, 256
Taxonomy, 17–22
Tetanus toxin fragment C (TTFC), 123
tetW, 276–280
Thermophiles, 6
Thermus sp. Z-1, 6
TH1 response, 126
Tight-junction, 149, 155
Toll-like receptors (TLRs), 232
Toxicity, 130
Transcriptomics, 131
Transfer RNA (tRNA), 15
Transgalactosyl reactions, 5, 6
Translocation, 157, 158
Trefoil factor families (TFFs), 125
T-lymphocytes, 235–237
Trials
  intervention, 258
  placebo-controlled, 257
Trophic effects, 41
Tumor necrosis factor (TNF)-α, 156
Typing, 250

U
Ulcerative colitis, 148–151, 159, 160
Urease A and B, 123
Urinary tract infections (UTI), 179
UVA, 232, 233
UVB, 233, 234
UV irradiation, 232, 233

V
Vaccination, 316
Vaginal candidiasis, 179
Vaginosis, 180
Various types of diarrhea, 40
Vegetable, 193, 208–212, 215
Viability, 193–201, 203, 205–211, 257
Viability of probiotics, 302
Viable state, 88, 105
Vibrio, 6, 8
Visceral hypersensitivity, 147
VSL#3, 145, 151, 157, 159

W
World Health Organization (WHO), 32

Y
Yarrowia lipolytica, 31, 34, 38
Yeast infection, 306